INEQUALITIES

Oh! the little more, and how much it is!
And the little less, and what worlds away!

ROBERT BROWNING

INEQUALITIES

By

G. H. HARDY
J. E. LITTLEWOOD
G. PÓLYA

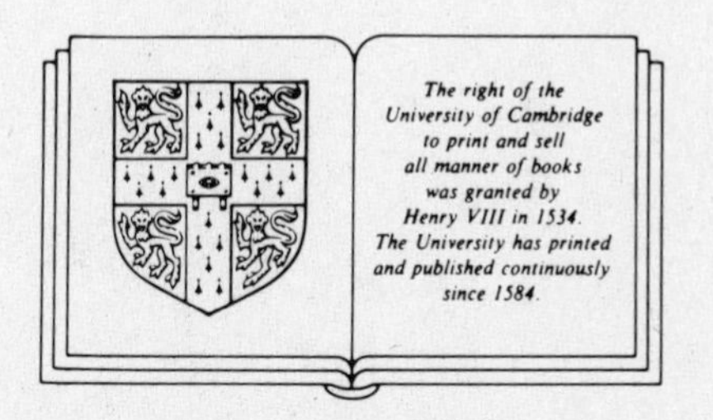

CAMBRIDGE UNIVERSITY PRESS
Cambridge
New York New Rochelle
Melbourne Sydney

Published by the Press Syndicate of the University of Cambridge
The Pitt Building, Trumpington Street, Cambridge CB2 1RP
32 East 57th Street, New York, NY 10022, USA
10 Stamford Road, Oakleigh, Melbourne 3166, Australia

First published 1934
Second edition 1952
Reprinted 1959, 1964, 1967, 1973, 1988 (twice)
First paperback edition 1988

Printed in Great Britain by Redwood Burn Limited,
Trowbridge, Wiltshire

ISBN 0 521 05260 8 hard covers
ISBN 0 521 35880 9 paperback

PREFACE TO FIRST EDITION

This book was planned and begun in 1929. Our original intention was that it should be one of the *Cambridge Tracts*, but it soon became plain that a tract would be much too short for our purpose.

Our objects in writing the book are explained sufficiently in the introductory chapter, but we add a note here about history and bibliography. Historical and bibliographical questions are particularly troublesome in a subject like this, which has applications in every part of mathematics but has never been developed systematically.

It is often really difficult to trace the origin of a familiar inequality. It is quite likely to occur first as an auxiliary proposition, often without explicit statement, in a memoir on geometry or astronomy; it may have been rediscovered, many years later, by half a dozen different authors; and no accessible statement of it may be quite complete. We have almost always found, even with the most famous inequalities, that we have a little new to add.

We have done our best to be accurate and have given all references we can, but we have never undertaken systematic bibliographical research. We follow the common practice, when a particular inequality is habitually associated with a particular mathematician's name; we speak of the inequalities of Schwarz, Hölder, and Jensen, though all these inequalities can be traced further back; and we do not enumerate explicitly all the minor additions which are necessary for absolute completeness.

We have received a great deal of assistance from friends. Messrs G. A. Bliss, L. S. Bosanquet, R. Courant, B. Jessen, V. Levin, R. Rado, I. Schur, L. C. Young, and A. Zygmund have all helped us with criticisms or original contributions. Dr Bosanquet, Dr Jessen, and Prof. Zygmund have read the proofs, and corrected many inaccuracies. In particular, Chapter III has been very largely rewritten as the result of Dr Jessen's suggestions. We hope that the book may now be reasonably free from error, in spite of the mass of detail which it contains.

Dr Levin composed the bibliography. This contains all the books and memoirs which are referred to in the text, directly or by implication, but does not go beyond them.

G. H. H.
J. E. L.
G. P.

Cambridge and Zürich
July 1934

PREFACE TO SECOND EDITION

The text of the first edition is reprinted with a few minor changes; three appendices are added.

J. E. L.
G. P.

Cambridge and Stanford
March 1951

TABLE OF CONTENTS

CHAPTER I. INTRODUCTION

CHAPTER II. ELEMENTARY MEAN VALUES

CHAPTER IV. VARIOUS APPLICATIONS OF THE CALCULUS

CHAPTER V. INFINITE SERIES

CHAPTER VI. INTEGRALS

Chapter VII. SOME APPLICATIONS OF THE CALCULUS OF VARIATIONS

Chapter VIII. SOME THEOREMS CONCERNING BILINEAR AND MULTILINEAR FORMS

Chapter X. REARRANGEMENTS

CHAPTER I

INTRODUCTION

1.1. Finite, infinite, and integral inequalities. It will be convenient to take some particular and typical inequality as a text for the general remarks which occupy this chapter; and we select a remarkable theorem due to Cauchy and usually known as 'Cauchy's inequality'.

Cauchy's inequality (Theorem 7) is

$$(1.1.1) \quad (a_1 b_1 + a_2 b_2 + \dots + a_n b_n)^2 \leqq (a_1{}^2 + a_2{}^2 + \dots + a_n{}^2)(b_1{}^2 + b_2{}^2 + \dots + b_n{}^2)$$

or

$$(1.1.2) \quad (\sum_1^n a_\nu b_\nu)^2 \leqq \sum_1^n a_\nu{}^2 \sum_1^n b_\nu{}^2,$$

and is true for all real values of $a_1, a_2, \dots, a_n, b_1, b_2, \dots, b_n$. We call $a_1, \dots, b_1, \dots$ the *variables* of the inequality. Here the number of variables is finite, and the inequality states a relation between certain finite sums. We call such an inequality an *elementary* or *finite* inequality.

The most fundamental inequalities are finite, but we shall also be concerned with inequalities which are not finite and involve generalisations of the notion of a sum. The most important of such generalisations are the infinite sums

$$(1.1.3) \quad \sum_1^\infty a_\nu, \quad \sum_{-\infty}^\infty a_\nu$$

and the integral

$$(1.1.4) \quad \int_a^b f(x)\,dx$$

(where a and b may be finite or infinite). The analogues of (1.1.2) corresponding to these generalisations are

$$(1.1.5) \quad (\sum_1^\infty a_\nu b_\nu)^2 \leqq \sum_1^\infty a_\nu{}^2 \sum_1^\infty b_\nu{}^2$$

(or the similar formula in which both limits of summation are infinite), and

$$(1.1.6) \qquad \left(\int_a^b f(x)\,g(x)\,dx\right)^2 \leqq \int_a^b f^2(x)\,dx \int_a^b g^2(x)\,dx.$$

We call (1.1.5) an *infinite*, and (1.1.6) an *integral*, inequality.

1.2. Notations. We have often to distinguish between different *sets* of the variables. Thus in (1.1.2) we distinguish the two sets $a_1, a_2, \ldots, a_n$ and $b_1, b_2, \ldots, b_n$. It is convenient to have a shorter notation for sets of variables, and often, instead of writing 'the set $a_1, a_2, \ldots, a_n$' we shall write 'the set (a)' or simply 'the a'.

We shall habitually drop suffixes and limits in summations, when there is no risk of ambiguity. Thus we shall write

$$\Sigma a$$

for any of $\sum_1^n a_\nu, \quad \sum_1^\infty a_\nu, \quad \sum_{-\infty}^{\infty} a_\nu;$

so that, for example,

$$(1.2.1) \qquad (\Sigma ab)^2 \leqq \Sigma a^2 \Sigma b^2$$

may mean either of (1.1.2) or (1.1.5), according to the context.

In integral inequalities, the *set* is replaced by a *function*; thus in passing from (1.1.2) to (1.1.6), (a) and (b) are replaced by f and g. We shall also often omit variables and limits in integrals, writing

$$\int f dx$$

for (1.1.4): so that (1.1.6), for example, will be written as

$$(1.2.2) \qquad \left(\int fg\,dx\right)^2 \leqq \int f^2 dx \int g^2 dx.$$

The ranges of the variables, whether in sums or integrals, are prescribed at the beginnings of chapters or sections, or may be inferred unambiguously from the context.

1.3. Positive inequalities. We are interested primarily in 'positive' inequalities[a]. A finite or infinite inequality is *positive* if all variables $a, b, \ldots$ involved in it are real and non-negative. An inequality of this type usually carries with it, as a trivial

[a] There are exceptions, as for example in §§ 8.8–8.17. There the 'positive' cases of the theorems discussed are relatively trivial.

for example in Dedekind's theory of real numbers, that we are concerned with such or such definite objects, we may say, as in projective geometry, that we are concerned with any system of objects which possesses certain properties specified in a set of axioms. We do not propose to consider the 'axiomatics' of different parts of the subject in detail, but it may be worth while to insert a few remarks concerning the axiomatic basis of those theorems which, like (1.1.2) and most of the theorems of Ch. II, belong properly to algebra.

We may take as the axioms of an algebra only the ordinary laws of addition and multiplication. All our theorems will then be true in many different fields, in real algebra, complex algebra, or the arithmetic of residues to any modulus. Or we may add axioms concerning the solubility of linear equations, axioms which secure the existence and uniqueness of difference and quotient. Our theorems will then be true in real or complex algebra or in arithmetic to a *prime* modulus.

In our present subject we are concerned with relations of *inequality*, a notion peculiar to *real* algebra. We can secure an axiomatic basis for theorems of inequality by taking, in addition to the 'indefinables' and axioms already referred to, one new indefinable and two new axioms. We take as indefinable the idea of a *positive* number, and as axioms the two propositions:

I. *Either a is 0 or a is positive or $-a$ is positive, and these possibilities are exclusive.*

II. *The sum and product of two positive numbers are positive.*

We say that a is *negative* if $-a$ is positive, and that a is *greater* (*less*) than b if $a-b$ is positive (negative). Any inequality of a purely algebraic type, such as (1.1.2), may be made to rest on this foundation.

1.6. Comparable functions. We may say that the functions

$$f(a)=f(a_1, a_2, \ldots, a_n), \quad g(a)=g(a_1, a_2, \ldots, a_n)$$

are *comparable* if there is an inequality between them valid for all non-negative real a, that is to say if either $f \leqq g$ for all such a or

$f \geqq g$ for all such a. Two given functions are not usually comparable. Thus two positive homogeneous polynomials of different degrees are certainly not comparable[a]; if $0 \leqq f \leqq g$ for all non-negative a, and both sides are homogeneous, then f and g are certainly of the same degree.

The definition may naturally be extended to functions $f(a, b, \ldots)$ of several sets of variables.

We shall be occupied throughout this volume with problems concerning the comparability of functions. Thus the arithmetic and geometric means of the a are comparable: $\mathfrak{G}(a) \leqq \mathfrak{A}(a)$ (Theorem **9**). The functions $\mathfrak{G}(a+b)$ and $\mathfrak{G}(a)+\mathfrak{G}(b)$ are comparable (Theorem **10**). The functions $\mathfrak{A}(ab)$ and $\mathfrak{A}(a)\,\mathfrak{A}(b)$ are not comparable; their relative magnitude depends upon the relations of magnitude of the a and b (Theorem **43**). The functions

$$\psi^{-1}(\Sigma p\psi(a)), \quad \chi^{-1}(\Sigma p\chi(a))$$

are comparable if and only if $\chi\psi^{-1}$ is convex or concave (Theorem **85**).

An important general theorem concerning the comparability of two functions of the form

$$\Sigma a_1^{\alpha_1} a_2^{\alpha_2} \ldots a_n^{\alpha_n},$$

due to Muirhead, will be found in § 2.18.

1.7. Selection of proofs. The methods of proof which we use in different parts of the book will depend on very different sets of ideas, and we shall often, particularly in Ch. II, give a number of alternative proofs of the same theorem. It may be useful to call attention here to certain broad distinctions between the methods which we employ.

In the first place, many of the proofs of Ch. II are 'strictly elementary', since they depend solely on the ideas and processes of finite algebra. We have made it a principle to give at any rate one such proof of any really important theorem whose character permits it.

Next we have, even in Ch. II, many proofs which are not elementary in this sense because they involve considerations of

[a] Compare § 2.19.

limits and continuity. We have also, particularly in Ch. IV, proofs which depend upon the standard properties of differential coefficients, as for example upon Rolle's Theorem. All these proofs belong to the elements of the theory of functions of one real variable.

Later, when dealing with integrals in Ch. VI, we naturally make use of the theory of measure and of the integral of Lebesgue. This we take for granted, but we give a summary in §§ 6.1–6.3 of the parts of the theory which we require.

Occasionally we appeal to the more remote parts of the theory of functions of real variables; but we do this only in alternative proofs or in the proofs of theorems of considerable intrinsic difficulty. Thus in Ch. IV (§ 4.6) we use the theory of the maxima and minima of functions of several variables; in Ch. VII we use the methods of the Calculus of Variations; and in Ch. IX we use the theory of double and repeated integration. We make no use of complex function theory, although, in the last chapters, we refer to it occasionally for purposes of illustration. The sections in which we do this do not belong properly to the main body of the book.

We add a few further remarks of a more detailed character.

(i) Cauchy's inequality (1.1.2) is a proposition of finite algebra, as defined in § 1.5. It is a recognised principle that the proof of such a theorem should involve only the methods of the theory to which it belongs.

(ii) We shall be continually meeting theorems, such as Hölder's inequality

$$(1.7.1) \qquad \Sigma ab \leq (\Sigma a^k)^{1/k} (\Sigma b^{k'})^{1/k'}$$

(Theorem **13**), whose status depends upon the value of a parameter k. If k is *rational*, the theorem is algebraical, and our remarks under (i) apply. If k is irrational, a^k is not an algebraical function, and it is obvious that there can be no strictly algebraical proof.

It is however reasonable to demand, when we are concerned with an inequality so fundamental as Hölder's, that our step outside algebra shall be the absolute minimum which the nature of the problem necessitates. It is plain that this step will depend upon our definition of a^k. We may define a^k as $\exp(k \log a)$, and in this case it is obviously legitimate and necessary to use the theory of the exponential and logarithmic functions. If, as is more usual, we define a^k as the limit of a^{k_n}, where k_n is an appropriate rational approximation to k, then *this* limiting process should be the *only* one to which we appeal.

(iii) Suppose that, adopting the last point of view, we have proved Hölder's inequality, for rational k, in the form (1.7.1.). We can infer its truth for irrational k by a passage to the limit.

Such a proof, however, is not usually sufficient for our purpose. We always wish to prove a theorem of a more precise type than (1.7.1), in which (as in Theorem **13**) we establish strict inequality except in certain specified special cases. When we pass to the limit, '$<$' becomes '$\leqq$', we lose touch with the cases of equality (though these are in fact the same as in the rational case), and our proof is incomplete. It is therefore necessary to arrange our proofs in such a manner as to avoid such passages to the limit wherever it is possible. The same point arises whenever we wish to pass from a finite inequality to the corresponding infinite or integral inequality. It recurs at intervals throughout the volume and has often determined our choice of a particular line of proof.

(iv) The general principles which have governed our choice of methods are as follows. When a theorem is simple and fundamental, like Theorems **7**, **9**, or **11**, we prove it by several different methods, and are careful that one of our methods at any rate shall conform to the canons laid down under (i) and (ii). When the theorem is subsidiary or difficult, or when a proof satisfying these conditions would be troublesome or long, we use whatever method seems to us simplest or most instructive.

1.8. Selection of subjects. The principles which have guided us in our selection of *subjects* may be summarised as follows.

(i) The first part of the book (Chs. II–VI[a]) contains a systematic treatment of a definite subject. Our object has been to

[a] Except perhaps some parts of Ch. IV.

discuss thoroughly (with their analogues and extensions) the simple inequalities which are 'in daily use' in analysis. Of these three are fundamental, viz.

(1) the theorem of the arithmetic and geometric means (Theorem **9**),

(2) Hölder's inequality (Theorem **11**),

(3) Minkowski's inequality (Theorem **24**);

and these three theorems dominate the first six chapters. We prove them in a variety of ways, in the finite case in Ch. II, in the infinite case in Ch. V, and in the integral case in Ch. VI; while Ch. III (which contains a general account of the theory of convex functions) is mainly occupied with their generalisations. In these chapters, of which the most important are II, III, and VI, we have aimed at a comprehensive and in some ways exhaustive treatment.

(ii) The rest of the book (Chs. VII–X) is written in a different spirit and must be judged by different standards. These chapters contain a series of essays on subjects suggested by the more systematic investigations which precede. In them there is very little attempt at system or completeness. They are intended as an introduction to certain fields of modern research, and we have allowed our personal interests to dominate our choice of topics.

In spite of this (or because of it) the chapters have a certain unity. There is much modern work, in real or complex function-theory, in the theory of Fourier series, or in the general theory of orthogonal developments, in which the 'Lebesgue classes L^k' occupy the central position. This work demands a considerable mastery of the technique of inequalities; Hölder's and Minkowski's inequalities, and other more modern and more sophisticated inequalities of the same general character, are required at every turn. Our object has been to write such an introduction to this field of analysis as may be made to hang naturally on the subject matter of the early chapters.

(iii) We are interested primarily in certain parts of *real analysis*, and not in arithmetic or in algebra for its own sake. The line between algebra and analysis is often difficult to draw, especially in the theory of quadratic or bilinear forms, and we have often doubted what to include or reject. We have however excluded all developments whose main interest seemed to us to be algebraical.

We have also excluded function-theory proper, real or complex. In the later chapters, however, we have sometimes tried to show the significance of our theorems by sketching the lines of some of their function-theoretic applications.

Thus (to give definite examples) our programme excludes

(1) inequalities of a definitely arithmetical character, such as those of the theory of primes, or those which give bounds for forms with integral variables;

(2) inequalities which belong properly to the algebraical theory of quadratic forms;

(3) inequalities, such as 'Bessel's inequality', which belong to the theory of orthogonal series;

(4) inequalities, such as 'Hadamard's three circle theorem', which belong to function-theory proper:

and there is no systematic discussion of geometrical inequalities, though we use them frequently for purposes of illustration.

It may be useful to end this introduction by a few words of advice to readers who are anxious to avoid unnecessary immersion in detail. The subject, attractive as it is, demands, for the writer at any rate, a great deal of attention to details of a rather tiresome kind. These details arise particularly in the exclusion of exceptional cases, the complete specification of cases of equality, and the conventional treatment of zero and infinite values. Such a reader as we have in mind may be content, in general, to simplify his task as follows. (1) He may ignore the distinction between *non-negative* and *positive*, so that the numbers and functions with which he is concerned are all positive in the narrow sense. (2) He may ignore our conventions concerning 'infinite values'. (3) He may assume that the parameter k or r of inequalities such as Hölder's and Minkowski's is greater than 1.

(4) He may take it for granted that 'what goes for sums goes, with the obvious modifications, for integrals' (or *vice versa*). He should then be able to master what is essential without undue trouble.

This advice for 'easy reading' must not be taken too literally. It is essential to understand the kind of exceptional cases which occur, and the general principles which govern the discrimination of cases of equality. It is not a mere academic exercise to pick out the cases of equality in such an inequality as Hölder's; a knowledge of these cases provides (as is shown very clearly in §§ 8.13–8.16) a powerful weapon for the discovery of deep and important theorems. Every reader should make it his business to explore this inequality at any rate to the end.

CHAPTER II

ELEMENTARY MEAN VALUES

2.1. Ordinary means. In what follows we are concerned with sets of n non-negative numbers a (or b, c, ...), say

(2.1.1) $$a_1, a_2, \ldots, a_\nu, \ldots, a_n \quad (a_\nu \geqq 0),$$

and a real parameter r, which we suppose for the present not to be zero.

We denote the ordered series (2.1.1) by (a). When we say that '(a) is proportional to (b)' we mean that there are two numbers λ and μ, not both zero, such that

(2.1.2) $$\lambda a_\nu = \mu b_\nu \quad (\nu = 1, 2, \ldots, n).$$

It will be observed that the null set, the set (a) in which every a is zero, is proportional to any (b). Proportionality, as we have defined it, is a symmetrical relation between sets but not a transitive one; it becomes transitive if we exclude the null set from consideration.

If (a) and (b) are proportional, and neither of them is null, then $b_\nu = 0$ whenever $a_\nu = 0$, and a_ν / b_ν is independent of ν for the remaining values of ν.

We write

(2.1.3) $$\mathfrak{M}_r = \mathfrak{M}_r(a) = \left(\frac{1}{n}\Sigma a^r\right)^{1/r} = \left(\frac{1}{n}\sum_{\nu=1}^{n} a_\nu^r\right)^{1/r},$$

except when (i) $r = 0$ or (ii) $r < 0$ and one or more of the a are zero. In the exceptional case (ii), when (2.1.3) has no meaning, we define $\mathfrak{M}_r$ as zero, so that

(2.1.4) $$\mathfrak{M}_r = 0 \quad (r < 0, \textit{ some } a \textit{ zero})^{\mathrm{a}}.$$

Here and elsewhere we shall omit the suffixes and limits of summation when it can be done without ambiguity.

In particular we write

(2.1.5) $$\mathfrak{A} = \mathfrak{A}(a) = \mathfrak{M}_1(a),$$

(2.1.6) $$\mathfrak{H} = \mathfrak{H}(a) = \mathfrak{M}_{-1}(a).$$

Finally, we write

(2.1.7) $$\mathfrak{G} = \mathfrak{G}(a) = \sqrt[n]{(a_1 a_2 \ldots a_n)} = \sqrt[n]{(\Pi a)}.$$

[a] If we admitted infinite values, there would be a corresponding case for positive r, viz. $r > 0$, some a infinite, $\mathfrak{M}_r = \infty$.

Thus $\mathfrak{A}(a)$, $\mathfrak{H}(a)$, $\mathfrak{G}(a)$ are the ordinary arithmetic, harmonic and geometric means.

We have excluded the case $r=0$, but we shall find later (§ 2.3) that we can interpret $\mathfrak{M}_0$ conventionally as $\mathfrak{G}$. We are not generally concerned with *negative* a, but it is sometimes convenient to use $\mathfrak{A}(a)$ without any restriction of sign. The definition is unchanged.

2.2. Weighted means. We shall however usually work with a more general system of mean values. We suppose that

$$p_\nu > 0 \quad (\nu = 1, 2, \ldots, n) \tag{2.2.1}$$

and write

$$\mathfrak{M}_r = \mathfrak{M}_r(a) = \mathfrak{M}_r(a, p) = \left(\frac{\Sigma p a^r}{\Sigma p}\right)^{1/r}, \tag{2.2.2}$$

$$\mathfrak{M}_r = 0 \quad (r < 0,\ \textit{some } a \textit{ zero}), \tag{2.2.3}$$

$$\mathfrak{G} = \mathfrak{G}(a) = \mathfrak{G}(a, p) = (\Pi a^p)^{1/\Sigma p}. \tag{2.2.4}$$

The equations (2.1.5) and (2.1.6) stand as before, with the addition of the symbols $\mathfrak{A}(a, p)$, $\mathfrak{H}(a, p)$. The last remark of § 2.1 applies also to the generalised $\mathfrak{A}$. The weighted means reduce to the ordinary means when $p_\nu = 1$ for every ν.

The means being homogeneous and of degree 0 in the p's, we may suppose, if we please, that $\Sigma p = 1$. In this case we shall replace p by q; thus

$$\mathfrak{M}_r(a) = \mathfrak{M}_r(a, q) = (\Sigma q a^r)^{1/r} \quad (\Sigma q = 1), \tag{2.2.5}$$

$$\mathfrak{G}(a) = \mathfrak{G}(a, q) = \Pi a^q \quad (\Sigma q = 1). \tag{2.2.6}$$

We shall not usually refer to the weights explicitly in our formulae, but it is always to be understood that *mean values which are compared with one another are formed with the same weights*.

Ordinary means are special cases of weighted means. On the other hand, weighted means *with commensurable weights* are special cases of ordinary means (with a different system of a); for we may suppose, on account of homogeneity, that the weights are *integral*, and we can derive means with integral weights from ordinary means by replacing every number by an appropriate set of equal numbers. Means with *incommensurable* weights may be regarded as limiting cases of ordinary means.

The following obvious formulae will be used repeatedly:

$$\mathfrak{M}_r(a) = \{\mathfrak{A}(a^r)\}^{1/r}, \tag{2.2.7}$$

$$\mathfrak{G}(a) = e^{\mathfrak{A}(\log a)}, \tag{2.2.8}$$

$$\mathfrak{M}_{-r}(a) = \frac{1}{\mathfrak{M}_r(1/a)}, \tag{2.2.9}$$

$$\mathfrak{M}_{rs}(a) = \{\mathfrak{M}_s(a^r)\}^{1/r}. \tag{2.2.10}$$

We suppose that $a > 0$ in (2.2.8), and in the other formulae if a suffix is negative; the formulae may be extended to cover the missing cases by appropriate conventions. Also

$$\mathfrak{A}(a+b) = \mathfrak{A}(a) + \mathfrak{A}(b), \tag{2.2.11}$$

$$\mathfrak{G}(ab) = \mathfrak{G}(a)\,\mathfrak{G}(b), \tag{2.2.12}$$

$$\mathfrak{M}_r(b) = k\mathfrak{M}_r(a) \quad \text{if } (b) = k(a) \tag{2.2.13}$$

(i.e. if $b_\nu = ka_\nu$, where k is independent of ν),

$$\mathfrak{G}(b) = k\mathfrak{G}(a) \quad \text{if } (b) = k(a), \tag{2.2.14}$$

$$\mathfrak{M}_r(a) \leqq \mathfrak{M}_r(b) \quad \text{if } a_\nu \leqq b_\nu \text{ for all } \nu. \tag{2.2.15}$$

2.3. Limiting cases of $\mathfrak{M}_r(a)$. We denote by

$$\operatorname{Min} a, \quad \operatorname{Max} a$$

the smallest and largest value of an a.

1. $\operatorname{Min} a < \mathfrak{M}_r(a) < \operatorname{Max} a$, *unless either all the a are equal, or else $r < 0$ and an a is zero.*

It is to be understood here, and in the enunciations of all later theorems, that, when we assert that inequalities hold unless some particular condition is satisfied, we imply that at least one of the inequalities degenerates into an equality in the case excluded. Here, for example, $\operatorname{Min} a = \mathfrak{M}_r(a) = \operatorname{Max} a$ if all a are equal, and $\operatorname{Min} a = \mathfrak{M}_r(a) \leqq \operatorname{Max} a$ in the other exceptional case.

We form our means with q. Since

$$\Sigma q(a - \mathfrak{A}) = 0,$$

every a is equal to $\mathfrak{A}$, or else $a - \mathfrak{A}$ is positive for at least one a and negative for another. This proves the theorem for $r = 1$.

In the general case we may suppose that either $a > 0$ or else $r > 0$, the cases excluded being trivial. It then follows that

$$\{\mathfrak{M}_r(a)\}^r = \mathfrak{A}(a^r)$$

lies between $(\text{Min}\, a)^r$ and $(\text{Max}\, a)^r$, which proves the theorem generally.

2. $\text{Min}\, a < \mathfrak{G}(a) < \text{Max}\, a$, *unless all the a are equal or an a is zero.*

In the second exceptional case $\mathfrak{G} = 0$. If $\mathfrak{G} > 0$ then

$$\Pi \left(\frac{a}{\mathfrak{G}}\right)^q = 1,$$

so that every a is $\mathfrak{G}$ or at least one is greater and one less than $\mathfrak{G}$.

3. $\lim_{r \to 0} \mathfrak{M}_r(a) = \mathfrak{G}(a)$.

If every a is positive

$$\mathfrak{M}_r(a) = \exp\left(\frac{1}{r} \log \Sigma q a^r\right)$$

$$= \exp\left\{\frac{1}{r} \log(1 + r\Sigma q \log a + O(r^2))\right\}$$

$$\to \exp(\Sigma q \log a) = \Pi a^q = \mathfrak{G}(a),$$

when $r \to 0$.

If there are some zero a, b denotes a positive a, and s is a q corresponding to a b, then

$$\mathfrak{M}_r(a, q) = (\Sigma q a^r)^{1/r} = (\Sigma s b^r)^{1/r} = (\Sigma s)^{1/r} \mathfrak{M}_r(b, s) \to 0$$

when $r \to +0$, since $\mathfrak{M}_r(b, s) \to \mathfrak{G}(b, s)$ and $\Sigma s < 1$. When $r < 0$, $\mathfrak{M}_r$ and $\mathfrak{G}$ are both zero, so that the result holds also when $r \to -0$.

Our proof depends on the theory of the exponential and logarithmic functions. We show in § 2.16 how a more elementary proof may be found if desired.

4. $\lim_{r \to \infty} \mathfrak{M}_r(a) = \text{Max}\, a$, $\lim_{r \to -\infty} \mathfrak{M}_r(a) = \text{Min}\, a$.

If a_k is the largest a, or one of the largest, and $r > 0$, we have

$$q_k^{1/r} a_k \leqq \mathfrak{M}_r(a) \leqq a_k;$$

from which the first equation follows at once. The second is trivial if any a is zero and follows from (2.2.9) otherwise.

We now agree to write

(2.3.1) $\quad \mathfrak{M}_0(a) = \mathfrak{G}(a), \quad \mathfrak{M}_\infty(a) = \text{Max}\, a, \quad \mathfrak{M}_{-\infty}(a) = \text{Min}\, a.$

With these conventions, we have

5. $\mathfrak{M}_{-\infty}(a) < \mathfrak{M}_r(a) < \mathfrak{M}_\infty(a)$ *for all finite r, unless the a are all equal, or* $r \leqq 0$ *and an a is zero.*

2.4. Cauchy's inequality. It is convenient to prove the next theorem here although it will be superseded later by a more complete theorem (Theorem **16**).

6. $\mathfrak{M}_r(a) < \mathfrak{M}_{2r}(a)$ $(r > 0)$, *unless all the a are equal.*

The inequality is
$$(\Sigma pa^r)^2 < \Sigma p \, \Sigma pa^{2r}$$
and is a special case of the very important theorem which follows.

7. $(\Sigma ab)^2 < \Sigma a^2 \Sigma b^2$, *unless (a) and (b) are proportional*[a].

For
$$\Sigma a^2 \Sigma b^2 - (\Sigma ab)^2 = \tfrac{1}{2} \sum_{\mu, \nu} (a_\mu b_\nu - a_\nu b_\mu)^2.$$

An alternative proof is as follows. The quadratic form
$$\Sigma (xa + yb)^2 = x^2 \Sigma a^2 + 2xy \Sigma ab + y^2 \Sigma b^2$$
is positive for all x, y, and therefore has a negative discriminant, unless $xa_\nu + yb_\nu = 0$ for some x, y, not both zero, and all ν.

To deduce Theorem **6**, take $\sqrt{p}$ and $a^r \sqrt{p}$ in place of a and b.

Theorem **7** may be generalised as follows:

8.
$$\begin{vmatrix} \Sigma a^2 & \Sigma ab & \dots & \Sigma al \\ \dots & \dots & \dots & \dots \\ \Sigma la & \Sigma lb & \dots & \Sigma l^2 \end{vmatrix} > 0,$$
unless the sets $(a), (b), \dots, (l)$ *are linearly dependent, i.e. unless there are numbers* $x, y, \dots, w$, *not all zero, such that* $xa_\nu + yb_\nu + \dots + wl_\nu = 0$ *for every* ν.

Either proof of Theorem **7** may be extended to prove Theorem **8**: we may either express the determinant as a sum of squares of determinants, or we may consider the non-negative quadratic form
$$\Sigma (xa + yb + \dots + wl)^2$$
in $x, y, \dots, w$. We do not go into details because any systematic discussion of inequalities connected with determinants and quadratic forms would carry us beyond the limits which we have imposed on the book.

2.5. The theorem of the arithmetic and geometric means. We come now to the most famous theorem of the subject.

[a] This is what is usually called Cauchy's inequality: see Cauchy (**1**, 373). The corresponding inequality for integrals (Theorem **181**) is usually called Schwarz's inequality, though it seems to have been stated first by Buniakowsky: see Buniakowsky (**1**, 4), Schwarz (**2**, 251).

9. $\mathfrak{G}(a) < \mathfrak{A}(a)$, *unless all the a are equal.*

The inequality to be proved may be written in either of the forms

$$(2.5.1) \qquad a_1^{p_1} a_2^{p_2} \ldots a_n^{p_n} < \left(\frac{p_1 a_1 + \ldots + p_n a_n}{p_1 + \ldots + p_n}\right)^{p_1 + \ldots + p_n},$$

$$(2.5.2) \qquad a_1^{q_1} a_2^{q_2} \ldots a_n^{q_n} < \Sigma q a$$

(where as usual $\Sigma q = 1$).

This theorem is so fundamental that we propose to give a number of proofs, of varying degrees of simplicity and generality. Of the two which we give in this section, the first is entirely elementary. The second depends on Theorem **3** and so, at present, on the theory of the exponential and logarithmic functions. We shall show later (§ 2.16) how this proof also may be made to conform more strictly to the canons of § 1.7.

(i)[a] We have[b]

$$a_1 a_2 = \left(\frac{a_1 + a_2}{2}\right)^2 - \left(\frac{a_1 - a_2}{2}\right)^2 < \left(\frac{a_1 + a_2}{2}\right)^2,$$

unless $a_1 = a_2$, and so

$$a_1 a_2 a_3 a_4 \leqq \left(\frac{a_1 + a_2}{2}\right)^2 \left(\frac{a_3 + a_4}{2}\right)^2 \leqq \left(\frac{a_1 + a_2 + a_3 + a_4}{4}\right)^4,$$

with inequality in one place or the other unless $a_1 = a_2 = a_3 = a_4$. Repeating the argument m times, we find

$$(2.5.3) \qquad a_1 a_2 \ldots a_{2^m} < \left(\frac{a_1 + a_2 + \ldots + a_{2^m}}{2^m}\right)^{2^m},$$

unless all the a are equal. This is (2.5.1) with unit weights and n a power of 2.

Suppose now that n is any number less than 2^m. Taking

$$b_1 = a_1, \quad b_2 = a_2, \quad \ldots, \quad b_n = a_n,$$

$$b_{n+1} = b_{n+2} = \ldots = b_{2^m} = \frac{a_1 + a_2 + \ldots + a_n}{n} = \mathfrak{A},$$

and applying (2.5.3) to the b, we find

$$a_1 a_2 \ldots a_n \mathfrak{A}^{2^m - n} < \left(\frac{b_1 + b_2 + \ldots + b_{2^m}}{2^m}\right)^{2^m} = \left(\frac{n\mathfrak{A} + (2^m - n)\mathfrak{A}}{2^m}\right)^{2^m} = \mathfrak{A}^{2^m},$$

or

$$a_1 a_2 \ldots a_n < \mathfrak{A}^n,$$

[a] Cauchy (**1**, 375). [b] Euclid (**1**: II 5, V 25).

unless all the b, and so all the a, are equal. This is (2.5.1) with unit weights. We deduce (2.5.1), with any commensurable weights, by the process explained in § 2.2.

When the weights are incommensurable, we can replace them by a set of commensurable approximations, prove (2.5.1) with the approximating weights, and proceed to the limit. In this process '$<$' is changed into '$\leqq$', so that we do not at first obtain a complete proof of the theorem. We may complete the proof as follows. Write

$$q_\nu = q'_\nu + q''_\nu \quad (\nu = 1, 2, \dots, n),$$

where $q'_\nu > 0$, $q''_\nu > 0$, and q'_ν is rational. Then

$$r' = \Sigma q'_\nu, \quad r'' = \Sigma q''_\nu$$

are rational and $r' + r'' = 1$. We have already proved (2.5.1) with '$<$' for rational p, and with '$\leqq$' in any case. Hence

$$\Pi a^{q'} < \left(\frac{\Sigma q' a}{\Sigma q'}\right)^{\Sigma q'}, \quad \Pi a^{q''} \leqq \left(\frac{\Sigma q'' a}{\Sigma q''}\right)^{\Sigma q''},$$

$$\Pi a^{q} = \Pi a^{q'} \Pi a^{q''} < \left(\frac{1}{r'} \Sigma q' a\right)^{r'} \left(\frac{1}{r''} \Sigma q'' a\right)^{r''}$$

$$\leqq \Sigma q' a + \Sigma q'' a = \Sigma q a.$$

Another way of completing the proof was shown us by R. E. A. C. Paley. This depends on Theorem **6**. From this theorem, the formula (2.2.10), and what has been proved before, it follows that

$$\mathfrak{A}(a) = \mathfrak{M}_1(a) > \mathfrak{M}_{\frac{1}{2}}(a) = \mathfrak{M}_1{}^2(a^{\frac{1}{2}}) \geqq \mathfrak{G}^2(a^{\frac{1}{2}}) = \mathfrak{G}(a).$$

(ii)[a] By Theorems **6** and **3**, we have

$$\mathfrak{A}(a) = \mathfrak{M}_1(a) > \mathfrak{M}_{\frac{1}{2}}(a) > \mathfrak{M}_{\frac{1}{4}}(a) > \dots > \lim_{m \to \infty} \mathfrak{M}_{2^{-m}}(a) = \mathfrak{G}(a).$$

This proof is very concise but not quite so elementary as the first. It may be observed that we require Theorem **3** only in the case in which the r of Theorem **3** tends to zero through the special sequence of values 2^{-m}.

2.6. Other proofs of the theorem of the means. We shall return to Theorem **9** in §§ 2.14–15 and again in § 2.21. We add here a few remarks about alternative proofs of the ordinary form of the theorem with unit weights.

[a] Schlömilch (**1**).

(i)[a] If the a are not all equal, let

$$a_1 = \text{Min}\, a < \text{Max}\, a = a_2.$$

If we replace each of a_1 and a_2 by $\frac{1}{2}(a_1+a_2)$, $\mathfrak{A}(a)$ is unaltered, but

$$\left(\frac{a_1+a_2}{2}\right)^2 > a_1 a_2,$$

so that $\mathfrak{G}(a)$ is increased.

Suppose now that we vary the a in such a manner that $\mathfrak{A}$ is constant, and that we assume the existence of a set (a^*) for which $\mathfrak{G}$ attains a maximum value. Then the a^* must be equal, since if not we can replace them as above by another system for which $\mathfrak{G}$ is greater. It follows that the maximum of $\mathfrak{G}$ is $\mathfrak{A}$, and that this maximum is attained only for equal a.

To prove the existence of (a^*), let

$$\phi(a_1, a_2, \ldots, a_{n-1}) = a_1 a_2 \ldots a_{n-1}(n\mathfrak{A} - a_1 - \ldots - a_{n-1}).$$

Then ϕ is continuous in the closed domain

$$a_1 \geqq 0, \ \ldots, \ a_{n-1} \geqq 0, \quad a_1 + a_2 + \ldots + a_{n-1} \leqq n\mathfrak{A}.$$

It therefore attains a maximum for some system of values $a_1^*, \ldots, a_{n-1}^*$ in the domain.

[a] This proof, the most familiar of all proofs of the theorem, is due (so far as we have been able to trace it) to Maclaurin (2). Maclaurin states the theorem in geometrical language, as follows: 'If the Line AB is divided into any Number of Parts AC, CD, DE, EB, the Product of all those Parts multiplied into one another will be a *Maximum* when the Parts are equal amongst themselves'. His proof is substantially that which follows. The proof has been rediscovered or reproduced by many later writers, for example by Grebe (**1**), Chrystal (**1**, 47).

Cauchy's proof (§ 2.5) may be regarded as a more sophisticated form of Maclaurin's, since he proves the theorem in the special case when $n = 2^m$ by a process similar to Maclaurin's. In general, Maclaurin's proof is not a 'finite' proof. As we have stated it, it depends on Weierstrass's theorem on the maximum of a continuous function. This would naturally have been taken for granted by Maclaurin (and has also been taken for granted by many of his modern followers, such as Grebe and Chrystal).

It is possible to avoid an appeal to Weierstrass's theorem, but at considerable cost. It is plain that if $a_1^1, a_2^1; a_1^2, a_2^2; \ldots$ are the smallest and largest of the sets resulting from 1, 2, ... repetitions of Maclaurin's process, then a_1^s increases and a_2^s decreases as s increases, so that

$$a_1^s \to \alpha_1, \quad a_2^s \to \alpha_2, \quad \alpha_2 \geqq \alpha_1.$$

A little consideration will show that n repetitions of the process diminish the greatest difference of the a by at least one-half, so that $a_2^n - a_1^n \leqq \frac{1}{2}(a_2 - a_1)$. Hence $a_2^s - a_1^s \to 0$, and $\alpha_1 = \alpha_2$. It follows that all the a tend to the same limit $\mathfrak{A}$. This gives a proof of the theorem, but one a good deal less simple than that in the text.

The reader should work out the analogous proof in which $\mathfrak{G}$ is kept constant and a_1 and a_2 are each replaced by $\sqrt{(a_1 a_2)}$.

(ii) There is a variation of Cauchy's proof which illustrates a point of some logical importance.

An ordinary inductive proof proceeds from n to $n+1$; the truth of a proposition $P(n)$ follows from the hypotheses

(*a*) $P(n)$ implies $P(n+1)$,

(*b*) $P(n)$ is true for $n=1$.

There is another mode of proof which may be called proof by 'backward induction'; the truth of $P(n)$ follows from

(*a*′) $P(n)$ implies $P(n-1)$,

(*b*′) $P(n)$ is true *for an infinity of n*.

Cauchy's proof may be arranged as a proof of this last type. First, Cauchy proves (*b*′) for $n=2^m$. Next, if the theorem is true for n, and if $\mathfrak{A}$ is the arithmetic mean of $a_1, a_2, \dots, a_{n-1}$, then an application of the theorem to the n numbers $a_1, \dots, a_{n-1}, \mathfrak{A}$ gives

$$\mathfrak{A}^n = \left(\frac{a_1 + \dots + a_{n-1} + \mathfrak{A}}{n}\right)^n > a_1 a_2 \dots a_{n-1} \mathfrak{A},$$

the result for $n-1$.

(iii)[a] Defining a_1 and a_2 as in (i), we may replace a_1 and a_2 by $\mathfrak{A}$ and $a_1 + a_2 - \mathfrak{A}$. Then $\mathfrak{A}$ is again unchanged, and

$$\mathfrak{A}(a_1 + a_2 - \mathfrak{A}) - a_1 a_2 = (\mathfrak{A} - a_1)(a_2 - \mathfrak{A}) > 0,$$

so that $\mathfrak{G}$ is increased. Repeating the process we arrive, after at most $n-1$ steps, at a system of a all equal to $\mathfrak{A}$. It follows that $\mathfrak{G} < \mathfrak{A}$.

This proof is a little more sophisticated but entirely elementary. There is an alternative, which we leave to the reader, in which a_1 and a_2 are replaced by $\mathfrak{G}$ and $a_1 a_2 / \mathfrak{G}$.

(iv) There are a number of inductive proofs of the theorem: see, for example, Chrystal (**1**, 46), Muirhead (**3**). One of the simplest runs as follows[b]. Suppose that $0 < a_1 \leqq a_2 \leqq \dots \leqq a_n$, $a_1 < a_n$, that $\mathfrak{A}_\nu$ and $\mathfrak{G}_\nu$ refer to the first ν of the a, and that it has been proved that $\mathfrak{A}_{n-1} \geqq \mathfrak{G}_{n-1}$. Then $a_n > \mathfrak{A}_{n-1}$, by Theorem **1**, and

$$\mathfrak{A}_n = \frac{(n-1)\mathfrak{A}_{n-1} + a_n}{n} = \mathfrak{A}_{n-1} + \frac{a_n - \mathfrak{A}_{n-1}}{n}.$$

[a] For these proofs see Sturm (**1**, 3), Crawford (**1**), Briggs and Bryan (**1**, 185), Muirhead (**3**), Hardy (**1**, 32).

[b] Another simple proof due to R. Rado is given at the end of the chapter (Theorem **60**).

Raising this equation to the nth power, and remembering that $n>1$, we obtain

$$\mathfrak{A}_n^n > \mathfrak{A}_{n-1}^n + n\mathfrak{A}_{n-1}^{n-1}\frac{a_n-\mathfrak{A}_{n-1}}{n} = a_n\mathfrak{A}_{n-1}^{n-1} \geqq a_n\mathfrak{G}_{n-1}^{n-1} = \mathfrak{G}_n^n.$$

(v) Another interesting proof was given very recently by Steffensen (**1**, **2**). It starts from the lemma: *if* $a_{\nu-1}\leqq a_\nu$, $b_{\nu-1}\leqq b_\nu$, *and* $a_\nu\leqq b_\nu$, *for all* ν, *then* $\Sigma a\,\Sigma b$ *is not decreased by exchanging* a_i *and* b_i, *and is increased except when* $a_i=b_i$ *or* $a_\nu=b_\nu$ *for* $\nu\neq i$. The lemma follows at once from the identity

$$\{\Sigma a+(b_i-a_i)\}\{\Sigma b+(a_i-b_i)\} = \Sigma a\,\Sigma b+(b_i-a_i)\{(\Sigma b-b_i)-(\Sigma a-a_i)\}.$$

To deduce the theorem of the means, we write it in the form

$$(a_1+a_1+\dots+a_1)\dots(a_n+a_n+\dots+a_n)\leqq(a_1+a_2+\dots+a_n)\dots(a_1+a_2+\dots+a_n).$$

If we suppose, as we may, that $a_1\leqq a_2\leqq\dots\leqq a_n$, and exchange $n-1$ terms of the first factor of the left-hand side against one term of each of the other factors, we obtain

$$(a_1+a_2+a_3+\dots+a_n)(a_1+a_2+a_2+\dots+a_2)\dots(a_1+a_n+a_n+\dots+a_n),$$

which is greater, by the lemma, unless all the a are equal. The theorem follows by repetition of the argument.

(vi) Further proofs of Theorem **9** (or of the special case considered in this section) are given in §§ 2.14, 2.15, 2.21, 3.11, and 4.2.

2.7. Hölder's inequality and its extensions.

Our next group of theorems centres round Theorem **11** (Hölder's inequality)[a].

10. *Suppose that* (a), (b), ..., (l) *are* m *sets each of* n *numbers. Then*

$$\mathfrak{G}(a)+\mathfrak{G}(b)+\dots+\mathfrak{G}(l)<\mathfrak{G}(a+b+\dots+l), \tag{2.7.1}$$

unless either (1) *every two of* (a), (b), ..., (l) *are proportional, or* (2) *there is a* ν *such that* $a_\nu=b_\nu=\dots=l_\nu=0$.

The theorem states that, if $\Sigma q=1$, then

$$a_1^{q_1}a_2^{q_2}\dots a_n^{q_n}+b_1^{q_1}b_2^{q_2}\dots b_n^{q_n}+\dots+l_1^{q_1}l_2^{q_2}\dots l_n^{q_n} < (a_1+b_1+\dots+l_1)^{q_1}(a_2+b_2+\dots+l_2)^{q_2}\dots,$$

unless every two columns of the array

$$\begin{matrix} a_1, & b_1, & \dots, & l_1 \\ a_2, & b_2, & \dots, & l_2 \\ \dots, & \dots, & \dots, & \dots \end{matrix}$$

[a] Strictly, 'Hölder's inequality' is Theorem **14**, or (2.8.3) of Theorem **13**. The inequality (2.7.1) was stated explicitly, for two sets and equal weights, by Minkowski (**1**, 117).

are proportional or there is a row containing only zeros. A necessary and sufficient condition that all columns should be proportional (i.e. that every pair of columns should be proportional) is that $a_\mu b_\nu - a_\nu b_\mu = 0$, $a_\mu c_\nu - a_\nu c_\mu = 0$, ..., for every μ and ν; and this condition is also necessary and sufficient for the proportionality of all rows. If we remember this, change our notation as between rows and columns of the array, and write $\alpha, \beta, \ldots, \lambda$ for $q_1, q_2, \ldots, q_n$, we see that Theorem **10** is equivalent to

11. *If* $\alpha, \beta, \ldots, \lambda$ *are positive and* $\alpha + \beta + \ldots + \lambda = 1$, *then*

$$\Sigma a^\alpha b^\beta \ldots l^\lambda < (\Sigma a)^\alpha (\Sigma b)^\beta \ldots (\Sigma l)^\lambda, \tag{2.7.2}$$

unless either (1) *the sets* $(a), (b), \ldots, (l)$ *are all proportional, or* (2) *one set is null.*

The conditions for equality might also be expressed by saying that *there is one set which is proportional to all the others* (the null set being proportional to all other sets). The case in which one set is null is trivial, and we may ignore it in the proof.

Here again we given two proofs.

(i) By Theorem **7**, $(\Sigma ab)^2 < \Sigma a^2 \Sigma b^2$

unless (a) and (b) are proportional. Hence

$$(\Sigma abcd)^4 \leqq (\Sigma a^2 b^2)^2 \Sigma (c^2 d^2)^2 \leqq \Sigma a^4 \Sigma b^4 \Sigma c^4 \Sigma d^4,$$

with inequality somewhere unless $(a), (b), (c), (d)$ are proportional[a]. Repeating the argument we see that

$$(\Sigma ab \ldots l)^{2^m} < \Sigma a^{2^m} \Sigma b^{2^m} \ldots \Sigma l^{2^m}, \tag{2.7.3}$$

with 2^m sets $(a), (b), \ldots$, unless all the sets are proportional. This is equivalent to (2.7.2) when every index is 2^{-m}.

Suppose next that M is any number less than 2^m, and let (g) be the Mth set. If $(ab \ldots g)$ is not null, we define $A, B, \ldots, L$ by

$$A^{2^m} = a^M, \ \ldots, \ G^{2^m} = g^M \quad (M \text{ sets}),$$

$$H^{2^m} = K^{2^m} = \ldots = L^{2^m} = ab \ldots g \quad (2^m - M \text{ sets}),$$

so that $AB \ldots L = ab \ldots g$, and apply (2.7.3) to $A, B, \ldots, L$. We thus obtain

$$(\Sigma ab \ldots g)^{2^m} < \Sigma a^M \ldots \Sigma g^M (\Sigma ab \ldots g)^{2^m - M}$$

or

$$(\Sigma ab \ldots g)^M < \Sigma a^M \Sigma b^M \ldots \Sigma g^M, \tag{2.7.4}$$

[a] The null set being excluded, proportionality is now transitive: see §2.1.

unless the sets $(A), (B), \ldots, (L)$, and so the sets $(a), (b), \ldots, (g)$, are proportional. This is equivalent to (2.7.2) with every index $1/M$. We have supposed $(ab \ldots g)$ not null; if it is null then (2.7.4) is obviously true, since none of $(a), (b), \ldots, (g)$ is null.

If now $\alpha, \beta, \ldots$ are rational, we can write

$$\alpha = \frac{\alpha'}{M}, \quad \beta = \frac{\beta'}{M}, \quad \ldots,$$

where $\alpha', \beta', \ldots$ are integers and $\Sigma\alpha' = M$. Applying (2.7.2), with every index $1/M$, to M sets formed by α' like sets of a, β' sets of b, and so on, we obtain (2.7.2) with indices $\alpha, \beta, \ldots$.

Finally, when $\alpha, \beta, \ldots$ are not all rational, we replace them by rational approximations whose sum is 1, form (2.7.2) for these rational indices, and proceed to the limit. In this process '$<$' degenerates into '$\leqq$' and, as in § 2.5 (i), we do not at first obtain a complete proof. We can complete the proof as follows. We can write $\alpha = \alpha_1 + \alpha_2$, $\beta = \beta_1 + \beta_2$, ..., where all the numbers are positive and those with suffix 1 are rational. If then $\Sigma\alpha_1 = \sigma_1$, $\Sigma\alpha_2 = \sigma_2$, so that $\sigma_1 + \sigma_2 = 1$, and $P_1^{\sigma_1} = a^{\alpha_1} b^{\beta_1} \ldots$, $P_2^{\sigma_2} = a^{\alpha_2} b^{\beta_2} \ldots$, we have

$$\Sigma a^\alpha b^\beta \ldots l^\lambda = \Sigma P_1^{\sigma_1} P_2^{\sigma_2} \leqq (\Sigma P_1)^{\sigma_1} (\Sigma P_2)^{\sigma_2}.$$

Since $\alpha_1, \beta_1, \ldots$ are rational

$$\Sigma P_1 = \Sigma a^{\alpha_1/\sigma_1} \ldots l^{\lambda_1/\sigma_1} < (\Sigma a)^{\alpha_1/\sigma_1} \ldots (\Sigma l)^{\lambda_1/\sigma_1};$$

while for ΣP_2 we have a similar inequality, but with '$\leqq$' only. Combining our results we obtain (2.7.2).

(ii) We may deduce Theorem **11** from Theorem **9**. We have in fact (since no set is null)

$$\frac{\Sigma a^\alpha b^\beta \ldots l^\lambda}{(\Sigma a)^\alpha (\Sigma b)^\beta \ldots (\Sigma l)^\lambda} = \Sigma \left(\frac{a}{\Sigma a}\right)^\alpha \left(\frac{b}{\Sigma b}\right)^\beta \ldots \left(\frac{l}{\Sigma l}\right)^\lambda$$
$$\leqq \Sigma \left(\alpha \frac{a}{\Sigma a} + \beta \frac{b}{\Sigma b} + \ldots + \lambda \frac{l}{\Sigma l}\right) = \alpha + \beta + \ldots + \lambda = 1.$$

There can be equality only if

$$\frac{a_\nu}{\Sigma a} = \frac{b_\nu}{\Sigma b} = \ldots = \frac{l_\nu}{\Sigma l} \quad (\nu = 1, 2, \ldots, n),$$

i.e. if $(a), (b), \ldots, (l)$ are proportional.

It will be observed that, whether α, β, ... are rational or not, no limiting processes are involved in the proof beyond those already present in the proof of Theorem **9**. The principle of the proof is the same as that of the proof of Theorem **13** below given independently by Francis and Littlewood[a] (**1**) and F. Riesz (**6**).

2.8. Hölder's inequality and its extensions (*continued*). If we suppose $r \neq 0$, and replace $a, b, \ldots, l$ in Theorem **11** by $qa^{r/\alpha}$, $qb^{r/\beta}, \ldots, ql^{r/\lambda}$, we obtain

12. *If $r, \alpha, \beta, \ldots, \lambda$ are positive and $\alpha + \beta + \ldots + \lambda = 1$, then*

$$\mathfrak{M}_r(ab\ldots l) < \mathfrak{M}_{r/\alpha}(a)\,\mathfrak{M}_{r/\beta}(b)\ldots\mathfrak{M}_{r/\lambda}(l)$$

unless $(a^{1/\alpha}), (b^{1/\beta}), \ldots, (l^{1/\lambda})$ are proportional or one of the factors on the right-hand side is zero. If $r < 0$, the inequality is reversed.

It is to be observed that, when $r > 0$, the second exceptional case occurs only if one of the sets $(a), (b), \ldots$ is null, whereas when $r < 0$ it occurs if any number of any set is zero. When $r = 0$ there is equality in any case.

We shall often find it convenient, when we are concerned with two sets of numbers only, to use the notation

$$k' = \frac{k}{k-1}, \tag{2.8.1}$$

k being any real number except 1. The relation (2.8.1) may also be written in the symmetrical forms

$$(k-1)(k'-1) = 1, \quad \frac{1}{k} + \frac{1}{k'} = 1 \tag{2.8.2}$$

(the last form failing when $k = 0$, $k' = 0$). We say that k and k' are *conjugate*.

13. *Suppose that $k \neq 0$, $k \neq 1$, and that k' is conjugate to k. Then*

$$\Sigma ab < (\Sigma a^k)^{1/k}(\Sigma b^{k'})^{1/k'} \quad (k > 1) \tag{2.8.3}$$

unless (a^k) and $(b^{k'})$ are proportional; and

$$\Sigma ab > (\Sigma a^k)^{1/k}(\Sigma b^{k'})^{1/k'} \quad (k < 1) \tag{2.8.4}$$

unless either (a^k) and $(b^{k'})$ are proportional or (ab) is null.

[a] See Hardy (**8**).

Cauchy's inequality (Theorem **7**) is the special case $k = k' = 2$, in which k is conjugate to itself.

(i) Suppose that $k > 1$. Then (2.8.3) is the special case of Theorem **11** in which there are two sets of letters and $\alpha = 1/k$, $\beta = 1/k'$. This is the ordinary form of Hölder's inequality[a].

(ii) Suppose that $0 < k < 1$, so that $k' < 0$. If any b is 0 then the second factor on the right-hand side of (2.8.4) is, as in § 2.1, to be interpreted as 0, so that (2.8.4) is true unless (ab) is null. If every b is positive, we define l, u, v by

$$l = 1/k,$$

so that $$l > 1, \quad k' = -kl'$$

and $$u = (ab)^k, \quad v = b^{-k},$$

so that $$ab = u^l, \quad a^k = uv, \quad b^{k'} = v^{l'}.$$

Then (2.8.4) reduces to (2.8.3) with u, v, l in place of a, b, k. The exceptional case is that in which (u^l) and $(v^{l'})$, i.e. (ab) and $(b^{k'})$, are proportional. If this is so then (since the b are now all positive) the sets (a) and $(b^{k'-1})$, and therefore the sets (a^k) and $(b^{k'})$, are proportional.

(iii) If $k < 0$, then $0 < k' < 1$. This case is reduced to (ii) by exchanging a and b, k and k'. Both (ii) and (iii) are included in (2.8.4).

The inequalities remain true in the excluded cases $k = 0$, $k = 1$ if we adopt appropriate conventions. If $k = 0$, $k' = 0$, we must interpret (2.8.4) as

$$a_1 b_1 + a_2 b_2 + \dots + a_n b_n > n (a_1 \dots a_n b_1 \dots b_n)^{1/n}.$$

If $k = 1$ we may interpret k' as $+\infty$ or as $-\infty$. In the first case we interpret (2.8.3) as $\Sigma ab < \operatorname{Max} b \, \Sigma a$, and in the second we interpret (2.8.4) as $\Sigma ab > \operatorname{Min} b \, \Sigma a$. We may leave it to the reader to pick out the cases of equality.

We can combine (2.8.3) and (2.8.4) in the single inequality

$$(2.8.5) \qquad (\Sigma ab)^{kk'} < (\Sigma a^k)^{k'} (\Sigma b^{k'})^k \quad (k \neq 0,\ k \neq 1).$$

In view of the extreme importance of Hölder's inequality, we

[a] Hölder (**1**). Hölder states the theorem in a less symmetrical form given a little earlier by Rogers (**1**).

depart from our usual practice here and state explicitly the derivative theorem for complex a, b.

14. *If $k > 1$, and k' is the conjugate of k, then*

$$|\Sigma ab| \leqq (\Sigma |a|^k)^{1/k} (\Sigma |b|^{k'})^{1/k'}.$$

There is equality if and only if $(|a_\nu|^k)$ and $(|b_\nu|^{k'})$ are proportional and $\arg a_\nu b_\nu$ *is independent of ν.*

The only additional remark needed for the proof is that

$$|\Sigma ab| < \Sigma |ab|$$

unless $\arg a_\nu b_\nu$ is independent of ν. We regard 0 as having any argument we please.

The following variant of the first part of Theorem **13** is sometimes called 'the converse of Hölder's inequality'.

15. *Suppose that $k > 1$, that k' is conjugate to k, and that $B > 0$. Then a necessary and sufficient condition that $\Sigma a^k \leqq A$ is that $\Sigma ab \leqq A^{1/k} B^{1/k'}$ for all b for which $\Sigma b^{k'} \leqq B$.*

The condition is necessary, by (2.8.3). If $\Sigma a^k > A$, we can choose the b so that $\Sigma b^{k'} = B$ and $(b^{k'})$ is proportional to (a^k), and then

$$\Sigma ab = (\Sigma a^k)^{1/k} (\Sigma b^{k'})^{1/k'} > A^{1/k} B^{1/k'}.$$

Hence the condition is also sufficient.

Theorem **15** is often useful for the purpose of determining an upper bound for Σa^k. Any argument based on it can be changed into one which involves only a special (b), but the form stated here, with arbitrary (b), is sometimes more convenient[a].

2.9. General properties of the means $\mathfrak{M}_r(a)$.

We can now prove a theorem which completes and supersedes some of those of §§ 2.3–4.

16.[b] *If $r < s$ then*

$$\mathfrak{M}_r(a) < \mathfrak{M}_s(a), \tag{2.9.1}$$

unless the a are all equal, or $s \leqq 0$ and an a is zero.

We have proved this already in the special cases (i) $r = -\infty$

[a] Compare §§ 6.9 (p. 142) and 6.13 (p. 149).

[b] Schlömilch (**1**). See also Reynaud and Duhamel (**1**, 155) and Chrystal (**1**, 48).

(Theorem **5**), (ii) $s=+\infty$ (Theorem **5**), (iii) $r=0, s=1$ (Theorem **9**), (iv) $s=2r$ (Theorem **6**).

Suppose first that $0<r<s$, and write $r=s\alpha$, so that $0<\alpha<1$, and

$$pa^s=u, \quad p=v,$$

so that $v>0$ and $\quad pa^{s\alpha}=(pa^s)^\alpha p^{1-\alpha}=u^\alpha v^{1-\alpha}.$

Then

$$\Sigma u^\alpha v^{1-\alpha}<(\Sigma u)^\alpha(\Sigma v)^{1-\alpha}, \tag{2.9.2}$$

by Theorem **11**, unless u_ν/v_ν is independent of ν, i.e. unless a_ν is independent of ν. Hence

$$\left(\frac{\Sigma pa^{s\alpha}}{\Sigma p}\right)^{1/s\alpha}<\left(\frac{\Sigma pa^s}{\Sigma p}\right)^{1/s},$$

which is (2.9.1).

The cases in which $r\leqq 0$ and an a is zero are trivial and we may ignore them. If every a is positive, and $r=0<s$, we have

$$(\mathfrak{M}_0(a))^s=(\mathfrak{G}(a))^s=\mathfrak{G}(a^s)<\mathfrak{A}(a^s)=(\mathfrak{M}_s(a))^s,$$

by Theorem **9** and (2.2.7). The two remaining cases, $r<s<0$ and $r<s=0$, reduce to those already discussed in virtue of (2.2.9).

17.[a] *If* $0<r<s<t$ *then*

$$\mathfrak{M}_s^s<(\mathfrak{M}_r^r)^{\frac{t-s}{t-r}}(\mathfrak{M}_t^t)^{\frac{s-r}{t-r}}, \tag{2.9.3}$$

unless all the a which are not zero are equal.

We restrict the parameters to be positive, the complications introduced by negative or zero values being hardly worth pursuing systematically.

We may write

$$s=r\alpha+t(1-\alpha) \quad (0<\alpha<1).$$

The inequality is then

$$\Sigma qa^s<(\Sigma qa^r)^\alpha(\Sigma qa^t)^{1-\alpha}$$

and reduces to a case of Theorem **11** when we write $u=qa^r$, $v=qa^t$. The condition for equality is that (u) and (v) should be proportional, and this is plainly equivalent to that stated in the enunciation. The reader should observe the difference between the conditions for equality in Theorems **16** and **17**.

[a] Liapounoff (**1**, **2**).

We shall see later (§3.6, Theorem **87**) that Theorem **17** may be stated in a more striking form.

2.10. The sums $\mathfrak{S}_r(a)$. (i) We write

$$\mathfrak{S}_r = \mathfrak{S}_r(a) = (\Sigma a^r)^{1/r} \quad (r > 0).$$

We confine our attention to positive r, leaving the construction of a theory of $\mathfrak{S}_r$ for $r \leqq 0$ as an exercise to the reader.

18. *If $0 < r < s < t$ then*

(2.10.1) $$\mathfrak{S}_s^s < (\mathfrak{S}_r^r)^{\frac{t-s}{t-r}} (\mathfrak{S}_t^t)^{\frac{s-r}{t-r}},$$

unless all the a which are not zero are equal.

This is essentially the same theorem as Theorem **17**. In fact

(2.10.2) $$\mathfrak{S}_r(a) = n^{1/r}\mathfrak{M}_r(a),$$

the mean $\mathfrak{M}_r(a)$ being formed with unit weights, and (2.10.1) reduces to (2.9.3), the powers of n disappearing.

The correspondence between Theorems **17** and **18** depends essentially on the fact that (2.9.3) and (2.10.1) are homogeneous in the second sense of §1.4, namely in the sign Σ. There is a theorem for sums corresponding to Theorem **16**, but in this theorem, which is expressed by (2.10.3) below, the sign of inequality is reversed; (2.10.3) is not homogeneous in Σ, and is not related to (2.9.1) as (2.10.1) is related to (2.9.3).

19.[a] *If $0 < r < s$ then*

(2.10.3) $$\mathfrak{S}_s(a) < \mathfrak{S}_r(a),$$

unless all the a but one are zero.

Since the inequality is homogeneous in the a, we may suppose $\Sigma a^r = 1$, i.e. $\mathfrak{S}_r = 1$.[b] Then $a_\nu \leqq 1$ for every ν, and so $a_\nu{}^s \leqq a_\nu{}^r$ and

$$\Sigma a^s \leqq \Sigma a^r = 1.$$

If more than one a is positive then at least one positive a is less than 1, and then there is inequality. Theorem **19** is usually quoted as Jensen's inequality.

(ii) We add the theorems for $\mathfrak{S}_r(a)$ corresponding to Theorems **4** and **3**.

[a] Pringsheim (**1**), Jensen (**2**). Pringsheim attributes his second proof to Lüroth.
[b] Compare the remarks on this proof in §1.4.

20. $\mathfrak{S}_r \to \operatorname{Max} a$ *when* $r \to \infty$.

21. $\mathfrak{S}_r \to \infty$ *when* $r \to 0$, *unless all the a but one are zero.*

Theorem **20** follows from (2.10.2) and Theorem **4**. To prove Theorem **21** we have only to observe that $\Sigma a^r = N + o(1)$, where N is the number of positive a.

(iii) Theorem **19**, combined with Theorem **11**, gives the following theorem of Jensen[a].

22. *If* $\alpha, \beta, \ldots, \lambda$ *are positive and* $\alpha + \beta + \ldots + \lambda > 1$, *then*

$$\Sigma a^\alpha b^\beta \ldots l^\lambda < (\Sigma a)^\alpha (\Sigma b)^\beta \ldots (\Sigma l)^\lambda,$$

unless every number of one set or all but one of each set is zero, and, in the latter case, those which are positive have the same rank.

We can write $\alpha = k\alpha'$, $\beta = k\beta'$, ..., where $k > 1$ and $\alpha' + \beta' + \ldots = 1$. If then $a^k = A$, $b^k = B$, ..., we have

$$\Sigma a^\alpha b^\beta \ldots l^\lambda = \Sigma A^{\alpha'} B^{\beta'} \ldots L^{\lambda'} \leqq (\Sigma A)^{\alpha'} (\Sigma B)^{\beta'} \ldots (\Sigma L)^{\lambda'}$$
$$= (\Sigma a^k)^{\alpha/k} \ldots (\Sigma l^k)^{\lambda/k} \leqq (\Sigma a)^\alpha \ldots (\Sigma l)^\lambda,$$

by Theorems **11** and **19**. There is inequality somewhere unless the conditions for equality in both theorems are satisfied.

(iv) It is natural to consider *weighted* sums

$$\mathfrak{T}_r = \mathfrak{T}_r(a) = \mathfrak{T}_r(a, p) = (\Sigma p a^r)^{1/r}.$$

It is plain that there can be no universal relation of the type (2.9.1) or (2.10.3), since $\mathfrak{T}_r$ is the $\mathfrak{S}_r$ of Theorem **19** when $p_\nu = 1$ and is $\mathfrak{M}_r$ when $\Sigma p_\nu = 1$. The possibilities in this direction are settled by the following theorem.

23. *A necessary and sufficient condition that*

$$\mathfrak{T}_r \leqq \mathfrak{T}_s \quad (0 < r < s), \tag{2.10.4}$$

for given weights p and all a, is that $\Sigma p \leqq 1$. *There is then inequality unless* (a) *is null, or* $\Sigma p = 1$ *and all the a are equal.*

A necessary and sufficient condition that

$$\mathfrak{T}_s \leqq \mathfrak{T}_r \quad (0 < r < s), \tag{2.10.5}$$

for given weights p and all a, is that $p_\nu \geqq 1$ *for every* ν. *There is then inequality unless* (a) *is null, or* $a_k > 0$, $p_k = 1$, *and the remaining a are zero.*

(i) If we take $a_\nu = 1$ for every ν, then $\mathfrak{T}_r = (\Sigma p)^{1/r}$, and (2.10.4) can be true only if $\Sigma p \leqq 1$. If this condition is satisfied, and $r = s\alpha$, so that $0 < \alpha < 1$, we have

$$\Sigma p a^r = \Sigma\, (p a^s)^\alpha\, p^{1-\alpha} \leqq (\Sigma p a^s)^\alpha\, (\Sigma p)^{1-\alpha} \leqq (\Sigma p a^s)^\alpha,$$

which is (2.10.4). The conditions for equality are plainly as stated.

[a] Jensen (**2**).

(ii) If we take $a_k = 1$ and the other a zero, $\mathfrak{S}_r = p_k^{1/r}$, and (2.10.5) can be true only if $p_k \geqq 1$. If we assume that this condition is satisfied, write $s = r\beta$, so that $\beta > 1$, and assume, as we may on grounds of homogeneity, that $\Sigma pa^r = 1$, then $pa^r \leqq 1$ for every ν and

$$\Sigma pa^s = \Sigma (pa^r)^\beta p^{1-\beta} \leqq \Sigma (pa^r)^\beta \leqq \Sigma pa^r,$$

which is (2.10.5). The conditions for equality are again plainly as stated.

2.11. Minkowski's inequality. Our next theorem is a generalisation of Theorem **10**.

24. *Suppose that r is finite and not equal to* 1. *Then*

(2.11.1)

$$\mathfrak{M}_r(a) + \mathfrak{M}_r(b) + \dots + \mathfrak{M}_r(l) > \mathfrak{M}_r(a+b+\dots+l) \quad (r > 1),$$

(2.11.2)

$$\mathfrak{M}_r(a) + \mathfrak{M}_r(b) + \dots + \mathfrak{M}_r(l) < \mathfrak{M}_r(a+b+\dots+l) \quad (r < 1),$$

unless (a), (b), ..., (l) *are proportional, or* $r \leqq 0$ *and*

$$a_\nu = b_\nu = \dots = l_\nu = 0$$

for some ν.

There is equality for any $a, b, \dots$ when $r = 1$. Theorem **10** is the special case $r = 0$. The main result remains true (and is trivial) when $r = \infty$ or $r = -\infty$, except that the conditions for equality require a restatement which may be left to the reader.

We take the means with q, and write

$$a + b + \dots + l = s, \quad \mathfrak{M}_r(s) = S.$$

Then
$$S^r = \Sigma q s^r = \Sigma q a s^{r-1} + \Sigma q b s^{r-1} + \dots + \Sigma q l s^{r-1}$$
$$= \Sigma (q^{1/r} a)(q^{1/r} s)^{r-1} + \dots + \Sigma (q^{1/r} l)(q^{1/r} s)^{r-1}.$$

Suppose first that $r > 1$. Applying (2.8.3) of Theorem **13** to each sum on the right, we obtain

$$(2.11.3) \quad S^r \leqq (\Sigma q a^r)^{1/r} (\Sigma q s^r)^{1/r'} + \dots = S^{r-1}((\Sigma q a^r)^{1/r} + \dots).$$

There is equality only if (qa^r), (qb^r), ... are all proportional to (qs^r), i.e. if (a), (b), ... are proportional. Since S is positive (except in the trivial case when every set is null), this establishes (2.11.1)[a].

Suppose next that $0 < r < 1$. Unless all the sets (a), (b), ... are null, $s_\nu > 0$ for some ν. If $s_\nu = 0$ for any particular ν, then

[a] This proof is due to F. Riesz (**1**, 45).

$a_\nu = b_\nu = \ldots = l_\nu = 0$, and we may omit that value of ν from consideration. We may therefore argue as if $s_\nu > 0$ for every ν. In that case (2.8.4) of Theorem **13** gives (2.11.3) with the sign of inequality reversed, and the proof may be completed as before.

Finally, suppose that $r < 0$. If any s_ν is zero, all the means are zero; we may therefore assume that $s_\nu > 0$ for every ν. If any a_ν is zero, $\mathfrak{M}_r(a) = 0$, and we may omit the letter a.[a] We may therefore argue on the assumption that every $a, b, \ldots$ is positive, and then again everything follows from (2.8.4) of Theorem **13**.

When the q are equal, we obtain

25. *If r is finite and not equal to* 0 *or* 1, *then*

(2.11.4)
$$(\Sigma(a+b+\ldots+l)^r)^{1/r} < (\Sigma a^r)^{1/r} + \ldots + (\Sigma l^r)^{1/r} \quad (r > 1),$$

(2.11.5)
$$(\Sigma(a+b+\ldots+l)^r)^{1/r} > (\Sigma a^r)^{1/r} + \ldots + (\Sigma l^r)^{1/r} \quad (r < 1),$$

unless $(a), (b), \ldots, (l)$ *are proportional, or* $r < 0$ *and* $a_\nu, b_\nu, \ldots, l_\nu$ *are all zero for some* ν.

It is (2.11.4) which is usually called Minkowski's inequality[b]. Theorem **24** is more general than Theorem **25** in appearance only, since it may be deduced from Theorem **25** by writing $p^{1/r}a, p^{1/r}b, \ldots$ for $a, b, \ldots$.

Theorem **24** may be given a very elegant symmetrical form[c].

26. *Suppose that* $\mathfrak{M}^{(\mu)}$ *denotes a mean taken with respect to the suffix* μ, *with weights* p_μ, *and* $\mathfrak{M}^{(\nu)}$ *one taken with respect to* ν *with weights* q_ν;[d] *and that* $0 < r < s < \infty$. *Then*

$$\mathfrak{M}_s^{(\nu)} \mathfrak{M}_r^{(\mu)}(a_{\mu\nu}) < \mathfrak{M}_r^{(\mu)} \mathfrak{M}_s^{(\nu)}(a_{\mu\nu}),$$

except when $a_{\mu\nu} = b_\mu c_\nu$.

The result holds generally for all r, s *such that* $r < s$, *except for the specification of the cases of equality.*

[a] Here we use (2.2.15).

[b] Minkowski (**1**, 115–117).

[c] Theorem **26** was communicated to us in 1929 by Mr A. E. Ingham. The same formulation of Minkowski's inequality was found independently by Jessen and published in his paper **1**. This and his later papers **2** and **3** contain many interesting generalisations: see Theorems **136** and **137**.

[d] We depart here from our usual convention about q; Σq is not necessarily 1 (though we prove the inequality by transforming it into one in which we may suppose $\Sigma q = 1$).

We prove the theorem for $0 < r < s < \infty$, leaving the other cases to the reader. There are various supplementary cases of equality when $r \leqq 0$ or one of r and s is infinite.

Let $s/r = k > 1$ and $p_\mu a_{\mu\nu}^r = A_{\mu\nu}$. Then the inequality to be proved is

$$\left\{\sum_{\nu=1}^{n} q_\nu \left(\sum_{\mu=1}^{m} p_\mu a_{\mu\nu}^r\right)^{s/r}\right\}^{1/s} < \left\{\sum_{\mu=1}^{m} p_\mu \left(\sum_{\nu=1}^{n} q_\nu a_{\mu\nu}^s\right)^{r/s}\right\}^{1/r}$$

or

$$\left\{\sum_{\nu=1}^{n} q_\nu \left(\sum_{\mu=1}^{m} A_{\mu\nu}\right)^{k}\right\}^{1/k} < \sum_{\mu=1}^{m} \left(\sum_{\nu=1}^{n} q_\nu A_{\mu\nu}^k\right)^{1/k}.$$

This reduces to (2.11.1) when $\Sigma q = 1$, and, being homogeneous in the q, is true without this restriction.

2.12. A companion to Minkowski's inequality.

The theorem which follows is an analogue of Theorem **25** of a simpler kind.

27. *If r is positive and not equal to 1 then*

$$\Sigma(a+b+\ldots+l)^r > \Sigma a^r + \Sigma b^r + \ldots + \Sigma l^r \quad (r > 1), \tag{2.12.1}$$

$$\Sigma(a+b+\ldots+l)^r < \Sigma a^r + \Sigma b^r + \ldots + \Sigma l^r \quad (0 < r < 1), \tag{2.12.2}$$

unless all numbers but one of each set $a_\nu, b_\nu, \ldots, l_\nu$ $(\nu = 1, 2, \ldots, n)$ are zero.

This follows at once from Theorem **19**, since for example

$$(a+b+\ldots+l)^r > a^r + b^r + \ldots + l^r$$

if $r > 1$, unless all of $a, b, \ldots, l$ but one are zero. It should be noticed that the sense of (2.12.1) and (2.12.2) is opposite to that of (2.11.4) and (2.11.5).

What is usually required in practice is a combination of (2.11.4) and (2.12.2), viz.

28. *If $r > 0$ then*

$$(\Sigma(a+b+\ldots+l)^r)^R \leqq (\Sigma a^r)^R + (\Sigma b^r)^R + \ldots + (\Sigma l^r)^R,$$

where $R = 1$ if $0 < r \leqq 1$ and $R = 1/r$ if $r > 1$.

2.13. Illustrations and applications of the fundamental inequalities.

(i) *Geometrical interpretations of Hölder's and Min-*

kowski's inequalities. Two particularly simple cases of Hölder's and Minkowski's inequalities are

$$(2.13.1)\quad (x_1x_2+y_1y_2+z_1z_2)^2 < (x_1^2+y_1^2+z_1^2)(x_2^2+y_2^2+z_2^2),$$

$$(2.13.2)\quad \sqrt{\{(x_1+x_2)^2+(y_1+y_2)^2+(z_1+z_2)^2\}} < \sqrt{(x_1^2+y_1^2+z_1^2)}+\sqrt{(x_2^2+y_2^2+z_2^2)}.$$

These hold for all real values of the variables, and express the facts that (1) the cosine of a real angle is numerically less than 1, and (2) the sum of two sides of a triangle is greater than the third side. The exceptional cases are those in which (1) the vectors (x_1, y_1, z_1) and (x_2, y_2, z_2) are parallel (with the same or opposite senses), and (2) the vectors are parallel and have the same sense.

The ordinary form of Minkowski's inequality is the extension of (2.13.2) to space of n dimensions with a generalised definition of distance, viz.

$$P_1P_2 = (|x_1-x_2|^r + |y_1-y_2|^r + \dots)^{1/r} \quad (r \geqq 1).$$

The most obvious extensions of (2.13.1) are connected not with Hölder's inequality for general r but with a generalisation of the case $r=2$ in a different direction.

29. *If* $\Sigma a_{\mu\nu} x_\mu x_\nu$, *where* $a_{\mu\nu} = a_{\nu\mu}$, *is a positive quadratic form (with real, but not necessarily positive, coefficients), then*

$$(\Sigma a_{\mu\nu} x_\mu y_\nu)^2 < \Sigma a_{\mu\nu} x_\mu x_\nu \, \Sigma a_{\mu\nu} y_\mu y_\nu,$$

unless (x) *and* (y) *are proportional.*

This is an immediate consequence of the fact that

$$\Sigma a_{\mu\nu}(\lambda x_\mu + \mu y_\mu)(\lambda x_\nu + \mu y_\nu)$$

is positive: compare the second proof of Theorem **7**. It represents geometrically an extension of (2.13.1) to n-dimensional space, with oblique coordinates or a non-Euclidean metric.

To illustrate Theorem **15**, take $k=2$, $A=l^2$, $B=1$, and rectangular coordinates. The theorem then asserts that, if the length of the projection of a vector along an arbitrary direction does not exceed l, the length of the vector does not exceed l.

(ii) *A theorem of Hadamard*[a]. In our next theorem also we are concerned with a set of numbers $a_{\mu\nu}$ real but not necessarily positive.

30. *If D is the determinant whose constituents are*

$$a_{\mu\nu} \quad (\mu, \nu = 1, 2, \ldots, n),$$

then

$$D^2 \leqq \Sigma a_{1\kappa}{}^2 \Sigma a_{2\kappa}{}^2 \ldots \Sigma a_{n\kappa}{}^2. \tag{2.13.3}$$

There is equality only when

$$a_{\mu 1} a_{\nu 1} + a_{\mu 2} a_{\nu 2} + \ldots + a_{\mu n} a_{\nu n} = 0 \tag{2.13.4}$$

for every distinct pair μ, ν, *or when one of the factors on the right-hand side of* (2.13.3) *vanishes.*

The geometrical significance of the theorem is that the volume of a parallelepiped in n-space does not exceed the product of the edges diverging from one corner, and that there is equality only when they are orthogonal or an edge vanishes.

Suppose that $\Sigma c_{\mu\nu} x_\mu x_\nu$, where $c_{\mu\nu} = c_{\nu\mu}$, is a positive quadratic form, and that Δ is the determinant whose constituents are $c_{\mu\nu}$. Then the equation

$$\begin{vmatrix} c_{11} - \lambda & c_{12} & \ldots \\ c_{21} & c_{22} - \lambda & \ldots \\ \ldots & \ldots & \ldots \end{vmatrix} = 0 \tag{2.13.5}$$

has n positive roots[b] whose sum is $\Sigma c_{\mu\mu}$ and whose product is Δ. Hence, by Theorem **9**,

$$\Delta \leqq \left(\frac{c_{11} + c_{22} + \ldots + c_{nn}}{n} \right)^n. \tag{2.13.6}$$

If $c_{\mu\mu} > 0$ for all μ, then the form

$$\Sigma \frac{c_{\mu\nu}}{\sqrt{(c_{\mu\mu} c_{\nu\nu})}} x_\mu x_\nu = \Sigma C_{\mu\nu} x_\mu x_\nu$$

is also positive; and if we apply (2.13.6) to this form, we obtain

$$\Delta \leqq c_{11} c_{22} \ldots c_{nn}. \tag{2.13.7}$$

[a] Hadamard (**1**) considers determinants with complex constituents. Theorem **30** was found earlier by Kelvin and proved by Muir (**1**).

[b] See Bôcher (**1**, 171).

This is substantially equivalent to Hadamard's theorem. For the form

$$\Sigma (a_{1\kappa} x_1 + a_{2\kappa} x_2 + \ldots + a_{n\kappa} x_n)^2 = \Sigma c_{\mu\nu} x_\mu x_\nu$$

is positive unless $D=0$. Also $\Delta = D^2$ and

$$c_{\mu\mu} = a_{\mu 1}{}^2 + a_{\mu 2}{}^2 + \ldots + a_{\mu n}{}^2,$$

so that (2.13.7) is (2.13.3).

For equality in (2.13.6), all the roots of (2.13.5) must be equal, which is only possible if $c_{\mu\nu}=0$ whenever $\mu \neq \nu$ and $c_{\mu\mu}$ is independent of μ. Hence, for equality in (2.13.7), we must have $C_{\mu\nu}=0$ for $\mu \neq \nu$, $C_{\mu\mu}$ independent of μ. The last condition is certainly satisfied, since $C_{\mu\mu}=1$, and $C_{\mu\nu}=0$ is $c_{\mu\nu}=0$, which is (2.13.4).

We can extend the theorem to determinants with complex constituents by using Hermitian instead of quadratic forms. Further extensions have been made by Schur (**2**)[a].

The following ingenious proof of (2.13.7) is due to Oppenheim[b]. Oppenheim's argument establishes not only (2.13.7), and so Hadamard's theorem, but also the inequalities (2.13.8) and (2.13.9) below, due to Minkowski[c] and Fischer[d] respectively.

Any two positive quadratic forms $\Sigma c_{ik} x_i x_k$, $\Sigma d_{ik} x_i x_k$ may be reduced simultaneously, by a linear transformation of determinant unity, to sums of squares[e], say $\Sigma c_\nu y_\nu{}^2$, $\Sigma d_\nu y_\nu{}^2$, where c_ν and d_ν are positive. Then $\Sigma (c_{ik}+d_{ik}) x_i x_k$ is reduced to $\Sigma (c_\nu + d_\nu) y_\nu{}^2$, and the determinants $|c_{ik}|$, ... of the forms satisfy

$$|c_{ik}| = \Pi c_\nu, \quad |d_{ik}| = \Pi d_\nu, \quad |c_{ik}+d_{ik}| = \Pi (c_\nu + d_\nu).$$

Hence, applying Theorem **10** to the sets (c_ν), (d_ν), we obtain

$$(2.13.8) \qquad |c_{ik}|^{1/n} + |d_{ik}|^{1/n} \leqq |c_{ik}+d_{ik}|^{1/n}.$$

Suppose now that the matrix of the d is formed from that of the c by multiplying, first the first r rows, and then the first r

[a] See also A. L. Dixon (**1**).
[b] Oppenheim (**2**).
[c] Minkowski (**2**).
[d] Fischer (**1**).
[e] See Bôcher (**1**, 171).

columns, by -1.[a] If then we divide (2.13.8) by 2, and raise to the nth power, we obtain

$$|c_{ik}| = |c_{11} \dots c_{nn}| \leqq |c_{11} \dots c_{rr}| \, |c_{r+1,r+1} \dots c_{nn}|, \tag{2.13.9}$$

where $|c_{11} \dots c_{rr}|$ denotes the north-west diagonal minor of r rows and columns in $|c_{ik}|$, and $|c_{r+1,r+1} \dots c_{nn}|$ denotes the complementary south-east minor. Repeating the argument, replacing each of the factors on the right-hand side of (2.13.9) by two factors, and so on, we ultimately obtain (2.13.7).

(iii) *The modulus of a matrix.* Suppose that A and B are the matrices of n rows and columns whose elements are $a_{\mu\nu}$ and $b_{\mu\nu}$; the elements may be complex. The matrices $A+B$ and BA are defined as the matrices whose elements are

$$a_{\mu\nu} + b_{\mu\nu}, \quad b_{\mu 1} a_{1\nu} + b_{\mu 2} a_{2\nu} + \dots + b_{\mu n} a_{n\nu}.$$

31.[b] *If $|A|$, the modulus of the matrix A, is defined by*

$$|A| = \sqrt{\{\Sigma |a_{\mu\nu}|^2\}},$$

then $$|A+B| \leqq |A| + |B|, \quad |BA| \leqq |B|\,|A|.$$

The first inequality is an immediate consequence of Theorem **25**, with $r=2$. The second follows from Theorem **7**, since

$$\sum_{\mu,\nu} |b_{\mu 1} a_{1\nu} + \dots + b_{\mu n} a_{n\nu}|^2 \leqq \sum_{\mu,\nu,p,q} |b_{\mu p}|^2 \, |a_{q\nu}|^2.$$

(iv) *Maxima and minima in elementary geometry.* We quote (as exercises for the reader) a few of the numerous applications of the fundamental inequalities to problems of elementary geometry.

32. *The area of a triangle of given perimeter $2p$ is a maximum if the sides a, b, c are equal.*

[Apply Theorem **9** to $p-a$, $p-b$, $p-c$.]

33. *If the surface of a rectangular parallelepiped is given, the volume is greatest when the parallelepiped is a cube.*

[Denote the edges diverging from a corner by a, b, c and apply Theorem **9** to bc, ca, ab. There is an analogous theorem for a parallelepiped in n dimensions; if $k<n$, and the surface of the k-dimensional boundary is given, the volume is greatest when the parallelepiped is rectangular and its edges are equal. This may be proved by combining Theorems **9** and **30** with identities between determinants.]

[a] Thus $\Sigma d_{ik} x_i x_k$ is formed from $\Sigma c_{ik} x_i x_k$ by replacing x_i, x_k $(i, k = 1, 2, \dots, r)$ by $-x_i$, $-x_k$ (and is therefore positive if $\Sigma c_{ik} x_i x_k$ is positive).

[b] See Wedderburn (**1**).

34. Définition. *Si la Base d'une Pyramide est circonscriptible à un Cercle; et si le Pié de la Hauteur est au Centre de ce Cercle: J'appelle cette Pyramide droite.*

Dans une Pyramide droite toutes les Faces ont une même Hauteur, et sont également inclinées au Plan de la Base.

Théorème. *Soient deux Pyramides de même Hauteur, dont les Bases sont égales tant en Surface qu'en Contour; que l'une soit droite et que l'autre ne le soit pas: j'affirme que la Surface de la première Pyramide est plus petite que la Surface de la seconde.* [Lhuilier (**1**, 116).]

[Let h be the height, b_ν a side of the base. and p_ν the perpendicular from the foot of the altitude on to b_ν. Then the lateral surface of the second pyramid is

$$\tfrac{1}{2}\Sigma b_\nu \sqrt{(h^2+p_\nu^2)} > \tfrac{1}{2}\sqrt{\{(\Sigma h b_\nu)^2+(\Sigma p_\nu b_\nu)^2\}},$$

by (2.11.4) of Theorem **25**, unless all the p_ν are equal.]

(v) *Some inequalities useful in elementary analysis.* The following theorems, which are easy deductions from Theorem **9**, are fundamental in the theory of the exponential and logarithmic functions.

35. *If* $\xi > -m$, $0 < m < n$, *then*

$$\left(1+\frac{\xi}{m}\right)^m < \left(1+\frac{\xi}{n}\right)^n.$$

If also $\xi < m$, *then*

$$\left(1-\frac{\xi}{m}\right)^{-m} > \left(1-\frac{\xi}{n}\right)^{-n}.$$

36. *If* $\xi > 0$, $\xi \neq 1$, $0 < m < n$, *then*

$$n(\xi^{1/n}-1) < m(\xi^{1/m}-1).$$

We have, by Theorem **9**,

$$\left(1+\frac{\xi}{m}\right)^{\frac{m}{n}} 1^{\frac{n-m}{n}} < \frac{m}{n}\left(1+\frac{\xi}{m}\right)+\frac{n-m}{n}1 = 1+\frac{\xi}{n}.$$

If $\xi < m$, we may write $-\xi$ for ξ. This proves Theorem **35**. Theorem **36** follows from Theorem **35** if we replace ξ in Theorem **36** by $\left(1 \pm \frac{\xi}{m}\right)^m$.

2.14. Inductive proofs of the fundamental inequalities.

Our fundamental theorems are Theorems **9**, **10** (or **11**), and **24** (or **25**), which we refer to shortly as G, H, M. We deduced H from G[a] and M from H; G is a limiting case of H, H a special case, or anticipation, of M.

The simplest case of G is

37. (G_0): $\quad a^\alpha b^\beta < a\alpha + b\beta \quad (\alpha+\beta=1).$

We show first that G can be deduced from G_0 by induction.

[a] Though giving also an independent proof of H.

Suppose that G has been proved for m letters $a, b, \ldots, k$ (or for any smaller number), and that

$$\alpha+\beta+\ldots+\kappa+\lambda=1, \quad \alpha+\beta+\ldots+\kappa=\sigma.$$

Then

$$a^{\alpha}b^{\beta}\ldots k^{\kappa}l^{\lambda}=(a^{\alpha/\sigma}b^{\beta/\sigma}\ldots k^{\kappa/\sigma})^{\sigma}l^{\lambda}$$
$$\leqq (a^{\alpha/\sigma}b^{\beta/\sigma}\ldots k^{\kappa/\sigma})\,\sigma+l\lambda \leqq a\alpha+b\beta+\ldots+k\kappa+l\lambda,$$

by G for 2 and for m letters. There is equality in the final result only if

$$a^{\alpha/\sigma}b^{\beta/\sigma}\ldots k^{\kappa/\sigma}=l, \quad a=b=\ldots=k,$$

i.e. if all letters are equal. Hence G is true for $m+1$ letters.

The simplest cases of H and M are

38. (H_0):

$$a_1{}^{\alpha}b_1{}^{\beta}+a_2{}^{\alpha}b_2{}^{\beta}<(a_1+a_2)^{\alpha}(b_1+b_2)^{\beta} \quad (\alpha+\beta=1).$$

39. (M_0):

$$\{(a_1+b_1)^r+(a_2+b_2)^r\}^{1/r}<(a_1{}^r+a_2{}^r)^{1/r}+(b_1{}^r+b_2{}^r)^{1/r} \quad (r>1)$$

(with a reversed inequality when $r<1$). We can deduce H_0 from G_0 and M_0 from H_0 by specialising our deductions of H from G and of M from H. We can also deduce H and M from H_0 and M_0 by induction, but, since these inductive proofs are not essential to our argument, we need only sketch them.

(i) We have

$$a_1{}^{\alpha}b_1{}^{\beta}+a_2{}^{\alpha}b_2{}^{\beta}+a_3{}^{\alpha}b_3{}^{\beta}\leqq(a_1+a_2)^{\alpha}(b_1+b_2)^{\beta}+a_3{}^{\alpha}b_3{}^{\beta}$$
$$\leqq(a_1+a_2+a_3)^{\alpha}(b_1+b_2+b_3)^{\beta}.$$

The process may be repeated, and there is no difficulty in picking out the cases of equality. We thus obtain (2.8.3) of Theorem **13** (H for two sets of n numbers).

Next, if $\alpha+\beta+\gamma=1$, $\alpha+\beta=\sigma$, we have

$$\Sigma a^{\alpha}b^{\beta}c^{\gamma}=\Sigma(a^{\alpha/\sigma}b^{\beta/\sigma})^{\sigma}c^{\gamma}\leqq(\Sigma a^{\alpha/\sigma}b^{\beta/\sigma})^{\sigma}(\Sigma c)^{\gamma}\leqq(\Sigma a)^{\alpha}(\Sigma b)^{\beta}(\Sigma c)^{\gamma}.$$

This process also may be repeated, and leads to the general form of H.

We may arrange the induction differently, increasing the number of sets first. The intermediate generalisation (H for any number of sets of two numbers) is worth separate statement.

40. *If* $\alpha+\beta+\ldots+\lambda=1$, *then*

$$\alpha_1{}^\alpha b_1{}^\beta \ldots l_1{}^\lambda + a_2{}^\alpha b_2{}^\beta \ldots l_2{}^\lambda < (a_1+a_2)^\alpha (b_1+b_2)^\beta \ldots (l_1+l_2)^\lambda,$$

unless $a_1/a_2=b_1/b_2=\ldots=l_1/l_2$ *or one of the sets is null.*

(ii) Similarly we can generalise $\mathfrak{M}_0$ in two directions. On the one hand

$$\begin{aligned}\{(a_1+b_1+c_1)^r + (a_2+b_2+c_2)^r\}^{1/r} \\ &\leqq (a_1{}^r+a_2{}^r)^{1/r} + \{(b_1+c_1)^r+(b_2+c_2)^r\}^{1/r} \\ &\leqq (a_1{}^r+a_2{}^r)^{1/r} + (b_1{}^r+b_2{}^r)^{1/r} + (c_1{}^r+c_2{}^r)^{1/r},\end{aligned}$$

and on the other

$$\begin{aligned}\{(a_1+b_1)^r + (a_2+b_2)^r + (a_3+b_3)^r\}^{1/r} \\ &\leqq [\{(a_1{}^r+a_2{}^r)^{1/r} + (b_1{}^r+b_2{}^r)^{1/r}\}^r + (a_3+b_3)^r]^{1/r} \\ &\leqq (a_1{}^r+a_2{}^r+a_3{}^r)^{1/r} + (b_1{}^r+b_2{}^r+b_3{}^r)^{1/r}.\end{aligned}$$

Repeating and combining these processes, we arrive at the general case.

2.15. Elementary inequalities connected with Theorem 37. We can write $\mathfrak{G}_0$ in the form

$$a^\alpha < \{a\alpha + b(1-\alpha)\} b^{\alpha-1}$$

or

$$a^\alpha - b^\alpha < \alpha b^{\alpha-1}(a-b) \quad (0<\alpha<1),$$

which is one case of a system of inequalities prominent in textbooks of analysis. The complete system is stated in Theorem 41 below. The theorem is so important that it is worth while to give a direct proof from first principles which conforms strictly to the criteria of § 1.7.

41. *If x and y are positive and unequal, then*

$$rx^{r-1}(x-y) > x^r - y^r > ry^{r-1}(x-y) \quad (r<0 \text{ or } r>1), \tag{2.15.1}$$

$$rx^{r-1}(x-y) < x^r - y^r < ry^{r-1}(x-y) \quad (0<r<1). \tag{2.15.2}$$

There is obviously equality when $r=0$, $r=1$, or $x=y$. We begin by reducing the theorem to one of its cases.

(i) We may suppose r positive. For let us assume that (2.15.1) has been proved when $r>1$, and that $r<0$, $r=-s$, so that $s+1>1$. Then

$$\begin{aligned} x^r-y^r &= x^{-s}-y^{-s}=x^{-s}y^{-s-1}(y^{s+1}-x^s y) \\ &= x^{-s}y^{-s-1}\{y^{s+1}-x^{s+1}-x^s(y-x)\}>x^{-s}y^{-s-1}sx^s(y-x) \\ &= ry^{r-1}(x-y). \end{aligned}$$

The other inequality in (2.15.1) can be treated similarly.

(ii) Let us denote the left- and right-hand inequalities in (2.15.1) by (1a) and (1b) respectively, and similarly for (2.15.2). If we interchange x and y, (1b) and (2b) become (1a) and (2a). It is therefore sufficient to prove (1b) and (2b).

(iii) We may now suppose, on grounds of homogeneity, that $y=1$.

The proof of Theorem **41** is now reduced to that of the next theorem.

42. *If x is positive and not equal to* 1, *then*

$$x^r-1>r(x-1) \quad (r>1), \tag{2.15.3}$$

$$x^r-1<r(x-1) \quad (0<r<1). \tag{2.15.4}$$

If in (2.15.3) we write $r=1/s$ and $x=y^{1/r}=y^s$, it becomes (2.15.4) with y, s for x, r. It is therefore sufficient to prove (2.15.3).

If q is an integer greater than 1,[a] and $y>1$, then

$$qy^q>1+y+\dots+y^{q-1}=\frac{y^q-1}{y-1}>q.$$

If $0<y<1$, the inequalities are reversed. Replacing y^q by x, we obtain in either case

$$\frac{x-1}{x}<q(x^{1/q}-1)<x-1. \tag{2.15.5}$$

Next, we have

$$\begin{aligned} \frac{y^{q+1}-1}{q+1}-\frac{y^q-1}{q} &= \frac{y-1}{q(q+1)}(qy^q-y^{q-1}-y^{q-2}-\dots-1) \\ &= \frac{(y-1)^2}{q(q+1)}\{y^{q-1}+(y^{q-1}+y^{q-2})+\dots+(y^{q-1}+y^{q-2}+\dots+1)\}. \end{aligned}$$

[a] We abandon here our usual convention concerning the meanings of q and p.

The curly bracket contains $\frac{1}{2}q(q+1)$ terms, all of which lie between y^q and 1, so that

(2.15.6)
$$\tfrac{1}{2}(y-1)^2 \lessgtr \frac{y^{q+1}-1}{q+1} - \frac{y^q-1}{q} \lessgtr \tfrac{1}{2}y^q(y-1)^2 \quad (y \gtrless 1);$$

and so, if p is any integer greater than q,

(2.15.7)
$$\tfrac{1}{2}(p-q)(y-1)^2 \lessgtr \frac{y^p-1}{p} - \frac{y^q-1}{q} \lessgtr \tfrac{1}{2}(p-q)y^p(y-1)^2 \quad (y \gtrless 1).$$

Now it follows from (2.15.5) that

$$\frac{(x-1)^2}{x^2} < q^2(x^{1/q}-1)^2 < (x-1)^2$$

if $x > 1$, while if $0 < x < 1$ the inequalities are reversed. Hence, replacing y^q by x in (2.15.7), we obtain

(2.15.8)
$$\frac{p-q}{2q}\frac{(x-1)^2}{x^2} \lessgtr \frac{x^{p/q}-1}{p/q} - (x-1) \lessgtr \frac{p-q}{2q}x^{p/q}(x-1)^2 \quad (x \gtrless 1).$$

Suppose now that $r > 1$. If r is rational, we write r for p/q; if r is irrational, we make $p/q \to r$. In either case we have

(2.15.9)
$$\tfrac{1}{2}(r-1)\frac{(x-1)^2}{x^2} \lesseqqgtr \frac{x^r-1}{r} - (x-1) \lesseqqgtr \tfrac{1}{2}(r-1)x^r(x-1)^2 \quad (r>1,\ x \gtrless 1),$$

which plainly includes (2.15.3).

This proves Theorems **42** and **41**, but it will be useful to have the inequalities corresponding to (2.15.9) when $r < 1$. We now replace y^p by x in (2.15.7), and use (2.15.5) with q replaced by p. We thus obtain

(2.15.10)
$$\frac{p-q}{2p}\frac{(x-1)^2}{x^2} \lessgtr x-1-\frac{x^{q/p}-1}{q/p} \lessgtr \frac{p-q}{2p}x(x-1)^2 \quad (x \gtrless 1),$$

(2.15.11)
$$\tfrac{1}{2}(1-r)\frac{(x-1)^2}{x^2} \lesseqqgtr x-1-\frac{x^r-1}{r} \lesseqqgtr \tfrac{1}{2}(1-r)x(x-1)^2 \quad (0<r<1,\ x \gtrless 1).$$

We have made the proof of (2.15.3) rather more elaborate than is necessary, in order to obtain the 'second order' inequalities (2.15.6)–(2.15.11), which are interesting in themselves. If we are concerned only to prove (2.15.3), we can argue as follows. Instead of (2.15.6) we write simply

$$\frac{y^{q+1}-1}{q+1} > \frac{y^q-1}{q},$$

whence

$$\frac{y^p-1}{p} > \frac{y^q-1}{q},$$

if p and q are integers and $p>q$. Hence we obtain (2.15.3) for *rational* r, and so, by a passage to the limit

$$x^r - 1 \geqq r(x-1),$$

for any $r>1$. If now r is irrational, we may write $r=\alpha s$, where α and s are both greater than 1 and α is rational. Then

$$x^r - 1 = (x^s)^\alpha - 1 > \alpha(x^s-1) \geqq \alpha s(x-1) = r(x-1),$$

so that (2.15.3) is true generally.

For other proofs of Theorem **41** which satisfy the requirements, see Stolz and Gmeiner (**1**, 202–208) and Pringsheim (**1**). Pringsheim uses the result to obtain an elementary proof of H. Radon (**1**, 1351) deduces H and M from Theorem **41**, but proves this by differential calculus. The proofs of Theorem **41** given in textbooks are usually limited to rational r; see for example Chrystal (**1**, 42–45), Hardy (**1**, 138).

2.16. Elementary proof of Theorem 3. We have proved incidentally in the last section a number of inequalities sharper than those stated in Theorems **41** and **42**. We lay no stress on these, since it is easy to find still more precise inequalities by the aid of the differential calculus (see § 4.2); but it may be interesting to show shortly how they enable us, if we desire, to 'elementarise' the proof of Theorem **3**.

We observe first that

$$a^r = 1 + O(r) \tag{2.16.1}$$

for fixed positive a and small (positive or negative) r;

$$(1+u)^q = 1 + qu + O(u^2) \tag{2.16.2}$$

for fixed q and small u; and

$$\{1+O(r^2)\}^{1/r} = 1 + O(r) \tag{2.16.3}$$

for small r. We leave the deduction of these formulae from those of the last section to the reader.

Supposing now that r is small, we have $a_\nu{}^r = 1 + u_\nu$, where $u_\nu = O(r)$, by (2.16.1), and

$$a_\nu{}^{q_\nu r} = (1+u_\nu)^{q_\nu} = 1 + q_\nu u_\nu + O(r^2),$$

by (2.16.2). Hence

$$\frac{\mathfrak{G}}{\mathfrak{M}_r}=\left(\frac{a_1^{q_1 r}\dots a_n^{q_n r}}{q_1 a_1^r+\dots+q_n a_n^r}\right)^{1/r}=\left\{\frac{(1+u_1)^{q_1}\dots(1+u_n)^{q_n}}{q_1(1+u_1)+\dots+q_n(1+u_n)}\right\}^{1/r}$$

$$=\left\{\frac{1+q_1u_1+\dots+q_nu_n+O(r^2)}{1+q_1u_1+\dots+q_nu_n}\right\}^{1/r}$$

$$=\{1+O(r^2)\}^{1/r}=1+O(r)\to 1.$$

2.17. Tchebychef's inequality. We know (Theorem **24**) that $\mathfrak{M}_r(a+b)$ is comparable (§ 1.6) with $\mathfrak{M}_r(a)+\mathfrak{M}_r(b)$. It is natural to ask whether $\mathfrak{M}_r(ab)$ is comparable with $\mathfrak{M}_r(a)\,\mathfrak{M}_r(b)$. Theorem **43** below shows that this is not so.

We say that (a) and (b) are *similarly ordered* if

$$(a_\mu-a_\nu)(b_\mu-b_\nu)\geqq 0,$$

for all μ, ν, and *oppositely ordered* if the inequality is always reversed. It is evident that (a) and (b) are similarly ordered if there is a permutation $\nu_1, \nu_2, \dots, \nu_n$ of the suffixes such that $a_{\nu_1}, a_{\nu_2}, \dots, a_{\nu_n}$ and $b_{\nu_1}, b_{\nu_2}, \dots, b_{\nu_n}$ are both non-decreasing sequences, and oppositely ordered if $a_{\nu_1}, \dots$ is non-increasing and $b_{\nu_1}, \dots$ non-decreasing; and that the converses of these propositions are also true.

43.[a] *If* $r>0$, *and* (a) *and* (b) *are similarly ordered, then*

(2.17.1) $$\mathfrak{M}_r(a)\,\mathfrak{M}_r(b)<\mathfrak{M}_r(ab),$$

unless all the a or all the b are equal. The inequality is reversed when the sets are oppositely ordered.

It is enough, after (2.2.7), to consider the case $r=1$. Then

$$\begin{aligned}\Sigma p\,\Sigma pab-\Sigma pa\,\Sigma pb&=\Sigma p_\mu\,\Sigma p_\nu a_\nu b_\nu-\Sigma p_\mu a_\mu\,\Sigma p_\nu b_\nu\\&=\Sigma\Sigma(p_\mu p_\nu a_\nu b_\nu-p_\mu p_\nu a_\mu b_\nu)=\Sigma\Sigma(p_\nu p_\mu a_\mu b_\mu-p_\nu p_\mu a_\nu b_\mu)\\&=\tfrac12\Sigma\Sigma(p_\mu p_\nu a_\nu b_\nu-p_\mu p_\nu a_\mu b_\nu+p_\nu p_\mu a_\mu b_\mu-p_\nu p_\mu a_\nu b_\mu)\\&=\tfrac12\Sigma\Sigma p_\mu p_\nu(a_\mu-a_\nu)(b_\mu-b_\nu)\geqq 0,\end{aligned}$$

or

$$\mathfrak{A}(a)\,\mathfrak{A}(b)\leqq\mathfrak{A}(ab),$$

if the series are similarly ordered.

We can determine the cases of equality as follows. Suppose, as

[a] The integral analogue is due to Tchebychef. See Hermite (**1**, 46–47), Franklin (**1**), Jensen (**1**), and Theorem **236**. When $r=1$, $\mathfrak{M}_r=\mathfrak{A}$, the inequality holds for any real and similarly ordered a, b.

we may in virtue of the remarks earlier in the section, that (a) and (b) are non-decreasing. The double sum contains a term

$$p_1 p_n (a_1 - a_n)(b_1 - b_n)$$

and this can vanish only if all a or all b are equal.

An immediate deduction is that

$$\mathfrak{M}_r(a)\mathfrak{M}_r(b)\ldots\mathfrak{M}_r(l) < \mathfrak{M}_r(ab\ldots l)$$

if r is positive and (a), (b), ..., (l) are all similarly ordered; in particular

$$\mathfrak{M}_r(a) < \mathfrak{M}_{mr}(a)$$

if m is an integer greater than 1. This includes Theorem **6** and is included in Theorem **16**.

The question asked at the beginning of this section is included in the more general question settled by the next theorem.

44. *A necessary and sufficient condition that* $\mathfrak{M}_r(ab\ldots l)$ *and* $\mathfrak{M}_s(a)\mathfrak{M}_t(b)\ldots\mathfrak{M}_v(l)$, *where* $r, s, \ldots, v$ *are positive, should be comparable, is that*

$$\frac{1}{r} \geqq \frac{1}{s} + \frac{1}{t} + \ldots + \frac{1}{v}; \tag{2.17.2}$$

in which case

$$\mathfrak{M}_r(ab\ldots l) \leqq \mathfrak{M}_s(a)\mathfrak{M}_t(b)\ldots\mathfrak{M}_v(l). \tag{2.17.3}$$

The sufficiency of the condition follows at once from Theorems **12** and **16**. If we take every set in (2.17.3) to be $(1, 0, 0, \ldots, 0)$, we see at once that (2.17.2) must be satisfied. A general inequality *opposite* to (2.17.3) is impossible for any $r, s, \ldots$, since $a_\nu b_\nu \ldots l_\nu$ may vanish for every ν and yet the right-hand side be positive.

2.18. Muirhead's theorem. In this and the four succeeding sections we suppose the a to be strictly positive. We denote by

$$\Sigma!\, F(a_1, a_2, \ldots, a_n)$$

the sum of the $n!$ terms obtained from $F(a_1, a_2, \ldots, a_n)$ by the possible permutations of the a. We shall be concerned only with the special case

$$F(a_1, a_2, \ldots, a_n) = a_1^{\alpha_1} a_2^{\alpha_2} \ldots a_n^{\alpha_n} \quad (a_\nu > 0,\ \alpha_\nu \geqq 0).$$

We write

$$[\alpha] = [\alpha_1, \alpha_2, \ldots, \alpha_n] = \frac{1}{n!}\Sigma!\, a_1^{\alpha_1} a_2^{\alpha_2} \ldots a_n^{\alpha_n}.$$

It is plain that $[\alpha]$ is unaltered by any permutation of the α, so that we may regard two sets of α as the same if they differ only in arrangement. We may describe a mean value of the type $[\alpha]$ as a *symmetrical mean.*

In particular

$$[1,0,0,\ldots,0]=\frac{(n-1)!}{n!}(a_1+a_2+\ldots+a_n)=\mathfrak{A}(a),$$

$$\left[\frac{1}{n},\frac{1}{n},\ldots,\frac{1}{n}\right]=\frac{n!}{n!}a_1^{1/n}a_2^{1/n}\ldots a_n^{1/n}=\mathfrak{G}(a),$$

the arithmetic and geometric means with unit weights. When $\alpha_1+\alpha_2+\ldots+\alpha_n=1$, $[\alpha]$ is a common generalisation of $\mathfrak{A}(a)$ and $\mathfrak{G}(a)$.

In general $[\alpha']$ is not comparable with $[\alpha]$ in the sense of § 1.6. The problem solved in this and the next two sections is that of determining conditions for comparability.

We say that (α') is *majorised* by (α), and write

$$(\alpha')\prec(\alpha),$$

when the (α) and (α') can be arranged so as to satisfy the following three conditions:

$$\alpha_1'+\alpha_2'+\ldots+\alpha_n'=\alpha_1+\alpha_2+\ldots+\alpha_n; \tag{2.18.1}$$

$$\alpha_1'\geqq\alpha_2'\geqq\ldots\geqq\alpha_n',\quad \alpha_1\geqq\alpha_2\geqq\ldots\geqq\alpha_n; \tag{2.18.2}$$

$$\alpha_1'+\alpha_2'+\ldots+\alpha_\nu'\leqq\alpha_1+\alpha_2+\ldots+\alpha_\nu\quad(1\leqq\nu<n). \tag{2.18.3}$$

The second condition is in itself no restriction, since we may rearrange (α') and (α) in any order, but it is essential to the statement of the third. It is plain that $(\alpha)\prec(\alpha)$.

45. *A necessary and sufficient condition that* $[\alpha']$ *should be comparable with* $[\alpha]$, *for all positive values of the a, is that one of* (α') *and* (α) *should be majorised by the other. If* $(\alpha')\prec(\alpha)$ *then*

$$[\alpha']\leqq[\alpha]. \tag{2.18.4}$$

There is equality only when (α') *and* (α) *are identical or when all the a are equal*[a].

[a] Theorem **45** is due substantially to Muirhead (2); but Muirhead considers only integral α.

2.19. Proof of Muirhead's theorem. (1) *The condition is necessary.* Suppose, as we may, that (2.18.2) is satisfied, and that (2.18.4) holds for all positive a. Taking all the a equal to x, we obtain

$$x^{\Sigma\alpha'} = [\alpha'] \leqq [\alpha] = x^{\Sigma\alpha}.$$

This can only be true both for large and for small x if (2.18.1) is true.

Next, take

$$a_1 = a_2 = \ldots = a_\nu = x, \quad a_{\nu+1} = \ldots = a_n = 1,$$

x being large. Since (α') and (α) are in descending order, the indices of the highest powers of x in $[\alpha']$ and $[\alpha]$ are

$$\alpha_1' + \alpha_2' + \ldots + \alpha_\nu', \quad \alpha_1 + \alpha_2 + \ldots + \alpha_\nu$$

respectively. It is plain that the first cannot exceed the second, and this proves (2.18.3).

(2) *The condition is sufficient.* The proof of this is rather more troublesome, and we require a new definition and two lemmas.

We define a special type of linear transformation of the α, which we call a transformation T, as follows. Suppose that α_k and α_l are two unequal α, the first being the greater; we may write

$$\alpha_k = \rho + \tau, \quad \alpha_l = \rho - \tau \quad (0 < \tau \leqq \rho). \tag{2.19.1}$$

If now

$$0 \leqq \sigma < \tau \leqq \rho \tag{2.19.2}$$

then a transformation T is defined by

$$\begin{cases} \alpha_k' = \rho + \sigma = \dfrac{\tau+\sigma}{2\tau}\alpha_k + \dfrac{\tau-\sigma}{2\tau}\alpha_l, \\ \alpha_l' = \rho - \sigma = \dfrac{\tau-\sigma}{2\tau}\alpha_k + \dfrac{\tau+\sigma}{2\tau}\alpha_l, \\ \alpha_\nu' = \alpha_\nu \quad (\nu \neq k, \ \nu \neq l). \end{cases} \tag{2.19.3}$$

If (α') arises from (α) by a transformation T, we write $\alpha' = T\alpha$. The definition does not necessarily imply that either the α or the α' are in decreasing order.

It is plain that the sufficiency of our condition for comparability will be established, and that we shall also have proved what is stated in Theorem **45** about the case of equality, if we have proved the two following lemmas.

Lemma 1. *If* $\alpha' = T\alpha$ *then* $[\alpha'] \leqq [\alpha]$, *with equality only when all a are equal.*

Lemma 2. *If* $(\alpha') \prec (\alpha)$, *but* (α') *is not identical with* (α), *then* (α') *can be derived from* (α) *by the successive application of a finite number of transformations* T.

Proof of Lemma 1. We may rearrange (α) and (α') so that $k=1$, $l=2$. Then

(2.19.4) $\quad n!\,2[\alpha]-n!\,2[\alpha']$

$$= n!\,2[\rho+\tau,\rho-\tau,\alpha_3,\ldots]-n!\,2[\rho+\sigma,\rho-\sigma,\alpha_3,\ldots]$$
$$= \Sigma!\,a_3^{\alpha_3}\ldots a_n^{\alpha_n}(a_1^{\rho+\tau}a_2^{\rho-\tau}+a_1^{\rho-\tau}a_2^{\rho+\tau}-a_1^{\rho+\sigma}a_2^{\rho-\sigma}-a_1^{\rho-\sigma}a_2^{\rho+\sigma})$$
$$= \Sigma!\,(a_1a_2)^{\rho-\tau}a_3^{\alpha_3}\ldots a_n^{\alpha_n}(a_1^{\tau+\sigma}-a_2^{\tau+\sigma})(a_1^{\tau-\sigma}-a_2^{\tau-\sigma}) \geqq 0,$$

with equality only when all the a are equal.

Proof of Lemma 2. We suppose that the condition (2.18.2) is satisfied, and call the number of the differences $\alpha_\nu-\alpha_\nu'$ which are not zero the *discrepancy* of (α) and (α'); if the discrepancy is zero the sets are identical. We prove the lemma by induction, assuming it to be true when the discrepancy is less than r and proving that it is then true when the discrepancy is r.

Suppose then that $(\alpha') \prec (\alpha)$ and that the discrepancy is $r>0$. Since, by (2.18.1), $\Sigma(\alpha_\nu-\alpha_\nu')=0$, and not all of these differences are zero, there must be positive and negative differences; and, by (2.18.3) the first which is not zero must be positive. We can therefore find k and l so that

(2.19.5) $\quad \alpha_k' < \alpha_k, \quad \alpha'_{k+1}=\alpha_{k+1}, \quad \ldots, \quad \alpha'_{l-1}=\alpha_{l-1}, \quad \alpha_l' > \alpha_l.$[a]

We take $\alpha_k=\rho+\tau$, $\alpha_l=\rho-\tau$, as in (2.19.1), and define σ by

(2.19.6) $$\sigma = \mathrm{Max}(|\alpha_k'-\rho|, |\alpha_l'-\rho|).$$

Then $0<\tau\leqq\rho$, since $\alpha_k>\alpha_l$. Also one or other[b] of

$$\alpha_l'-\rho=-\sigma, \quad \alpha_k'-\rho=\sigma,$$

[a] $\alpha_l-\alpha_l'$ is the first negative difference, $\alpha_k-\alpha_k'$ the last positive difference which precedes it. The text assumes $l-k>1$; the case $l-k=1$ is easier.

[b] Possibly both.

is true, since $\alpha_k' \geqq \alpha_l'$; and $\sigma < \tau$, since $\alpha_k' < \alpha_k$ and $\alpha_l' > \alpha_l$. Hence

$$0 \leqq \sigma < \tau \leqq \rho,$$

as in (2.19.2).

We now write

$$\alpha_k'' = \rho + \sigma, \quad \alpha_l'' = \rho - \sigma, \quad \alpha_\nu'' = \alpha_\nu \quad (\nu \neq k, \ \nu \neq l). \tag{2.19.7}$$

If $\alpha_k' - \rho = \sigma$, $\alpha_k'' = \alpha_k'$; if $\alpha_l' - \rho = -\sigma$, $\alpha_l'' = \alpha_l'$.[a] Since the pairs α_k, α_k' and α_l, α_l' each contribute a unit to the discrepancy r between (α') and (α), the discrepancy between (α') and (α'') is smaller, being $r-1$ or $r-2$.

Next, comparing (2.19.7) with (2.19.3), and observing that (2.19.2) is satisfied, we see that (α'') arises from (α) by a transformation T.

Finally, (α') is majorised by (α''). To prove this we must verify that the conditions corresponding to (2.18.1), (2.18.2) and (2.18.3), with α'' for α, are satisfied. For the first, we have

$$\alpha_k'' + \alpha_l'' = 2\rho = \alpha_k + \alpha_l, \quad \Sigma\alpha' = \Sigma\alpha = \Sigma\alpha''. \tag{2.19.8}$$

For the second, we observe first that

$$\alpha_k' \leqq \rho + |\alpha_k' - \rho| \leqq \rho + \sigma = \alpha_k'',$$
$$\alpha_l' \geqq \rho - |\alpha_l' - \rho| \geqq \rho - \sigma = \alpha_l''$$

and so, by (2.19.5),

$$\alpha''_{k-1} = \alpha_{k-1} \geqq \alpha_k = \rho + \tau > \rho + \sigma = \alpha''_k \geqq \alpha'_k \geqq \alpha'_{k+1} = \alpha_{k+1} = \alpha''_{k+1},$$
$$\alpha''_{l-1} = \alpha_{l-1} = \alpha'_{l-1} \geqq \alpha'_l \geqq \alpha''_l = \rho - \sigma > \rho - \tau = \alpha_l \geqq \alpha_{l+1} = \alpha''_{l+1};$$

and the inequalities affecting the α'' are those required. Finally, we have to prove that

$$\alpha_1' + \alpha_2' + \ldots + \alpha_\nu' \leqq \alpha_1'' + \alpha_2'' + \ldots + \alpha_\nu''.$$

Now this is true if $\nu < k$ or $\nu \geqq l$, by (2.19.7) and (2.18.3); it is true for $\nu = k$, because it is true for $\nu = k-1$ and $\alpha_k' \leqq \alpha_k''$; and it is true for $k < \nu < l$ because it is true for $\nu = k$ and the intervening α' and α'' are identical.

We have thus proved that (α') is majorised by (α''), a set arising from (α) by a transformation T and having a discrepancy from (α') less than r. This proves Lemma 2 and so completes the proof of Theorem 45.[b]

[a] Again, both these equations may be true.

[b] For another proof, see Theorems **74** and **75**.

2.20. An alternative theorem. We shall say that (α') *is an average of* (α) if there are n^2 numbers $p_{\mu\nu}$ such that

$$(2.20.1)\qquad p_{\mu\nu} \geqq 0, \quad \sum_{\mu=1}^{n} p_{\mu\nu} = 1, \quad \sum_{\nu=1}^{n} p_{\mu\nu} = 1$$

and

$$(2.20.2)\qquad \alpha_\mu' = p_{\mu 1}\alpha_1 + p_{\mu 2}\alpha_2 + \dots + p_{\mu n}\alpha_n.$$

Since the conditions (2.20.1) are not affected by a permutation of the μ or the ν, the definition is, like that of § 2.18, independent of the order of the α or the α'. The equations (2.19.3) show that $(\rho+\sigma, \rho-\sigma, \alpha_3, \dots)$ is an average of $(\rho+\tau, \rho-\tau, \alpha_3, \dots)$ when (2.19.2) is satisfied.

The last two conditions (2.20.1) may also be stated as follows: $\Sigma\alpha'$, when expressed as a function of the α, is identical with $\Sigma\alpha$, and every α' is 1 if every α is 1. From this it follows that the relationship is transitive; if (α') is an average of (α), and (α'') of (α'), then (α'') is an average of (α). And from this and Lemma 2 of § 2.19 it follows that *if* $(\alpha') \prec (\alpha)$ *then* (α') *is an average of* (α).

The converse is also true. For suppose that (2.20.1) and (2.20.2) are satisfied. Then (2.18.1) follows by addition of the equations (2.20.2). Finally, if we suppose (α) and (α') in descending order, and write

$$p_{1\nu} + p_{2\nu} + \dots + p_{m\nu} = k_\nu,$$

we have $k_\nu \leqq 1$ and $\Sigma k_\nu = m$, by (2.20.1); and so

$$\alpha_1' + \alpha_2' + \dots + \alpha_m' \leqq k_1\alpha_1 + \dots + k_{m-1}\alpha_{m-1} + (m - k_1 - \dots - k_{m-1})\alpha_m$$
$$\leqq (\alpha_1 - \alpha_m) + \dots + (\alpha_{m-1} - \alpha_m) + m\alpha_m = \alpha_1 + \alpha_2 + \dots + \alpha_m,$$

which is (2.18.3).

We have therefore proved the two following theorems.

46. *A necessary and sufficient condition that* (α') *should be an average of* (α) *is that* $(\alpha') \prec (\alpha)$.

47. *A necessary and sufficient condition that* $[\alpha']$ *should be comparable with* $[\alpha]$ *is that one of* (α') *and* (α) *should be an average of the other. If* (α') *is an average of* (α) *then* $[\alpha'] \leqq [\alpha]$, *with equality only as in Theorem* **45**.

2.21. Further theorems on symmetrical means. (1) Theorems **45** and **47** fulfil two purposes. First, either theorem gives a simple criterion for deciding whether two means $[\alpha]$ and $[\alpha']$ are or are not comparable. Secondly, the proof of Theorem **45** shows us how, by repeated application of the transformation (2.19.3) and the formula (2.19.4), to decompose the difference of two comparable means into a sum of obviously positive terms. We obtain, for example, a new and interesting proof of the theorem of the arithmetic and geometric means (with unit weights); in fact

$$\begin{aligned}\mathfrak{A}(a^n)-\mathfrak{G}(a^n) &= [n,0,0,\dots,0]-[1,1,\dots,1]\\ &= ([n,0,0,\dots,0]-[n-1,1,0,\dots,0])\\ &\quad +([n-1,1,0,\dots,0]-[n-2,1,1,0,\dots,0])\\ &\quad +([n-2,1,1,0,\dots,0]-[n-3,1,1,1,0,\dots,0])+\dots\\ &= \frac{1}{2(n!)}\{\Sigma!\,(a_1^{n-1}-a_2^{n-1})(a_1-a_2)+\Sigma!\,(a_1^{n-2}-a_2^{n-2})(a_1-a_2)a_3\\ &\quad +\Sigma!\,(a_1^{n-3}-a_2^{n-3})(a_1-a_2)a_3a_4+\dots\}.\end{aligned}$$

Since
$$(a_r-a_s)(a_r^\nu-a_s^\nu)>0$$
unless $a_r=a_s$, the theorem follows[a].

(2) **48.** *If* $\alpha_1+\alpha_2+\dots+\alpha_n=1$, *then*

$$\mathfrak{G}(a)<[\alpha]<\mathfrak{A}(a),$$

unless $[\alpha]$ *is* $\mathfrak{G}(a)$ *or* $\mathfrak{A}(a)$, *or all the* a *are equal.*

This theorem[b] shows that all the $[\alpha]$ of degree of homogeneity 1 are comparable with $\mathfrak{G}(a)$ and $\mathfrak{A}(a)$, though not in general comparable among themselves. To prove it we apply Theorem **47**; since

$$\frac{1}{n}=\frac{\alpha_1}{n}+\frac{\alpha_2}{n}+\dots+\frac{\alpha_n}{n}$$

and $\quad \alpha_\mu=\alpha_\mu.1+\alpha_{\mu+1}.0+\dots+\alpha_n.0+\alpha_1.0+\dots+\alpha_{\mu-1}.0,$

$(1/n,\ 1/n,\ \dots,\ 1/n)$ is an average of (α) and (α) an average of $(1,0,\dots,0)$. Or we may deduce Theorem **48** directly from Theorem **45**.

(3) We add two further theorems of a similar character, with indications only of the proofs.

49. *If* $0<\sigma\leqq 1$, *then a necessary and sufficient condition that* $[\alpha']\leqq[\alpha]^\sigma$ *is that* $(\alpha')\prec(\sigma\alpha)$. *If* $\sigma>1$, *the condition is necessary but not sufficient.*

[a] This proof was known before Muirhead's work; see Hurwitz (**1**).

[b] Communicated to us by Prof. I. Schur.

[To prove the condition necessary, follow the line of § 2.19 (1). To prove it sufficient, combine Theorems **45** and **11**. As an example

$$[r, 0, 0, \ldots] \leqq [s, 0, 0, \ldots]^{r/s} \quad (0 < r < s);$$

this is $\mathfrak{M}_r(a) \leqq \mathfrak{M}_s(a)$ (Theorem **16**), with unit weights. The same example shows that the condition is no longer sufficient when $\sigma > 1$.]

50. *If r, p, a are positive and*

$$T_r = \Sigma p_\nu a_\nu^r = \mathfrak{T}_r^r$$

(in the notation of § 2.10 (iv)), *then a necessary and sufficient condition that*

$$T_{\alpha_1'} T_{\alpha_2'} \ldots T_{\alpha_n'} \leqq T_{\alpha_1} T_{\alpha_2} \ldots T_{\alpha_n}$$

for all a and p is that $(\alpha') \prec (\alpha)$.

[The necessity of the condition may be established as before. To prove it sufficient we use Theorem **46** and Hölder's inequality, which give

$$T_{\alpha_\mu'} = T_{s_{\mu 1}\alpha_1 + s_{\mu 2}\alpha_2 + \cdots + s_{\mu m}\alpha_m}$$
$$\leqq (T_{\alpha_1})^{s_{\mu 1}} (T_{\alpha_2})^{s_{\mu 2}} \ldots (T_{\alpha_m})^{s_{\mu m}}:$$

we have changed the notation slightly in order to avoid conflict with that of § 2.10. The result follows by multiplication.]

2.22. The elementary symmetric functions of n positive numbers. If

$$(x+a_1)(x+a_2)\ldots(x+a_n) = x^n + c_1 x^{n-1} + c_2 x^{n-2} + \ldots + c_n$$
$$= x^n + \binom{n}{1} p_1 x^{n-1} + \binom{n}{2} p_2 x^{n-2} + \ldots + p_n,$$

then c_r is the rth elementary symmetric function of the a, i.e. the sum of the products, r at a time, of different a, and p_r the average of these products. In this section we consider two well-known theorems concerning the p_r. We write $c_0 = p_0 = 1$.

In the notation of § 2.18

$$c_r = \frac{1}{r!\,(n-r)!} \Sigma !\, a_1 a_2 \ldots a_r,$$

$$p_r = \frac{r!\,(n-r)!}{n!} c_r = [1, 1, \ldots, 1, 0, 0, \ldots, 0],$$

there being r 1's and $n-r$ 0's. Also $p_1 = \mathfrak{A}(a)$ and $p_n = \mathfrak{G}^n(a)$, with unit weights. The different p_r, being of different degrees, are not comparable[a]; but they are connected by non-linear inequalities.

[a] This is a trivial case of Theorem **45**.

51. $p_{r-1}p_{r+1} < p_r^2$ $(1 \leqq r < n)$, *unless all the a are equal.*

52. $p_1 > p_2^{\frac{1}{2}} > p_3^{\frac{1}{3}} > \ldots > p_n^{1/n}$, *unless all the a are equal.*

Theorem **51**, which was stated by Newton[a], is actually true for real, not necessarily positive, a; and we shall give a proof of the more general theorem, depending on the methods of the differential calculus, in § 4.3. Theorem **52** is due to Maclaurin[b].

Theorem **52** is a corollary of Theorem **51**, since

$$(p_0 p_2)(p_1 p_3)^2 (p_2 p_4)^3 \ldots (p_{r-1} p_{r+1})^r < p_1^2 p_2^4 p_3^6 \ldots p_r^{2r}$$

gives $p_{r+1}^r < p_r^{r+1}$ or

$$p_r^{1/r} > p_{r+1}^{1/(r+1)}.$$

This remark, together with the proof of § 4.3, disposes of the theorems, but it is interesting to consider proofs of them by the methods of this chapter.

(i) *Proof of Theorem* **52** *by the method of* § 2.6 (iii). We begin by proving a theorem similar to but weaker than Theorem **51**.

53.[c] $c_{r-1}c_{r+1} < c_r^2$.

This theorem is weaker than Theorem **51**, since $p_{r-1}p_{r+1} < p_r^2$ is

$$\frac{(r+1)(n-r+1)}{r(n-r)} c_{r-1}c_{r+1} < c_r^2.$$

To prove it we observe that a typical term in $c_{r-1}c_{r+1} - c_r^2$ is

$$a_1^2 a_2^2 \ldots a_{r-s}^2 a_{r-s+1} \ldots a_{r+s}$$

and that this occurs with the coefficient

$$\binom{2s}{s-1} - \binom{2s}{s} < 0.$$

From Theorem **53** it follows that

$$(2.22.1) \qquad c_{r-1}c_s < c_r c_{s-1}$$

if $r < s$.

[a] Newton (**1**, 173). See also Maclaurin (**2**).

[b] Maclaurin (**2**). See also Schlömilch (**1**). The inequality $p_1 > p_n^{1/n}$ is a case of Theorem **9**.

[c] The theorem is stated, like Theorems **51** and **52**, for positive a. It remains true, as the proof shows, for non-negative a, unless $c_r = 0$ (i.e. unless all but $r-1$ of the a are 0).

We can now prove Theorem **52** as follows. If the a are not all equal, let $a_1 = \operatorname{Min} a$, $a_2 = \operatorname{Max} a$. Then

$$a_1 < \alpha_1 < a_2, \tag{2.22.2}$$

where

$$\alpha_1 = p_\mu^{1/\mu}.$$

We replace a_1 and a_2 by α_1 and α_2, choosing α_2 so that p_μ shall be unaltered, and prove that any p_ν for which $\nu > \mu$ is increased by the substitution. The result will then follow as in § 2.6 (iii).

We have

$$\binom{n}{\mu} p_\mu = c_\mu = a_1 a_2 c'_{\mu-2} + (a_1 + a_2) c'_{\mu-1} + c'_\mu,$$

where c_r' is the c_r formed from the $n-2$ numbers other than a_1 and a_2. Since p_μ is to be unchanged

$$a_1 a_2 c'_{\mu-2} + (a_1 + a_2) c'_{\mu-1} + c'_\mu = \alpha_1 \alpha_2 c'_{\mu-2} + (\alpha_1 + \alpha_2) c'_{\mu-1} + c'_\mu,$$

$$(\alpha_1 \alpha_2 - a_1 a_2) c'_{\mu-2} = -(\alpha_1 + \alpha_2 - a_1 - a_2) c'_{\mu-1}, \tag{2.22.3}$$

$$(\alpha_1 c'_{\mu-2} + c'_{\mu-1}) \alpha_2 = a_1 a_2 c'_{\mu-2} + (a_1 + a_2 - \alpha_1) c'_{\mu-1}. \tag{2.22.4}$$

The value of α_2 defined by (2.22.4) is positive because of (2.22.2).

Also, if p_ν becomes p_ν^*,

$$\binom{n}{\nu} (p_\nu^* - p_\nu) = (\alpha_1 \alpha_2 - a_1 a_2) c'_{\nu-2} + (\alpha_1 + \alpha_2 - a_1 - a_2) c'_{\nu-1};$$

and so $p_\nu^* - p_\nu$ has the sign of

$$(\alpha_1 + \alpha_2 - a_1 - a_2) \left(\frac{c'_{\nu-1}}{c'_{\nu-2}} - \frac{c'_{\mu-1}}{c'_{\mu-2}} \right).$$

The second factor is negative, by (2.22.1); and, by (2.22.3),

$$\begin{aligned} \operatorname{sgn}(\alpha_1 + \alpha_2 - a_1 - a_2) &= \operatorname{sgn}(a_1 a_2 - \alpha_1 \alpha_2) \\ &= \operatorname{sgn}\{\alpha_1(\alpha_1 + \alpha_2 - a_1 - a_2) + a_1 a_2 - \alpha_1 \alpha_2\} \\ &= \operatorname{sgn}\{(\alpha_1 - a_1)(\alpha_1 - a_2)\} = -1, \end{aligned}$$

by (2.22.2). Hence $p_\nu^* > p_\nu$, which proves the theorem.

(ii) *Proof of Theorem* **51** *by induction*[a]. Suppose that Theorem **51** has been proved for $n-1$ numbers, $a_1, a_2, \ldots, a_{n-1}$, and that

[a] This proof was communicated to us independently by Messrs A. L. Dixon, A. E. Jolliffe, and M. H. A. Newman.

c'_r, p'_r are the c_r, p_r formed from these $n-1$ numbers, which we shall at first suppose not all equal. Then

$$c_r = c'_r + a_n c'_{r-1}$$

and so
$$p_r = \frac{n-r}{n} p'_r + \frac{r}{n} a_n p'_{r-1}.$$

Hence we deduce

$$n^2 (p_{r-1} p_{r+1} - p_r^2) = A + Ba_n + Ca_n^2,$$

where
$$A = \{(n-r)^2 - 1\} p'_{r-1} p'_{r+1} - (n-r)^2 p'^2_r,$$

$$B = (n-r+1)(r+1) p'_{r-1} p'_r + (n-r-1)(r-1) p'_{r-2} p'_{r+1} - 2r(n-r) p'_{r-1} p'_r,$$

$$C = (r^2-1) p'_{r-2} p'_r - r^2 p'^2_{r-1}.$$

Since $a_1, a_2, \ldots, a_{n-1}$ are not all equal, we have, by the inductive hypothesis,

$$p'_{r-1} p'_{r+1} < p'^2_r, \quad p'_{r-2} p_r < p'^2_{r-1}, \quad p'_{r-2} p'_{r+1} < p'_{r-1} p'_r,$$

so that
$$A < -p'^2_r, \quad B < 2p'_{r-1} p'_r, \quad C < -p'^2_{r-1},$$

and
$$n^2 (p_{r-1} p_{r+1} - p_r^2) < -(p'_r - a_n p'_{r-1})^2 \leqq 0.$$

This proves the theorem. The result is still true when

$$a_1 = a_2 = \ldots = a_{n-1},$$

because then $a_n \neq a_1 = p'_r / p'_{r-1}$.

It is also possible to prove Theorem **51** by means of identities of the type considered in § 2.21 (1).

54.
$$p_r^2 - p_{r-1} p_{r+1} = \frac{1}{r(r+1)\binom{n}{r}\binom{n}{r+1}} \sum_{i=0}^{r-1} \binom{2i}{i} \frac{(r,i)}{i+1},$$

where
$$(r,i) = \Sigma a_1^2 \ldots a^2_{r-i-1} a_{r-i} \ldots a_{r+i-1} (a_{r+i} - a_{r+i+1})^2,$$

the summation extending over all products formed from the a and of the type shown.

55.
$$\left\{\frac{n!}{(r-1)!\,(n-r-1)!}\right\}^2 (p_r^2 - p_{r-1} p_{r+1}) = (n-1) \Sigma (a_1 - a_2)^2 (c^{n-2}_{r-1})^2$$
$$+ \frac{2!\,(n-3)}{(r-1)(n-r-1)} \Sigma (a_1-a_2)^2 (a_3-a_4)^2 (c^{n-4}_{r-2})^2$$
$$+ \frac{3!\,(n-5)}{(r-1)(r-2)(n-r-1)(n-r-2)} \Sigma (a_1-a_2)^2 (a_3-a_4)^2 (a_5-a_6)^2 (c^{n-6}_{r-3})^2$$
$$+ \ldots,$$

where c^{n-2}_{r-1} *is the sum of the products,* $r-1$ *at a time, of the* $n-2$ *a other than* a_1, a_2, *and so on.*

Theorem **54** is due to Muirhead (**1**), and Theorem **55** to Joliffe (**1**). Theorem 55 gives an 'intuitive' proof of the more general form of Theorem **51** (for real a of either sign) referred to at the beginning of this section.

2.23. A note on definite forms. The identity of Hurwitz and Muirhead proved in § 2.21 (1) shows that, when $a > 0$,

$$a_1{}^n + a_2{}^n + \ldots + a_n{}^n - na_1 a_2 \ldots a_n$$

can be expressed as a sum in which every term is obviously non-negative.

If we write $a_1 = x_1{}^2$, $a_2 = x_2{}^2$, ..., we obtain

$$(2.23.1)\quad x_1{}^{2n} + \ldots + x_n{}^{2n} - nx_1{}^2 x_2{}^2 \ldots x_n{}^2$$
$$= \frac{1}{2(n-1)!}\{\Sigma!\,(x_1{}^{2n-2} - x_2{}^{2n-2})(x_1{}^2 - x_2{}^2) + \ldots\}.$$

Now

$$(x_1{}^{2n-2} - x_2{}^{2n-2})(x_1{}^2 - x_2{}^2)$$
$$= (x_1{}^2 - x_2{}^2)^2 (x_1{}^{2n-4} + x_1{}^{2n-6} x_2{}^2 + \ldots + x_2{}^{2n-4})$$

is a sum of squares of polynomials such as $(x_1{}^2 - x_2{}^2)\, x_1{}^{n-2}$; and so the right-hand side of (2.23.1) is a sum of squares. Finally, since

$$x_1^{2n} + \ldots + x_{2n}^{2n} - 2nx_1 x_2 \ldots x_{2n} = x_1^{2n} + \ldots + x_n^{2n} - nx_1{}^2 \ldots x_n{}^2$$
$$+ x_{n+1}^{2n} + \ldots + x_{2n}^{2n} - nx_{n+1}^2 \ldots x_{2n}^2 + n\,(x_1 \ldots x_n - x_{n+1} \ldots x_{2n})^2,$$

it follows that

$$(2.23.2)\qquad F = x_1{}^{2n} + \ldots + x_{2n}{}^{2n} - 2nx_1 x_2 \ldots x_{2n} = \sum_i P_i{}^2,$$

where the P_i are real polynomials of degree n. For example

$$x^6 + y^6 + z^6 + u^6 + v^6 + w^6 - 6xyzuvw$$
$$= \tfrac{1}{2}(x^2 + y^2 + z^2)\{(y^2 - z^2)^2 + (z^2 - x^2)^2 + (x^2 - y^2)^2\}$$
$$+ \tfrac{1}{2}(u^2 + v^2 + w^2)\{(v^2 - w^2)^2 + (w^2 - u^2)^2 + (u^2 - v^2)^2\} + 3\,(xyz - uvw)^2$$

is a sum of $9 + 9 + 1 = 19$ squares of real polynomials.

A *real form* is a homogeneous polynomial $F(x_1, x_2, \ldots, x_m)$, with real coefficients, in the m real variables $x_1, x_2, \ldots, x_m$. A form F is said to be *definite*, in a certain region of the variables, if it does not change sign in that region, for example if $F \geqq 0$. We may divide definite forms into positive and negative forms, and it is plainly sufficient to consider *positive* forms. Thus the form (2.23.2) is positive in the region of all real values of the variables. It is plain that a form which has this property must be of even degree.

If $F > 0$ in a certain region, then F is said to be *strictly positive* in that region.

The form (2.23.2), and the forms considered in Theorems **7** and **55** (with x for a), can be expressed as sums of squares of real polynomials, and it is natural to ask whether this is a general property of definite forms. *Is it true that, if $F \geqq 0$ for all real x, then*

$$F = \sum_i P_i^2,$$

where the P_i are real polynomials?

This problem was solved completely by Hilbert[a]. Here we have space only for a few fragmentary remarks. We begin by observing that there are two cases in which the answer is immediate. We denote the degree of F by $2n$, and the number of variables by m.

If $m = 2$, so that $F = F(x, y)$ and n is arbitrary, then any real factor $ax + by$ of F must occur in even multiplicity, and the complex factors must occur in conjugate pairs $ax + by$, $\bar{a}x + \bar{b}y$. Hence, grouping the factors appropriately, we obtain

$$F = p^2(q + ir)(q - ir) = (pq)^2 + (pr)^2,$$

where p, q, r are real polynomials.

It is a familiar theorem of algebra[b] that any definite quadratic form in any number m of variables may be expressed as a sum of at most m squares of real linear forms. Thus the answer is affirmative in the two cases

(1) $m = 2$, n arbitrary,

(2) m arbitrary, $2n = 2$.

Hilbert found a third case

(3) $m = 3$, $2n = 4$,

and proved that any positive biquadratic form in three variables is representable as the sum of three squares of real quadratics. He also proved that in all other cases the answer is negative, there being definite forms of degree $2n$ in m variables which cannot be represented in the manner proposed.

[a] Hilbert (**1**). [b] See, for example, Bôcher (**1**, 144–154).

Hilbert then suggested the following theorem: *any positive F can be expressed as*

$$F = \sum_i R_i^2,$$

where R_i is a real ***rational*** *function.* An equivalent theorem is: *any positive F may be expressed as a quotient of sums of squares of real forms*[a].

Hilbert[b] gave a very difficult proof of these theorems for ternary forms in (x, y, z). The general theorems were first proved by Artin[c]. Artin's proof is very remarkable and comparatively simple, but depends upon the ideas of modern abstract algebra in a manner which makes it impossible for us to reproduce it here.

2.24. A theorem concerning strictly positive forms. The rather fragmentary remarks of § 2.23 form a natural introduction to the simpler problem which we consider here. We are concerned now with forms which are *strictly* positive in the region of *positive x*. The theorem which we shall prove resembles those of § 2.23 in asserting that a positive form can be represented in a manner which renders its positive character intuitive. It is no longer necessary that the degree of the form should be even.

56.[d] *If the form $F(x_1, x_2, \ldots, x_m)$ is strictly positive for*

$$x \geqq 0, \quad \Sigma x > 0,$$

then F may be expressed as

$$F = \frac{G}{H},$$

where G and H are forms with positive coefficients. In particular we may suppose that

$$H = (x_1 + x_2 + \ldots + x_m)^p$$

for a suitable p.

[a] It is evident that the first theorem implies the second (with one square only in the denominator). And since

$$\frac{\Sigma g_i^2}{\Sigma h_j^2} = \sum_{i,j} \left(\frac{g_i h_j}{\Sigma h_j^2}\right)^2,$$

the second theorem implies the first.

[b] Hilbert (2).

[c] Artin (1).

[d] Pólya (3). The theorem had been proved before (apart from the last clause) by Poincaré (1) when $m=2$ and by Meissner (1) when $m=3$. Meissner's method is applicable in principle in the general case, but does not lead to so simple a result.

For simplicity of writing we suppose $m=3$; no new point of principle arises for general m.

The function $F(x, y, z)$ is positive and continuous in the closed region

(2.24.1) $$x \geqq 0, \quad y \geqq 0, \quad z \geqq 0, \quad x+y+z=1,$$

and has a positive minimum μ in that region. We write

(2.24.2) $$F(x,y,z)=\Sigma_n A_{\alpha\beta\gamma}\frac{x^\alpha y^\beta z^\gamma}{\alpha!\,\beta!\,\gamma!},$$

the summation being over

(2.24.3) $$\alpha \geqq 0, \quad \beta \geqq 0, \quad \gamma \geqq 0, \quad \alpha+\beta+\gamma=n;$$

and

(2.24.4) $$\phi(x,y,z;t)=t^n\,\Sigma_n A_{\alpha\beta\gamma}\binom{xt^{-1}}{\alpha}\binom{yt^{-1}}{\beta}\binom{zt^{-1}}{\gamma},$$

where $t>0$ and $\binom{xt^{-1}}{\alpha}$, ... are the usual binomial coefficients, so that

$$\binom{xt^{-1}}{0}=1$$

and $$t^\alpha\binom{xt^{-1}}{\alpha}=\frac{x(x-t)(x-2t)\ldots\{x-(\alpha-1)t\}}{1.2.3\ldots\alpha}$$

for $\alpha=1, 2, 3, \ldots$.

It is plain that $\phi(x,y,z;t)\to F(x,y,z)$ when $t\to 0$; and if we write

$$\phi(x,y,z;0)=F(x,y,z),$$

then ϕ is continuous in

$$x \geqq 0, \quad y \geqq 0, \quad z \geqq 0, \quad x+y+z=1, \quad 0 \leqq t \leqq 1.$$

There is therefore an ϵ such that

(2.24.5)
$$\phi(x,y,z;t)>\phi(x,y,z;0)-\tfrac{1}{2}\mu=F(x,y,z)-\tfrac{1}{2}\mu \geqq \tfrac{1}{2}\mu>0$$

for $0<t<\epsilon$ and all x, y, z in (2.24.1).

We have also

(2.24.6) $$(x+y+z)^{k-n}=(k-n)!\,\Sigma_{k-n}\frac{x^\kappa y^\lambda z^\mu}{\kappa!\,\lambda!\,\mu!},$$

the summation being over

$$\kappa \geqq 0, \quad \lambda \geqq 0, \quad \mu \geqq 0, \quad \kappa+\lambda+\mu=k-n.$$

Multiplying (2.24.2) and (2.24.6), we obtain

$$(x+y+z)^{k-n}F=(k-n)!\,\Sigma_n\,\Sigma_{k-n}\,A_{\alpha\beta\gamma}\frac{x^{\alpha+\kappa}y^{\beta+\lambda}z^{\gamma+\mu}}{\alpha!\,\kappa!\,\beta!\,\lambda!\,\gamma!\,\mu!}.$$

Here we write

$$\alpha+\kappa=a,\quad \beta+\lambda=b,\quad \gamma+\mu=c,$$

so that a, b, c vary over

(2.24.7) $$a\geqq 0,\quad b\geqq 0,\quad c\geqq 0,\quad a+b+c=k,$$

and α, β, γ over

(2.24.8) $$0\leqq\alpha\leqq a,\quad 0\leqq\beta\leqq b,\quad 0\leqq\gamma\leqq c,\quad \alpha+\beta+\gamma=n.$$

This gives

(2.24.9)

$$(x+y+z)^{k-n}F=(k-n)!\,\Sigma_k\frac{x^a y^b z^c}{a!\,b!\,c!}\,\Sigma'\,A_{\alpha\beta\gamma}\binom{a}{\alpha}\binom{b}{\beta}\binom{c}{\gamma}.$$

In (2.24.9), Σ' implies summation with respect to α, β, γ over (2.24.8); but, since $\binom{a}{\alpha}=0$, $\binom{b}{\beta}=0$, ... if $\alpha>a$, $\beta>b$, ..., we may replace this summation by summation over (2.24.3), i.e. by Σ_n. We thus obtain

(2.24.10)

$$(x+y+z)^{k-n}F=(k-n)!\,\Sigma_k\frac{x^a y^b z^c}{a!\,b!\,c!}\,\Sigma_n\,A_{\alpha\beta\gamma}\binom{a}{\alpha}\binom{b}{\beta}\binom{c}{\gamma}$$

$$=(k-n)!\,k^n\,\Sigma_k\,\phi\left(\frac{a}{k},\frac{b}{k},\frac{c}{k};\frac{1}{k}\right)\frac{x^a y^b z^c}{a!\,b!\,c!}.$$

The ϕ here is positive, by (2.24.5), if k is sufficiently large, and this proves the theorem.

(1) The theorem gives a systematic process for deciding whether a given form F is strictly positive for positive x. We multiply repeatedly by Σx, and, if the form is positive, we shall sooner or later obtain a form with positive coefficients.

It is instructive to consider the working of the process for

$$F=x_1{}^n+x_2{}^n+\dots+x_n{}^n-(n-\epsilon)\,x_1x_2\dots x_n,$$

where ϵ is positive and small. The coefficient of

$$x_1{}^{i_1}x_2{}^{i_2}\dots x_n{}^{i_n},$$

where $i_1+i_2+\dots+i_n=n(q+1)$, in

$$\phi=(x_1+\dots+x_n)^{nq}F$$

is certainly positive if one of the i is 0. If $i_1 \geqq 1, \ldots, i_n \geqq 1$ it is

$$\frac{(nq)!}{(i_1-n)!\,i_2!\,i_3!\ldots}+\frac{(nq)!}{i_1!\,(i_2-n)!\,i_3!\ldots}+\ldots-(n-\epsilon)\frac{(nq)!}{(i_1-1)!\,(i_2-1)!\ldots},$$

and has the sign of

$$\psi(i_1, i_2, \ldots, i_n)=i_1(i_1-1)\ldots(i_1-n+1)$$
$$+i_2(i_2-1)\ldots(i_2-n+1)+\ldots-(n-\epsilon)\,i_1 i_2\ldots i_n.$$

We require ψ to be positive for all $i_1, i_2, \ldots, i_n$.

If not all of $i_1, i_2, \ldots$ are equal to $q+1$, there is one, say i_1, less than $q+1$, and one, say i_2, greater than $q+1$; and changing i_1, i_2 into i_1+1, i_2-1 changes ψ by

$$n\{i_1(i_1-1)\ldots(i_1-n+2)-(i_2-1)(i_2-2)\ldots(i_2-n+1)\}$$
$$-(n-\epsilon)\,i_3\ldots i_n(i_2-i_1-1)<0.$$

Hence ψ will be positive for all i if it is positive when every i is $q+1$. It will be positive in this case if

$$n(q+1)q(q-1)\ldots(q-n+2)>(n-\epsilon)(q+1)^n$$

or

$$\left(1-\frac{1}{q+1}\right)\left(1-\frac{2}{q+1}\right)\ldots\left(1-\frac{n-1}{q+1}\right)>1-\frac{\epsilon}{n},$$

and *a fortiori*[a] if

$$q+1>\frac{n^2(n-1)}{2\epsilon}.$$

If this condition is satisfied, all the coefficients in ϕ are positive.

It follows that $F>0$ for $x>0$, $\Sigma x>0$. Making $\epsilon\to 0$, we obtain yet another proof of the theorem of the means, in the form $\Sigma x^n \geqq n\Pi x$.

(2) If we write

$$x_m=1-x_1-\ldots-x_{m-1}$$

we obtain a theorem concerning general non-homogeneous polynomials in $m-1$ variables.

57. *If a (non-homogeneous) polynomial $f(x_1, x_2, \ldots, x_{m-1})$ is positive in the region*

$$x_1 \geqq 0,\ \ldots,\ x_{m-1}\geqq 0,\quad x_1+x_2+\ldots+x_{m-1}\leqq 1,$$

then $f(x)$ can be expressed in the form

$$f(x)=\Sigma c x_1^{a_1}\ldots x_{m-1}^{a_{m-1}}(1-x_1-\ldots-x_{m-1})^{a_m},$$

where the a are non-negative integers and the c are positive.

The theorem is a generalisation of one due to Hausdorff[b].

MISCELLANEOUS THEOREMS AND EXAMPLES[c]

58. If $\alpha, \beta, \gamma, \ldots, \lambda$ are greater than -1, and are all positive or all negative, then

$$(1+\alpha)(1+\beta)\ldots(1+\lambda)>1+\alpha+\beta+\ldots+\lambda.$$

[For the case $\alpha=\beta=\ldots=\lambda$, see James Bernoulli (**1**, **5**, 112).]

[a] See Theorem 58. [b] Hausdorff (**1**). Hausdorff has $m=2$.

[c] Some of the theorems which follow here are mere exercises for the reader, but most have some independent interest.

59. If $c>0$ then

$$|a+b|^2 \leqq (1+c)|a|^2 + \left(1+\frac{1}{c}\right)|b|^2$$

for all (real or complex) a, b.

[See Bohr (**1**, 78).]

60. If $\mathfrak{A}_n$, $\mathfrak{G}_n$ are the arithmetic and geometric means of $a_1, a_2, \ldots, a_n$, with unit weights, and $\mathfrak{A}_{n+1}$, $\mathfrak{G}_{n+1}$ those of $a_1, a_2, \ldots, a_n, a_{n+1}$, then

$$n(\mathfrak{A}_n - \mathfrak{G}_n) < (n+1)(\mathfrak{A}_{n+1} - \mathfrak{G}_{n+1})$$

unless $$a_{n+1} = \mathfrak{G}_n.$$

[This theorem, communicated to us by Dr R. Rado, embodies another proof of the theorem of the means. If we write $a_{n+1} = x^{n+1}$, $\mathfrak{G}_n = y^{n+1}$, then the inequality to be proved is

$$x^{n+1} - (n+1)xy^n + ny^{n+1} > 0$$

and this follows from Theorem **41**.]

61. $$ab \leqq \frac{a^r}{r} + \frac{b^{r'}}{r'} \quad (a>0,\ b>0,\ r>1),$$

with equality when $b = a^{r-1}$.

[Another form of Theorem **37**. For this and the next two theorems see Young (**1**, **5**, **6**).]

62. $$uv \leqq u\frac{u^p - 1}{p} + \left(\frac{1+pv}{1+p}\right)^{(1+p)/p} \quad (u>0,\ v>-1/p,\ p>0).$$

[Replace r in Theorem **61** by $1+p$ and a, b by u, $(1+pv)/(1+p)$.]

63. $$uv \leqq u \log u + e^{v-1} \quad (u>0).$$

[Make $p \to 0$ in Theorem **62**. See also § 4.4(5).]

64. If $a>0$, $a_1 a_2 \ldots a_n = l^n$, then

$$(1+a_1)(1+a_2)\ldots(1+a_n) > (1+l)^n,$$

unless all the a are equal.

[Chrystal (**1**, 51). Example of Theorem **40**.]

65. If a and b are positive and $p>1$ or $p<0$, then

$$\Sigma \frac{a^p}{b^{p-1}} > \frac{(\Sigma a)^p}{(\Sigma b)^{p-1}},$$

unless (a) and (b) are proportional. The inequality is reversed if $0<p<1$.

[Radon (**1**, 1351): transformation of Theorem **13**.]

66. If $a>0$ then $\Sigma a\, \Sigma a^{-1} > n^2$, unless all the a are equal.

[From Theorems **7** or **9** or **16** or **43**.]

67. $$\Sigma(a+b)\,\Sigma\frac{ab}{a+b} < \Sigma a\, \Sigma b,$$

unless (a) and (b) are proportional.

[Milne (**1**).]

68. $$\Sigma(a+b+c)\,\Sigma\frac{bc+ca+ab}{a+b+c}\,\Sigma\frac{abc}{bc+ca+ab}<\Sigma a\,\Sigma b\,\Sigma c,$$

unless (a), (b), (c) are proportional.

69. If $0<r<s$ and

$$M_r(a,b)=\left(\frac{a^r+b^r}{2}\right)^{1/r},$$

then $$\Sigma M_r(a,b)\,\Sigma M_{-r}(a,b)<\Sigma M_s(a,b)\,\Sigma M_{-s}(a,b)<\Sigma a\,\Sigma b,$$

unless (a) and (b) are proportional.

70. If $0<k<1$ and

$$\Sigma ab \geqq A\,(\Sigma b^{k'})^{1/k'}$$

for all b, then $\Sigma a^k \geqq A^k$.

[This is an analogue of Theorem **15** for the case $0<k<1$ (when $k'<0$). If all a are positive, define b by

$$ab=a^k,\quad b=a^{k-1},\quad b^{k'}=a^k,$$

when

(i) $$\Sigma a^k \geqq A\,(\Sigma a^k)^{1/k'},\quad \Sigma a^k \geqq A^k.$$

If all a vanish, A must be 0 and there is nothing to prove. If some but not all vanish, suppose that $a>0$ in a set E of μ members and $a=0$ in the complementary set CE of $\nu=n-\mu$ members; and define b as above in E, and by $b=G$ in CE. Then

$$\Sigma a^k=\Sigma ab \geqq A\,(\sum_E a^k+\nu G^{k'})^{1/k'}=A\,(\Sigma a^k+\nu G^{k'})^{1/k'}.$$

Making $G\to\infty$ we obtain (i) again.]

71. If $0<h\leqq a_\nu\leqq H$, $0<k\leqq b_\nu\leqq K$, then

$$1\leqq\frac{\Sigma a^2\,\Sigma b^2}{(\Sigma ab)^2}\leqq\tfrac{1}{4}\left\{\sqrt{\left(\frac{HK}{hk}\right)}+\sqrt{\left(\frac{hk}{HK}\right)}\right\}^2.$$

[See Pólya and Szegö (**1**, I, 57, 213), where the conditions for equality are given.]

72. $$\lim_{r\to\infty}\frac{\mathfrak{M}_{r+1}^{r+1}(a)}{\mathfrak{M}_r^r(a)}=\mathfrak{M}_\infty(a).$$

If all the a are positive, there is a similar theorem for $-\infty$.

73. If $$\frac{\mathfrak{A}(a)-\mathfrak{G}(a)}{\mathfrak{A}(a)}\leqq\epsilon<1,$$

the means being formed with unit weights, then

$$1+\xi<\frac{a_\nu}{\mathfrak{A}}<1+\xi',$$

where ξ is the negative and ξ' the positive root of the equation

$$(1+x)\,e^{-x}=(1-\epsilon)^n.$$

[See Pólya and Szegö (**1**, I, 58, 215).]

74. If $[\gamma_1', \ldots, \gamma_k'] \leqq [\gamma_1, \ldots, \gamma_k]$, $[\delta_1', \ldots, \delta_l'] \leqq [\delta_1, \ldots, \delta_l]$
for all values of the variables involved, then

$$[\gamma_1', \ldots, \gamma_k', \delta_1', \ldots, \delta_l'] \leqq [\gamma_1, \ldots, \gamma_k, \delta_1, \ldots, \delta_l].$$

[It follows from the first hypothesis, and the definition of the sums, that

$$[\gamma_1', \ldots, \gamma_k', \delta_1] \leqq [\gamma_1, \ldots, \gamma_k, \delta_1];$$

and hence, by repetition of the argument, that

$$\text{(i)} \qquad [\gamma_1', \ldots, \gamma_k', \delta_1, \ldots, \delta_l] \leqq [\gamma_1, \ldots, \gamma_k, \delta_1, \ldots, \delta_l].$$

Similarly, using the second hypothesis,

$$\text{(ii)} \qquad [\gamma_1', \ldots, \gamma_k', \delta_1', \ldots, \delta_l'] \leqq [\gamma_1', \ldots, \gamma_k', \delta_1, \ldots, \delta_l].$$

The result follows from (i) and (ii).]

75. If $(\alpha') \prec (\alpha)$ and the α and α' are in decreasing order, then there is a greatest non-negative δ for which

$$\text{(i)} \qquad (\beta) = (\alpha_1' + \delta, \alpha_2', \ldots, \alpha'_{n-1}, \alpha_n' - \delta) \prec (\alpha).$$

If δ has this value then

$$\text{(ii)} \qquad (\alpha_1', \alpha_n') \prec (\beta_1, \beta_n)$$

and

$$\text{(iii)} \qquad (\beta_1, \ldots, \beta_k) \prec (\alpha_1, \ldots, \alpha_k), \ (\beta_{k+1}, \ldots, \beta_n) \prec (\alpha_{k+1}, \ldots, \alpha_n)$$

for some k between 1 and $n-1$ inclusive.

[It is plain from the definitions (*a*) that (i) is true for $\delta = 0$, (*b*) that the set of δ for which it is true is closed, (*c*) that, if it is true for a positive δ, then it is true for any smaller positive δ. Hence there is a maximum non-negative δ for which (i) and (ii) are true.

If δ has this value then either (*a*) $\beta_n = \alpha_n' - \delta = 0$, or

$$(b) \ \beta_1 + \ldots + \beta_k = \alpha_1' + \ldots + \alpha_k' + \delta = \alpha_1 + \ldots + \alpha_k$$

for a $k < n$; for otherwise we could increase δ without disturbing (i). In case (*a*)

$$\sum_1^{n-1} \alpha_\nu \leqq \sum_1^n \alpha_\nu = \sum_1^n \beta_\nu = \sum_1^{n-1} \beta_\nu \leqq \sum_1^{n-1} \alpha_\nu$$

and so $\beta_1 + \ldots + \beta_{n-1} = \alpha_1 + \ldots + \alpha_{n-1}$, which is (*b*) with $k = n-1$. Hence (*b*) is true in any case; and then (iii) follows from the definitions.

Dr R. Rado, who communicated Theorems **74** and **75** to us, uses them to obtain a new and elegant proof of the sufficiency of Muirhead's criterion (Theorem **45**). The result is true for $n = 2$ by Lemma 1 of p. 47; let us then suppose that $n > 2$, that the conditions (2.18.1), (2.18.2), and (2.18.3) are satisfied, and that the result is true for any number of variables less than n. Then, by the inductive hypothesis,

$$[\alpha_1', \alpha_n'] \leqq [\beta_1, \beta_n], \quad [\beta_1, \ldots, \beta_k] \leqq [\alpha_1, \ldots, \alpha_k], \quad [\beta_{k+1}, \ldots, \beta_n] \leqq [\alpha_{k+1}, \ldots, \alpha_n].$$

Hence, using Theorem **74** twice, we obtain

$$\begin{aligned} [\alpha_1', \ldots, \alpha_n'] &= [\alpha_1', \beta_2, \ldots, \beta_{n-1}, \alpha_n'] \leqq [\beta_1, \ldots, \beta_n] \\ &= [\beta_1, \ldots, \beta_k, \beta_{k+1}, \ldots \beta_n] \leqq [\alpha_1, \ldots, \alpha_k, \alpha_{k+1}, \ldots \alpha_n] = [\alpha_1, \ldots, \alpha_n].] \end{aligned}$$

76. If $a>0$, and r and s are positive integers, then

$$\Sigma a_1{}^r a_2{}^r \dots a_s{}^r \Sigma \frac{1}{a_1{}^s a_2{}^s \dots a_r{}^s} > \binom{n}{r}\binom{n}{s},$$

unless all the a are equal.

[Generalisation of Theorem **66**. By Theorem **45**,

$$[r, r, \dots (s \text{ times}), 0, 0, \dots, 0] > \left[\frac{rs}{n}, \frac{rs}{n}, \dots, \frac{rs}{n}\right].$$

Form the corresponding inequality with r and s interchanged and a replaced by $1/a$, and multiply.]

77. A necessary and sufficient condition that

$$p_1{}^{\alpha_1'} p_2{}^{\alpha_2'} \dots p_n{}^{\alpha_n'} \leqq p_1{}^{\alpha_1} p_2{}^{\alpha_2} \dots p_n{}^{\alpha_n},$$

where $p_1, \dots$ are defined as in § 2.22, and the α are positive, for all positive a, is that

$$\alpha_m' + 2\alpha'_{m+1} + \dots + (n-m+1)\,\alpha_n' \geqq \alpha_m + 2\alpha_{m+1} + \dots + (n-m+1)\,\alpha_n$$

for $1 \leqq m \leqq n$, with equality when $m=1$.

[The sufficiency follows from Theorem **51**; the necessity may be proved on the lines of § 2.19 (1). Dougall (**1**) gives a proof for integral α based on an identity. For certain special cases, such as

$$p_{\mu-\lambda} p_{\mu+\lambda} \leqq p_{\mu-\kappa} p_{\mu+\kappa} \quad (0 \leqq \kappa < \lambda < \mu),$$

$$p_{\mu_1+\mu_2+\dots+\mu_r} \leqq p_{\mu_1} p_{\mu_2} \dots p_{\mu_r},$$

see Kritikos (**1**).]

78. The means $[\frac{1}{2}, \frac{1}{2}, 0, 0, \dots, 0]$ and $[\frac{3}{5}, \frac{1}{5}, \frac{1}{5}, 0, \dots, 0]$ are not comparable.

[Example of Theorem **45** and illustration of Theorem **48**.]

79. If $a>0$, and P_μ is the arithmetic mean of the μth roots of the products of μ different a, then

$$P_1 > P_2 > \dots > P_n,$$

unless all the a are equal.

[Smith (**1**, 440). Example of Theorem **45**:

$$[1, 0, 0, \dots, 0] > [\tfrac{1}{2}, \tfrac{1}{2}, 0, \dots, 0] > [\tfrac{1}{3}, \tfrac{1}{3}, \tfrac{1}{3}, \dots, 0] > \dots.]$$

80. If $\mu \geqq 0$, and x, y, z are positive, then

$$x^\mu (x-y)(x-z) + y^\mu (y-z)(y-x) + z^\mu (z-x)(z-y) > 0,$$

unless $x=y=z$.

81. If $\nu \geqq 0$, $\delta > 0$, and the a are positive and not all equal, then

$$[\nu+2\delta, 0, 0, \alpha_4, \dots] - 2[\nu+\delta, \delta, 0, \alpha_4, \dots] + [\nu, \delta, \delta, \alpha_4, \dots] > 0.$$

[This result, communicated to us by Prof. I. Schur, is not a consequence of Theorem **45**, but follows from Theorem **80**, with $\mu = \nu/\delta$.]

CHAPTER III

MEAN VALUES WITH AN ARBITRARY FUNCTION AND THE THEORY OF CONVEX FUNCTIONS

3.1. Definitions. The means $\mathfrak{M}_r(a)$ and $\mathfrak{G}(a)$ are of the form

(3.1.1) $$\mathfrak{M}_\phi(a) = \phi^{-1}\{\Sigma q\phi(a)\},$$

where $\phi(x)$ is one of the functions

$$x^r, \quad \log x$$

and $\phi^{-1}(x)$ the inverse function. It is natural to consider more general means of the type (3.1.1), formed with an arbitrary function ϕ subject to appropriate conditions. The most obvious conditions to be imposed upon ϕ are that it should be continuous and strictly monotonic, in which case it has an inverse ϕ^{-1} which satisfies the same conditions.

We require the following preliminary theorem.

82. *If* (i) $\phi(x)$ *is continuous and strictly monotonic in* $H \leqq x \leqq K$,

(ii) $H \leqq a_\nu \leqq K \quad (\nu = 1, 2, \ldots, n)$,

(iii) $q_\nu > 0$, $\Sigma q_\nu = 1$,

then (1) *there is a unique* $\mathfrak{M}$ *in* (H, K) *for which*

(3.1.2) $$\phi(\mathfrak{M}) = \Sigma q\phi(a),$$

(2) $\mathfrak{M}$ *is greater than some and less than others of the* a, *unless the* a *are all equal.*

Since $\phi(x)$ is continuous and increases or decreases from $\phi(H)$ to $\phi(K)$ when x increases from H to K, and $\Sigma q\phi(a)$ lies between these limits, there is just one $\mathfrak{M}$ which satisfies (3.1.2). Also

$$\Sigma q\{\phi(\mathfrak{M}) - \phi(a)\} = 0$$

and some terms must be positive and some negative, unless all are zero. Hence $\mathfrak{M} - a$ is sometimes positive and sometimes negative, unless it is always zero.

We have assumed $\phi(x)$ continuous in the closed interval (H, K). The argument is still valid if $\phi(x)$ is continuous and strictly increasing in $H < x < K$, while $\phi(x) \to -\infty$ when $x \to H$, or $\phi(x) \to +\infty$ when $x \to K$, provided that we then interpret $\phi(H)$ as meaning $-\infty$, or $\phi(K)$ as meaning $+\infty$, and $\mathfrak{M}$ as being H when $\Sigma q\phi(a) = -\infty$ or K when $\Sigma q\phi(a) = +\infty$. Here H may be $-\infty$ or K may be $+\infty$; a particularly important case is that in which $H=0$, $K=+\infty$. In the definition which follows, and in all the discussion of the properties of $\mathfrak{M}_\phi$ later in this chapter, it is assumed that ϕ is strictly monotonic and is either continuous in the closed interval or behaves in the manner just explained.

We write[a]

(3.1.3) $\mathfrak{M}_\phi = \mathfrak{M}_\phi(a) = \mathfrak{M}_\phi(a, q) = \phi^{-1}\{\Sigma q\phi(a)\} = \phi^{-1}\{\mathfrak{A}[\phi(a)]\}$.

The weights q are arbitrary positive numbers whose sum is 1, and when we compare two means it is to be understood that the weights of the means are the same. For $\phi(x) = x$, $\log x$ and x^r, $\mathfrak{M}_\phi$ reduces to $\mathfrak{A}$, $\mathfrak{G}$ and $\mathfrak{M}_r$ respectively.

3.2. Equivalent means. The mean $\mathfrak{M}_\phi$ is determined when the function ϕ is given. We may ask whether the converse is true: if $\mathfrak{M}_\psi = \mathfrak{M}_\chi$ for all a and q, is ψ necessarily the same function as χ? This question is answered by the theorem which follows.

83.[b] *In order that*

(3.2.1) $$\mathfrak{M}_\psi(a) = \mathfrak{M}_\chi(a)$$

for all a and q, it is necessary and sufficient that

(3.2.2) $$\chi = \alpha\psi + \beta,$$

where α and β are constants and $\alpha \neq 0$.

In the argument which follows we assume ψ and χ continuous in the closed interval (H, K). It is easy to see that it applies with trivial variations in the exceptional cases mentioned in § 3.1. We shall actually prove more than we have stated, viz. that (3.2.2) is a sufficient condition for (3.2.1) to be true for all a and q, and that it is a necessary condition for (3.2.1) to be true for all sets of

[a] In this chapter we define $\mathfrak{M}_\phi$ directly, and deduce its properties from the definition. In Ch. VI (§§ 6.19–6.22) we shall show how $\mathfrak{M}_\phi$ may be defined 'axiomatically', that is to say by prescription of its characteristic properties.

[b] Knopp (2), Jessen (2).

two variables and weights. Later (§ 3.7) we shall prove still more, viz. that (3.2.2) is necessary for the truth of (3.2.1) for all sets of two variables and a fixed pair of weights.

(i) If (3.2.2) is satisfied

$$\begin{aligned}\chi\{\mathfrak{M}_\chi(a)\} &= \Sigma q\,\chi(a) = \Sigma q\,\{\alpha\psi(a)+\beta\}\\ &= \alpha\,\Sigma q\psi(a)+\beta\\ &= \alpha\psi\{\mathfrak{M}_\psi(a)\}+\beta = \chi\{\mathfrak{M}_\psi(a)\},\end{aligned}$$

and so $\mathfrak{M}_\chi = \mathfrak{M}_\psi$. Hence the condition is *sufficient*.

(ii) In proving the condition *necessary*, we assume only that (3.2.1) is true for all sets of two variables and weights.

In (3.2.1) take

$$n=2,\quad a_1=H,\quad a_2=K,\quad q_1=\frac{K-t}{K-H},\quad q_2=\frac{t-H}{K-H},$$

where $H<t<K$. Then

(3.2.3)

$$\psi^{-1}\left\{\frac{K-t}{K-H}\psi(H)+\frac{t-H}{K-H}\psi(K)\right\}=\chi^{-1}\left\{\frac{K-t}{K-H}\chi(H)+\frac{t-H}{K-H}\chi(K)\right\}$$

for $H<t<K$; and this is also true for $t=H$ and $t=K$. If we denote the common value by x then, as t varies from H to K, x assumes all values in (H, K) and

$$\frac{K-t}{K-H}\psi(H)+\frac{t-H}{K-H}\psi(K)=\psi(x),$$

$$\begin{aligned}\frac{K-t}{K-H}\chi(H)+\frac{t-H}{K-H}\chi(K) &= \frac{\psi(K)-\psi(x)}{\psi(K)-\psi(H)}\chi(H)+\frac{\psi(x)-\psi(H)}{\psi(K)-\psi(H)}\chi(K)\\ &= \alpha\psi(x)+\beta,\end{aligned}$$

where α and β are independent of x. Hence

$$x=\chi^{-1}\{\alpha\psi(x)+\beta\}$$

for all x in (H, K); and this is (3.2.2). This completes the proof of Theorem **83**.

One corollary of Theorem **83** which is sometimes useful is this. Since $-\phi$ is a linear function of ϕ, and $-\phi$ increases if ϕ decreases, we may always suppose, if we please, that the ϕ involved in $\mathfrak{M}_\phi(x)$ is an *increasing* function.

Theorem **83** also enables us to elucidate the apparently exceptional status of $\mathfrak{M}_0 = \mathfrak{G}$ among the means $\mathfrak{M}_r$ of Ch. II. Since

$$\phi_r(x) = \int_1^x t^{r-1}\,dt = \frac{x^r - 1}{r}$$

is a linear function of x^r, for $r \neq 0$, we have, by Theorem **83**,

$$\mathfrak{M}_r(a) = \mathfrak{M}_{\phi_r}(a).$$

This equation is still valid for $r = 0$, since $\phi_0(x) = \log x$.

3.3. A characteristic property of the means $\mathfrak{M}_r$. It is natural to ask whether there is any simple property of the means of Ch. II which characterises them among the more general means considered here.

84.[a] *Suppose that $\phi(x)$ is continuous in the open interval* $(0, \infty)$, *and that*

$$\mathfrak{M}_\phi(ka) = k\mathfrak{M}_\phi(a) \tag{3.3.1}$$

for all positive a, q, and k. Then $\mathfrak{M}_\phi(a)$ is $\mathfrak{M}_r(a)$. In other words, the means $\mathfrak{M}_r$ are the only homogeneous means $\mathfrak{M}_\phi$.

Naturally (3.3.1) does not imply $\phi = x^r$ (or $\log x$); for, by Theorem **83**, we can replace ϕ by $\alpha\phi + \beta$ without changing $\mathfrak{M}_\phi$.

That (3.3.1) is true when $\phi = x^r$ or $\phi = \log x$ is obvious. We now assume (3.3.1) and deduce the form of ϕ. After Theorem **83**, we may suppose that

$$\phi(1) = 0; \tag{3.3.2}$$

for we may replace $\phi(x)$ by $\phi(x) - \phi(1)$.

We write (3.3.1) in the form

$$\mathfrak{M}_\phi(a) = k^{-1}\mathfrak{M}_\phi(ka) = k^{-1}\phi^{-1}\{\Sigma q\phi(ka)\} = \mathfrak{M}_\psi(a),$$

where
$$\psi(x) = \phi(kx).$$

It follows from Theorem **83**[b] that

$$\phi(kx) = \alpha(k)\phi(x) + \beta(k), \tag{3.3.3}$$

where $\alpha(k)$ and $\beta(k)$ are functions of k, and $\alpha(k) \neq 0$; and from (3.3.2) and (3.3.3) that

$$\phi(k) = \beta(k). \tag{3.3.4}$$

[a] Nagumo (**1**), de Finetti (**1**), Jessen (**4**). The following simple version of de Finetti's proof was communicated to us by Dr Jessen.

[b] If we used one of the more precise forms of Theorem **83**, referred to after its enunciation in § 3.2, we should obtain a more precise form of Theorem **84**, in which homogeneity was only assumed for restricted classes of variables or weights.

If we substitute from (3.3.4) into (3.3.3), and write y for k, we find that

$$\phi(xy) = \alpha(y)\phi(x) + \phi(y) \tag{3.3.5}$$

for all positive x and y.

Similarly

$$\phi(xy) = \alpha(x)\phi(y) + \phi(x); \tag{3.3.6}$$

and (3.3.5) and (3.3.6) give

$$\frac{\alpha(x)-1}{\phi(x)} = \frac{\alpha(y)-1}{\phi(y)}.$$ [a]

Each of these functions must reduce to a constant c, so that $\alpha(y) = 1 + c\phi(y)$. It then follows from (3.3.5) that

$$\phi(xy) = c\phi(x)\phi(y) + \phi(x) + \phi(y). \tag{3.3.7}$$

In discussing this functional equation, we must distinguish two cases.

(1) If $c = 0$, (3.3.7) reduces to the classical equation

$$\phi(xy) = \phi(x) + \phi(y).$$

The most general solution, continuous for $x > 0$, is[b] $\phi = C \log x$.

(2) If $c \neq 0$, we put $c\phi(x) + 1 = f(x)$,

and the equation reduces to

$$f(xy) = f(x)f(y),$$

whose general solution is $f = x^r$. Hence

$$\phi(x) = \frac{x^r - 1}{c}.$$

3.4. Comparability. Our general remarks on the 'comparability' of functions of the a (§ 1.6) suggest the following problem: *given two functions ψ and χ, each continuous and strictly monotonic in (H, K), are $\mathfrak{M}_\psi$ and $\mathfrak{M}_\chi$ comparable; i.e. is there an inequality*

$$\mathfrak{M}_\psi \leqq \mathfrak{M}_\chi \tag{3.4.1}$$

[a] Provided $x \neq 1$, $y \neq 1$. Since (3.3.7) is plainly true when x or y is 1, the exception is irrelevant.

[b] Cauchy (**1**, 103–105).

(*or one in the opposite sense*) *valid for all a and q*? Theorem **16** tells us that the answer is affirmative when ψ and χ are powers.

We write $$\chi\{\psi^{-1}(x)\}=\phi(x).$$
Then ϕ is continuous and strictly monotonic, and has the inverse $\phi^{-1}=\psi\chi^{-1}$. We also write
$$x=\psi(a), \quad a=\psi^{-1}(x).$$
Then the x are arbitrary numbers between $\psi(H)$ and $\psi(K)$; and (3.4.1) takes the form

$$\phi(\Sigma qx)\leqq\Sigma q\phi(x) \tag{3.4.2}$$

(for all q) if χ is increasing, the reversed form if χ is decreasing.

We thus obtain

85. *If ψ and χ are continuous and strictly monotonic, then a necessary and sufficient condition that $\mathfrak{M}_\psi$ and $\mathfrak{M}_\chi$ should be comparable is that $\phi=\chi\psi^{-1}$ should satisfy* (3.4.2), *or the reversed inequality.*

In what follows, we examine this class of functions ϕ in detail.

For arbitrary weights p, (3.4.2) becomes

$$\phi\left(\frac{\Sigma px}{\Sigma p}\right)\leqq\frac{\Sigma p\phi(x)}{\Sigma p}. \tag{3.4.3}$$

3.5. Convex functions. The function ϕ of §3.4 was the resultant of two monotonic functions, and therefore itself monotonic; but now we consider a ϕ subject to (3.4.2) only.

The simplest case of (3.4.2) is

$$\phi\left(\frac{x+y}{2}\right)\leqq\frac{\phi(x)+\phi(y)}{2}. \tag{3.5.1}$$

A function which satisfies (3.4.2) satisfies (3.5.1), but the class of functions satisfying (3.5.1) is more general. We shall however show that the two inequalities are equivalent for functions subject to certain not very restrictive conditions.

A function which satisfies (3.5.1) in a certain interval is called *convex* in that interval. If $-\phi$ is convex, ϕ is *concave*. We may also define convexity or concavity in an open interval. It is often convenient to admit infinite values at the ends of the interval; it is obvious that such values must be positive for convex and negative for concave functions, if the interval is finite.

The foundations of the theory of convex functions are due to Jensen (**2**).[a] Geometrically, (3.5.1) means that the middle point of any chord of the curve $y=\phi(x)$ lies above or on the curve; here a curve means any, not necessarily continuous, graph. The inequality

$$\phi(q_1x_1+q_2x_2)\leqq q_1\phi(x_1)+q_2\phi(x_2) \tag{3.5.2}$$

(for all q) asserts that the whole chord lies above or on the curve, and the general inequality (3.4.2) asserts that the centre of gravity of any number of arbitrarily weighted points of the curve lies above or on the curve. It is geometrically intuitive that, when the curve is continuous, the weakest condition implies the stronger, and we shall find that much more than this is confirmed by our analysis. We might have taken (3.4.2) or (3.5.2) as our definition of convexity, but we have followed Jensen in starting from the weakest definition. The most *natural* definitions are perhaps (3.5.2) and another which we discuss in § 3.19. There is some logical interest in assuming as little as possible.

It is sometimes useful to have a definition of the convexity or concavity of a finite or enumerably infinite set of numbers. We shall say that the set $a_1,\dots,a_n$ is *convex* if

$$2a_\nu\leqq a_{\nu-1}+a_{\nu+1}\quad(\nu=2,3,\dots,n-1),$$

i.e. if the second differences of the set are non-negative.

Thus we can state Theorem **51**, in the less exact form with '$\leqq$', by saying that the set $\log p$ is concave; the full theorem is that $\log p$ is strictly concave (see § 3.8) unless the a are equal. When two products of powers of the p are comparable, the inequality which holds between them may be deduced (substantially as Theorem **52** was deduced from Theorem **51**) from the concavity of $\log p$. This is the kernel of Theorem **77**.

3.6. Continuous convex functions. We now proceed to investigate the simplest case in which (3.4.2) and (3.5.1) are equivalent.

If $\phi(x)$ satisfies (3.5.1), we have

$$4\phi\left(\frac{x_1+x_2+x_3+x_4}{4}\right)\leqq 2\phi\left(\frac{x_1+x_2}{2}\right)+2\phi\left(\frac{x_3+x_4}{2}\right)$$
$$\leqq\phi(x_1)+\phi(x_2)+\phi(x_3)+\phi(x_4),$$

[a] Though Hölder (**1**) had considered the inequality (3.4.2) before Jensen.

and so on. We thus prove

$$\phi\left(\frac{x_1+x_2+\ldots+x_n}{n}\right) \leqq \frac{\phi(x_1)+\phi(x_2)+\ldots+\phi(x_n)}{n} \tag{3.6.1}$$

for a particular sequence of values of n, viz. $n=2^m$.

To prove (3.6.1) true generally, it is enough to prove also that, if it is true for n, it is true for $n-1$.[a] Suppose then that (3.6.1) has been proved for n numbers, and that $x_1, x_2, \ldots, x_{n-1}$ are given. Taking x_n to be the arithmetic mean $\mathfrak{A}$ (with equal weights) of the $n-1$ numbers, and applying (3.6.1), we obtain

$$\phi(\mathfrak{A}) = \phi\left\{\frac{(n-1)\mathfrak{A}+\mathfrak{A}}{n}\right\} = \phi\left(\frac{x_1+x_2+\ldots+x_{n-1}+\mathfrak{A}}{n}\right)$$

$$\leqq \frac{\phi(x_1)+\phi(x_2)+\ldots+\phi(x_{n-1})+\phi(\mathfrak{A})}{n},$$

and so $$\phi(\mathfrak{A}) \leqq \frac{\phi(x_1)+\phi(x_2)+\ldots+\phi(x_{n-1})}{n-1},$$

which is (3.6.1) with $n-1$ for n. Hence (3.6.1) is true generally.

Next, supposing that, in (3.6.1), the x form appropriate groups of equal numbers, we obtain (3.4.2) for any *commensurable* q.

Finally, if $\phi(x)$ is *continuous*, we can prove (3.4.2) without restriction on the q; for we may replace the q by commensurable approximations and proceed to the limit. We thus obtain

86. *Any continuous convex function satisfies* (3.4.2).

As an application, we may consider Theorem **17**. If $s=\frac{1}{2}(r+t)$ we have, by Theorem **7**,

$$(\Sigma pa^s)^2 \leqq \Sigma pa^r \, \Sigma pa^t,$$

or $$\{\mathfrak{M}_s{}^s(a)\}^2 \leqq \mathfrak{M}_r{}^r(a)\,\mathfrak{M}_t{}^t(a),$$

or $$\log \mathfrak{M}_s{}^s(a) \leqq \tfrac{1}{2}\{\log \mathfrak{M}_r{}^r(a) + \log \mathfrak{M}_t{}^t(a)\}.$$

In other words

87. $\log \mathfrak{M}_r{}^r(a) = r \log \mathfrak{M}_r(a)$ *is a convex function of* r.

From this, by appealing to Theorem **86** (or repeating the argument by which this theorem was proved), we deduce Theorem **17** (apart from the specification of the cases of equality).

[a] Here we follow the lines of § 2.6 (ii). For a proof following Cauchy's argument more directly, see Jensen (2).

3.7. An alternative definition[a]. We characterised a convex function $\phi(x)$ in § 3.5 by the fact that the middle point of a chord of the curve $y=\phi(x)$ lies above or on the curve. Riesz and Jessen have made an observation which is interesting and sometimes important in applications[b], viz. that, when $\phi(x)$ is continuous, it is sufficient to require that *some* point of the chord should lie above or on the curve.

88. *If $\phi(x)$ is continuous, and there is at least one point of every chord of the curve $y=\phi(x)$, besides the end points of the chord, which lies above or on the curve, then every point of every chord lies above or on the curve, so that $\phi(x)$ is convex.*

Suppose that PQ is a chord, and R a point on the chord below the curve. Then there is a last point S on PR and a first point T on RQ in which the curve meets the chord: S may be P and T may be Q. The chord ST lies entirely below the curve, contradicting the hypothesis.

This remark gives us an alternative proof of Theorem **86**. If $\phi(x)$ is convex, the middle point of any chord lies above or on the curve. Hence, as we have proved, every point of the chord lies above or on the curve. That is to say

$$\phi(q_1x_1+q_2x_2)\leqq q_1\phi(x_1)+q_2\phi(x_2)$$

if $q_1>0$, $q_2>0$, $q_1+q_2=1$, but q_1 and q_2 are otherwise arbitrary. We may then proceed by induction. If $q_1+q_2+q_3=1$, then

$$(3.7.1)\quad \begin{aligned}\phi(q_1x_1+q_2x_2+q_3x_3)&=\phi\left\{q_1x_1+(q_2+q_3)\frac{q_2x_2+q_3x_3}{q_2+q_3}\right\}\\ &\leqq q_1\phi(x_1)+(q_2+q_3)\phi\left(\frac{q_2x_2+q_3x_3}{q_2+q_3}\right)\\ &\leqq q_1\phi(x_1)+(q_2+q_3)\frac{q_2\phi(x_2)+q_3\phi(x_3)}{q_2+q_3}\\ &=q_1\phi(x_1)+q_2\phi(x_2)+q_3\phi(x_3),\end{aligned}$$

and so generally.

A corollary of Theorem 88 is

89. *If $\phi(x)$ is continuous, and every chord of $y=\phi(x)$ meets the curve in a point distinct from its end-points, then $\phi(x)$ is linear.*

[a] M. Riesz (**1**), Jessen (**2**).

[b] See, for example, § 8.13.

By Theorem **88**, every point of every chord lies on or above the curve. But Theorem **88** remains true if 'above' is replaced by 'below' in hypothesis and conclusion. Hence every chord of the curve coincides with the curve.

From Theorem **89** we can deduce the refinement on Theorem **83** referred to in § 3.2. Suppose that

$$\psi^{-1}\{q_1\psi(a_1)+q_2\psi(a_2)\}=\chi^{-1}\{q_1\chi(a_1)+q_2\chi(a_2)\}$$

for fixed q_1, q_2 and arbitrary a. Writing $\chi\psi^{-1}=\phi$, $\psi(a)=x$, $a=\psi^{-1}(x)$, we obtain

$$\phi(q_1x_1+q_2x_2)=q_1\phi(x_1)+q_2\phi(x_2),$$

so that one point at any rate of every chord of $y=\phi(x)$ lies on the curve. It follows from Theorem **89** that ϕ is linear.

3.8. Equality in the fundamental inequalities. We now suppose $\phi(x)$ continuous and convex, and consider when equality can occur in (3.5.1), (3.5.2), or (3.4.2).

Suppose that $x_1<x_3<x_2$, that $x_3=q_1x_1+q_2x_2$, and that P_1, P_2, ... are the points on the curve $y=\phi(x)$ corresponding to $x_1, x_2, \ldots$. If $\phi(x)$ is not linear in (x_1, x_2), there is an x_4 in (x_1, x_2) such that P_4 lies below the line P_1P_2. Suppose for example that x_4 lies in (x_1, x_3). Then x_3 lies in (x_4, x_2), and P_3 lies on or below P_4P_2, and therefore below P_1P_2. Hence (3.5.2) holds with inequality. It follows that *equality can occur in* (3.5.2) *only when* $\phi(x)$ *is linear in* (x_1, x_2).

This conclusion is easily extended to the general inequality (3.4.2). Suppose, for example, that there is equality when $n=3$, and that $x_1<x_2<x_3$. Then all the signs of inequality in (3.7.1) must reduce to equality, and $\phi(x)$ must be linear in each of the intervals

$$\left(x_1, \frac{q_2x_2+q_3x_3}{q_2+q_3}\right), \quad (x_2, x_3)$$

and therefore over (x_1, x_3).

We have thus proved

90. *If $\phi(x)$ is continuous and convex, then*

$$\phi(\Sigma qx)<\Sigma q\phi(x), \tag{3.8.1}$$

$$\phi\left(\frac{\Sigma px}{\Sigma p}\right)<\frac{\Sigma p\phi(x)}{\Sigma p}, \tag{3.8.2}$$

unless either (i) *all the x are equal or* (ii) $\phi(x)$ *is linear in an interval including all the x.*

91. *Any chord of a continuous convex curve lies entirely above the curve, except for its end-points, or coincides with it.*

We may say that $\phi(x)$ is *strictly convex* if

$$\phi\left(\frac{x+y}{2}\right) < \tfrac{1}{2}\{\phi(x)+\phi(y)\} \tag{3.8.3}$$

for every unequal pair x, y. Since a strictly convex function cannot be linear in any interval, any such function, if continuous, satisfies (3.8.1) and (3.8.2), unless all the x are equal.

3.9. Restatements and extensions of Theorem 85.[a] We may restate Theorem **85** in the form

92. *If ψ and χ are continuous and strictly monotonic, and χ is increasing, then a necessary and sufficient condition that $\mathfrak{M}_\psi \leqq \mathfrak{M}_\chi$ for all a and q is that $\phi = \chi\psi^{-1}$ should be convex*[b].

We shall say in these circumstances that χ is *convex with respect to ψ*. Thus t^s is convex with respect to t^r when $s \geqq r > 0$.

The curve $y = \phi(x)$ has the parametric representation

$$x = \psi(t), \quad y = \chi(t).$$

The chord through the points on the curve corresponding to $t = t_1$ and $t = t_2$ is

$$x = \psi(t), \quad y = \psi^*(t),$$

where
$$\psi^*(t) = \frac{\psi(t_2)-\psi(t)}{\psi(t_2)-\psi(t_1)}\chi(t_1) + \frac{\psi(t)-\psi(t_1)}{\psi(t_2)-\psi(t_1)}\chi(t_2)$$

is the function
$$\alpha\psi(t)+\beta$$

which assumes the values $\chi(t_1)$ and $\chi(t_2)$ for $t = t_1$ and $t = t_2$. We may call $y = \psi^*(x)$ the ψ-chord of $y = \chi(x)$. In order that χ should be convex with respect to ψ, it is necessary and sufficient that $\chi \leqq \psi^*$, i.e. that every point of any ψ-chord of χ should lie on or above the curve.

[a] Jessen (**2**, **3**).

[b] We have actually proved more in regard to the necessity of the condition: see our remarks on Theorem **83**.

Theorem **92** may be generalised as follows. Suppose that

$$a = a_{\nu_1, \nu_2, \ldots, \nu_m}$$

is a function of m variables $\nu_1, \nu_2, \ldots, \nu_m$, and that

$$\mathfrak{M}_{\psi_m}^{\nu_m} \ldots \mathfrak{M}_{\psi_2}^{\nu_2} \mathfrak{M}_{\psi_1}^{\nu_1}(a)$$

is the result of taking means with respect to $\nu_1, \nu_2, \ldots \nu_m$ in succession.

93. *Suppose that ψ_μ and χ_μ are continuous and strictly monotonic, and that χ_μ is increasing. Then, in order that*

$$\mathfrak{M}_{\psi_m}^{\nu_m} \ldots \mathfrak{M}_{\psi_1}^{\nu_1}(a) \leqq \mathfrak{M}_{\chi_m}^{\nu_m} \ldots \mathfrak{M}_{\chi_1}^{\nu_1}(a)$$

for all a and q, it is necessary and sufficient that every χ_μ should be convex with respect to the corresponding ψ_μ.

It is understood, of course, that the weights involved in the operations $\mathfrak{M}_{\psi_\mu}$ and $\mathfrak{M}_{\chi_\mu}$ are the same, though they will generally vary with μ. That the conditions are sufficient follows at once from Theorem **92**. To see that they are necessary, we have only to suppose a to be a function of a single ν_μ.

3.10. Twice differentiable convex functions. We postpone to § 3.18 any further discussion of the general properties of convex functions, and consider now a particularly important sub-class of such functions, viz. those which possess a second differential coefficient.

94. *Suppose that $\phi(x)$ possesses a second derivative $\phi''(x)$ in the open interval (H, K). Then a necessary and sufficient condition that $\phi(x)$ should be convex in the interval is*

$$\phi''(x) \geqq 0.\text{[a]} \tag{3.10.1}$$

(i) *The condition is necessary.* Replacing $\frac{1}{2}(x+y)$ and $\frac{1}{2}(x-y)$ by t and h in (3.5.1), and supposing that $x > y$, so that $h > 0$, we obtain

[a] The important case in practice is that in which (as stated in the theorem) ϕ'' exists in the *open* interval. We usually wish, however, to assert convexity in the *closed* interval. Since $\phi'' \geqq 0$, ϕ' and ϕ are monotonic near the ends of the interval and tend to finite or infinite limits; ϕ' may tend to $-\infty$ at the left-hand end and to $+\infty$ at the right-hand end, and ϕ may tend to $+\infty$ at either. The function will be convex in the closed interval if its value at each end is not less than its limit at that end.

(3.10.2) $$\phi(t+h)+\phi(t-h)-2\phi(t) \geqq 0$$

for all t, h such that the arguments lie in the interval.

Suppose now that $\phi''(t)<0$. Then there are positive numbers δ and h such that

$$\phi'(t+u)-\phi'(t-u) < -\delta u$$

for $0<u \leqq h$. Integrating this inequality from $u=0$ to $u=h$, we obtain

$$\phi(t+h)+\phi(t-h)-2\phi(t) < -\tfrac{1}{2}\delta h^2,$$

in contradiction with (3.10.2).

(ii) *The condition is sufficient.* We prove that ϕ satisfies (3.4.2). In fact, if $X=\Sigma qx$, we have

$$\phi(x_\nu)=\phi(X)+(x_\nu-X)\phi'(X)+\tfrac{1}{2}(x_\nu-X)^2\phi''(\xi_\nu)$$

for some ξ_ν between X and x_ν, and so

$$\Sigma q\phi(x) \geqq \phi(X)=\phi(\Sigma qx).$$

If $\phi''(x)>0$, there can be equality only if every x is equal to X. We have therefore proved

95.[a] *If* $\phi''(x)>0$, *then* $\phi(x)$ *is strictly convex and satisfies* (3.8.1) *and* (3.8.2), *unless all the* x *are equal.*

3.11. Applications of the properties of twice differentiable convex functions. Combining the proofs of Theorems **85** and **95** (part (ii)) we obtain the following theorem, which will be found particularly useful in applications.

96. *If* ψ *and* χ *are strictly monotonic,* χ *increasing,* $\phi=\chi\psi^{-1}$, *and* $\phi''>0$, *then* $\mathfrak{M}_\psi<\mathfrak{M}_\chi$, *unless all the* a *are equal.*

Examples. (1) If $\psi=\log x$, $\chi=x$, then $\phi=\chi\psi^{-1}=e^x$. Theorem **96** reduces to Theorem **9**.

(2) If $\psi=x^r$, $\chi=x^s$, where $0<r<s$, then $\phi=x^{s/r}$, $\phi''>0$. Theorem **96** gives Theorem **16** (for positive indices). The other cases of Theorem **16** may be derived similarly.

(3) Suppose that $\phi=x^k$, where k is not 0 or 1. Then ϕ is convex in $(0,\infty)$ if $k<0$ or $k>1$, concave if $0<k<1$. Supposing $k>1$, and

[a] Hölder (**1**).

applying Theorem **95**, we find

$$\Sigma qx < (\Sigma qx^k)^{1/k}$$

or
$$\Sigma px < (\Sigma px^k)^{1/k} (\Sigma p)^{1/k'},$$

unless all the x are equal. If we write $px = ab$, $px^k = a^k$, we obtain Theorem **13**, for $k > 1$. The other cases follow similarly.

(4) Suppose that $\phi = \log(1+e^x)$, so that

$$\phi''(x) = \frac{e^x}{(1+e^x)^2} > 0,$$

and that the abscissae and weights in (3.8.1) are $\log(a_2/a_1)$, $\log(b_2/b_1)$, ..., and α, β, ... We obtain

$$a_1^\alpha b_1^\beta \ldots l_1^\lambda + a_2^\alpha b_2^\beta \ldots l_2^\lambda < (a_1+a_2)^\alpha (b_1+b_2)^\beta \ldots (l_1+l_2)^\lambda,$$

unless $a_2/a_1 = b_2/b_1 = \ldots$ (Theorem **40**: H for any number of sets of two numbers).

(5) Suppose that $\phi = (1+x^r)^{1/r}$, where r is not 0 or 1, and that the abscissae and weights in (3.8.2) are a_2/a_1, b_2/b_1, ... and a_1, b_1, ... In this case ϕ is convex if $r > 1$ and concave if $r < 1$. We find, for example, that

$$\{(a_1+b_1+\ldots+l_1)^r + (a_2+b_2+\ldots+l_2)^r\}^{1/r}$$
$$< (a_1^r+a_2^r)^{1/r} + (b_1^r+b_2^r)^{1/r} + \ldots + (l_1^r+l_2^r)^{1/r}$$

if $r > 1$ and a_2/a_1, b_2/b_1, ... are not all equal (M for any number of sets of two numbers). It will be remembered that both H and M can be extended to sets of more numbers inductively.

(6) **97.** *If $a > 0$, $p > 0$, then*

$$\exp\left(\frac{\Sigma p \log a}{\Sigma p}\right) < \frac{\Sigma pa}{\Sigma p} < \exp\left(\frac{\Sigma pa \log a}{\Sigma pa}\right),$$

unless all the a are equal.

We write this with p instead of q for the sake of symmetry. The first inequality is (3.8.2) reversed, with $\phi(x) = \log x$, a concave function. It is equivalent to G (Theorem **9**). The second inequality is (3.8.2), with $\phi(x) = x \log x$, a convex function.

3.12. Convex functions of several variables. Suppose that D is a convex domain in the plane of (x, y), that is to say, a domain which contains the whole of the segment of a straight line which connects any two of its points[a]. A function $\Phi(x, y)$ is said to be *convex in D* if it is defined everywhere in D and

$$\Phi\left(\frac{x_1+x_2}{2}, \frac{y_1+y_2}{2}\right) \leqq \tfrac{1}{2}\{\Phi(x_1, y_1)+\Phi(x_2, y_2)\}, \tag{3.12.1}$$

for all (x_1, y_1) and (x_2, y_2) of D.[b] The definition asserts more than convexity in x and y separately; thus xy is a convex function of x for every y, and a convex function of y for every x, but it is not a convex function of x and y.

It is often convenient to use an alternative form of the definition just given. Suppose that x, y, u, v are given, and consider the values of t (if any) for which $(x+ut, y+vt)$ belongs to D. Since D is convex, these values form an interval (which may be null). Then we say that $\Phi(x, y)$ is convex in D if

$$\chi(t)=\Phi(x+ut, y+vt) \tag{3.12.2}$$

is, for every x, y, u, v, a convex function of t throughout the interval of t in question. The definition is equivalent to that which we gave before, since, if

$$x+ut_1=x_1, \quad y+vt_1=y_1, \quad x+ut_2=x_2, \quad y+vt_2=y_2,$$

(3.12.1) becomes

$$\chi\left(\frac{t_1+t_2}{2}\right) \leqq \tfrac{1}{2}\{\chi(t_1)+\chi(t_2)\}.$$

Φ is said to be concave if $-\Phi$ is convex.

If $z=\Phi(x, y)$ is the equation of a surface in rectangular Cartesian coordinates, (3.12.1) asserts that the middle point of any chord of the surface lies above or on the corresponding point of the

[a] It would be sufficient to consider rectangular domains, but convexity is the natural limitation to impose on D. It is not part of our programme to consider questions of analysis situs connected with convex or general domains.

[b] There is a wider generalisation of the notion of a convex function of a single variable which is important in the theory of functions but with which we shall not be concerned. The function $\Phi(x, y)$ is *subharmonic* if its value at the centre of any circle does not exceed its average over the circumference. In particular Φ is subharmonic if it is twice differentiable and

$$\nabla^2\Phi=\Phi_{xx}+\Phi_{yy}\geqq 0.$$

For the theory of subharmonic functions see F. Riesz (**5**, **9**), Montel (**1**).

surface. If the surface is continuous we can deduce that the whole chord lies above or on the surface, and that the centre of gravity of any number of arbitrarily weighted points of the surface lies above or on the surface. This is what is asserted in the following theorem.

98. *If $\Phi(x,y)$ is convex and continuous, then*

(3.12.3) $$\Phi(\Sigma qx, \Sigma qy) \leqq \Sigma q \Phi(x,y).$$

The proof is the same as that of Theorem **86**, except for the obvious changes of notation.

There is also a theorem corresponding to Theorem **88**; it is sufficient to assert that no chord of the surface lies (except for its end-points) entirely below the surface. All the other remarks of § 3.7 remain true with the obvious changes.

A theorem corresponding to Theorems **94** and **95** is

99. *If $\Phi(x,y)$ is twice differentiable in an open D, then a necessary and sufficient condition that it should be convex in D is that the quadratic form*

$$Q = \Phi_{xx}u^2 + 2\Phi_{xy}uv + \Phi_{yy}v^2$$

should be positive[a] *for all u, v and all (x, y) of D.*

If Q is strictly positive[b], *then* (3.12.3) *holds with inequality, unless all the x and all the y are equal.*

(1) *The condition is necessary.* If (x, y) is in D, then $\chi(t)$, defined by (3.12.2), is convex in a neighbourhood of $t=0$. Hence, by Theorem **94**, $\chi''(0) \geqq 0$, i.e. $Q \geqq 0$.

(2) *The condition is sufficient.* If

$$\Sigma q = 1, \quad X = \Sigma qx, \quad Y = \Sigma qy$$

then

$$\Phi(x_\nu, y_\nu) = \Phi(X, Y) + (x_\nu - X)\Phi_x{}^0 + (y_\nu - Y)\Phi_y{}^0$$
$$+ \tfrac{1}{2}\{(x_\nu - X)^2 \Phi_{xx}{}^1 + 2(x_\nu - X)(y_\nu - Y)\Phi_{xy}{}^1 + (y_\nu - Y)^2 \Phi_{yy}{}^1\},$$

where the index 0 indicates the point (X, Y) and the index 1 some point on the line joining this point to (x_ν, y_ν). It follows that

$$\Sigma q \Phi(x,y) \geqq \Phi(X,Y) = \Phi(\Sigma qx, \Sigma qy).$$

If Q is strictly positive, and there is equality, then $x_\nu = X$, $y_\nu = Y$ for all ν.

[a] $Q \geqq 0$. [b] $Q > 0$ except for $u = v = 0$.

We notice that Q is positive if and only if

(3.12.4) $\quad \Phi_{xx} \geqq 0, \quad \Phi_{yy} \geqq 0, \quad \Phi_{xx}\Phi_{yy} - \Phi_{xy}^2 \geqq 0,$

and strictly positive if and only if

(3.12.5) $\quad \Phi_{xx} > 0, \quad \Phi_{xx}\Phi_{yy} - \Phi_{xy}^2 > 0.$

It is negative if (3.12.4) holds with the signs of the first two inequalities reversed.

The extension of the definitions and theorems of this section to functions of more than two variables may be left to the reader.

3.13. Generalisations of Hölder's inequality. We may write Hölder's inequality in the form

(3.13.1) $$\mathfrak{A}(ab) \leqq \mathfrak{M}_F(a)\,\mathfrak{M}_G(b)$$

or

(3.13.2) $$\Sigma qab \leqq F^{-1}\{\Sigma q F(a)\}\, G^{-1}\{\Sigma q G(b)\},$$

where $F(x) = x^r$ $(r > 1)$ and $G(x) = x^{r'}$, r' being as usual the index conjugate to r in the sense of § 2.8. If we write

$$\phi = F^{-1}, \quad \psi = G^{-1}, \quad F(a) = x, \quad G(b) = y, \quad a = \phi(x), \quad b = \psi(y),$$

we obtain

(3.13.3) $$\Sigma q \phi(x)\psi(y) \leqq \phi(\Sigma qx)\,\psi(\Sigma qy).$$

The simplest case of this is

$$\tfrac{1}{2}\{\phi(x_1)\psi(y_1) + \phi(x_2)\psi(y_2)\} \leqq \phi\{\tfrac{1}{2}(x_1+x_2)\}\,\psi\{\tfrac{1}{2}(y_1+y_2)\},$$

which expresses the fact that $\phi(x)\psi(y)$ is a concave function of x and y. When, as here, ϕ and ψ are continuous, it is equivalent to the more general inequality (3.13.3). Hence, reversing the argument (with general ϕ and ψ), we obtain

100. *If F and G are continuous and strictly monotonic, then a necessary and sufficient condition that $\mathfrak{A}(ab)$ should be comparable with $\mathfrak{M}_F(a)\,\mathfrak{M}_G(b)$ is that $F^{-1}(x)\,G^{-1}(y)$ should be a concave or convex function of the two variables x and y; in the first case* (3.13.1) *is true, in the second the reverse inequality.*

As an example we may take $F(x) = x^r$, $G(y) = y^s$. It then follows from Theorems **100** and **99** that

$$\mathfrak{A}(ab) \leqq \mathfrak{M}_r(a)\,\mathfrak{M}_s(b)$$

if $r > 1$, $s > 1$ and $(r-1)(s-1) \geqq 1$. If $r < 1$, $s < 1$, $(1-r)(1-s) \geqq 1$, the inequality is reversed. These are the only cases of comparability[a]. The argument excludes the cases $r=0$ and $s=0$, but they may be included by using an exponential instead of a power.

We might look for a more straightforward generalisation of Hölder's inequality. Hölder's inequality asserts that (3.13.1) holds if $f(x)$ and $g(x)$ are inverse positive powers of x, and either

(*a*) $$F(x)=xf(x), \qquad G(x)=xg(x),$$

or

(*b*) $$F(x)=\int_0^x f(t)\,dt, \quad G(x)=\int_0^x g(t)\,dt;$$

and we might expect that it would hold for other pairs of inverse f and g. The theorem which follows shows that no such extension is possible.

101. *Suppose that $f(x)$ is a continuous and strictly increasing function which vanishes for $x=0$ and has a second derivative continuous for $x>0$, and that $g(x)$ is the inverse function (which has necessarily the same properties). Suppose further that $F(x)$ and $G(x)$ are defined either by* (*a*) *or by* (*b*), *and that* (3.13.1) *is true for all positive a, b. Then f is a power of x and* (3.13.1) *is Hölder's inequality.*

We consider case (*a*).[b] If, as in the proof of Theorem **100**, we write ϕ and ψ for F^{-1} and G^{-1}, then $\phi(x)\psi(y)$ must be a concave function of x and y. It follows from Theorem **99** and (3.12.4)[c] that $\phi''\leqq 0$, $\psi''\leqq 0$ and

(3.13.4) $$\{\phi'(x)\psi'(y)\}^2 \leqq \phi(x)\psi(y)\phi''(x)\psi''(y)$$

for all positive x and y.

If $\phi(x)=u$, $\psi(x)=v$, we have

$$x=F(u)=uf(u), \quad \frac{x}{u}=f(u), \quad u=g\left(\frac{x}{u}\right),$$

$$x=\frac{x}{u}g\left(\frac{x}{u}\right)=G\left(\frac{x}{u}\right), \quad \frac{x}{u}=\psi(x)=v,$$

and so

(3.13.5) $$\phi(x)\psi(x)=x.$$

[a] Compare Theorem **44**. [b] See Cooper (**4**) for case (*b*).

[c] With the appropriate changes of sign.

Hence $\phi''(x)\psi(x)+2\phi'(x)\psi'(x)+\phi(x)\psi''(x)=0,$

and so, by (3.13.4),

$$(\phi''\psi+\phi\psi'')^2=4\phi'^2\psi'^2\leqq 4\phi\psi\phi''\psi'',$$

all the arguments being x. This is possible only if

$$\phi''\psi=\phi\psi''=-\phi'\psi', \quad \phi''\psi+\phi'\psi'=0,$$

or $\phi'\psi$ is constant. Hence, by (3.13.5), $x\phi'/\phi$ is constant, in which case ϕ and the other functions are powers of x.

3.14. Some theorems concerning monotonic functions. We collect here some simple theorems which will be useful later. The first characterises monotonic functions as (3.4.2) characterises continuous convex functions.

102.[a] *A necessary and sufficient condition that*

$$(\Sigma p)\,\phi(\Sigma x)\leqq \Sigma p\,\phi(x), \tag{3.14.1}$$

for all positive x and p, is that $\phi(x)$ should decrease (in the wide sense) for $x>0$. The opposite inequality is similarly characteristic of increasing functions.

There is strict inequality if $\phi(x)$ decreases strictly and there is more than one x.

(i) If ϕ decreases, $\phi(\Sigma x)\leqq\phi(x)$, whence (3.14.1) follows.

(ii) If in (3.14.1) we take $n=2$, $x_1=x$, $x_2=h$, $p_1=1$, $p_2=p$, we obtain

$$(1+p)\,\phi(x+h)\leqq\phi(x)+p\,\phi(h).$$

Making $p\to 0$, we see that $\phi(x+h)\leqq\phi(x)$.

The case $\phi(x)=x^{\alpha-1}$ $(0<\alpha<1)$, $p=x$, gives Theorem **19.**

103. *A sufficient condition that*

$$f(\Sigma x)\leqq\Sigma f(x), \tag{3.14.2}$$

for all positive x, is that $x^{-1}f(x)$ should decrease. There is strict inequality if $x^{-1}f(x)$ decreases strictly and there is more than one x.

For if we write $f(x)=x\phi(x)$, then (3.14.2) becomes (3.14.1) with $p=x$. The condition is not necessary, since (3.14.2) is satisfied by any $f(x)$ for which

$$f(x)>0, \quad \operatorname{Max} f(x)\leqq 2\operatorname{Min} f(x);$$

for example $f(x)=3+\cos x$.

[a] Jensen (**2**): Jensen does not refer to the necessity of the condition.

104. *If*

$$\phi(\Sigma px) \leqq \Sigma p\phi(x) \tag{3.14.3}$$

for all positive x and p, then $\phi(x)$ is a multiple of x.

If we take $n=2$, $x_1=x$, $x_2=y$, $p_1=y/2x$ and $p_2=\frac{1}{2}$ in (3.14.3), we obtain

$$\frac{\phi(y)}{y} \leqq \frac{\phi(x)}{x}.$$

Since we can interchange x and y, ϕ/x is constant.

3.15. Sums with an arbitrary function: generalisations of Jensen's inequality. We may define 'sums' involving an arbitrary function ϕ as well as means. We write

$$\mathfrak{S}_\phi(a) = \phi^{-1}\{\Sigma\phi(a)\}.$$

Here $\phi(x)$ is continuous and strictly monotonic, as in § 3.1; but it is necessary now to assume rather more, since $\Sigma\phi(a)$ is not, like $\Sigma q\phi(a)$, a mean of the values of $\phi(a)$, and so necessarily a possible value of $\phi(x)$. We therefore assume that $\phi(x)$ is positive for all positive x and tends to ∞ either when $x \to 0$ or when $x \to \infty$. We shall also assume that the a are all positive, leaving the reader to make the modifications appropriate when any a is zero[a].

105.[b] *If ψ and χ are continuous, positive, and strictly monotonic, then $\mathfrak{S}_\psi$ and $\mathfrak{S}_\chi$ are comparable whenever* (1) *ψ and χ vary in opposite directions, or* (2) *ψ and χ vary in the same direction and χ/ψ is monotonic.*

In case (1),

$$\mathfrak{S}_\psi \leqq \mathfrak{S}_\chi \tag{3.15.1}$$

if ψ decreases and χ increases. In case (2), (3.15.1) *is true if χ/ψ decreases. There is equality in case* (1), *and, if χ/ψ is strictly monotonic, also in case* (2), *only when there is only one a.*

[a] Suppose, for example, that $\phi(x)=x^r$, where $r>0$ (the case of § 2.10). Then $\phi(0)=0$, and we need make no distinction between two such systems of the a as (1, 1) and (1, 1, 0). If $\phi(0)$ were positive it would be necessary to distinguish, and the discrimination of the cases of equality in Theorem **105** would become tedious. If $\phi(x)\to\infty$ when $x\to 0$, then $\mathfrak{S}_\phi(a)=0$ whenever any a is zero.

[b] The substance of this theorem is due to Cooper (**2**).

In case (1), when χ increases,

$$\mathfrak{S}_\chi(a) \geqq \chi^{-1}\{\chi(\operatorname{Max} a)\} = \operatorname{Max} a$$

and similarly $\mathfrak{S}_\psi(a) \leqq \operatorname{Min} a$.

In case (2), suppose that ψ and χ increase, and write

$$\psi(a) = x, \quad a = \psi^{-1}(x), \quad \chi\psi^{-1} = f.$$

Then (3.15.1) reduces to (3.14.2), and is true if $x^{-1}f(x)$ decreases, i.e. if

$$\frac{f\{\psi(x)\}}{\psi(x)} = \frac{\chi(x)}{\psi(x)}$$

decreases. If ψ and χ decrease, (3.15.1) reduces to (3.14.2) reversed, and is true if f/x increases, or if χ/ψ decreases.

The reader will have no difficulty in distinguishing the cases of equality. The case $\psi = x^s$, $\chi = x^r$ gives Theorem **19**.

We may also define weighted sums analogous to those of § 2.10 (iv), viz.

$$\mathfrak{T}_\phi(a) = \phi^{-1}\{\Sigma p\,\phi(a)\},$$

where the p are arbitrary positive numbers. $\mathfrak{T}_\phi$ reduces to $\mathfrak{M}_\phi$ if $\Sigma p = 1$, to $\mathfrak{S}_\phi$ if every p is 1.

3.16. Generalisations of Minkowski's inequality. If $\phi(x) = x^r$, where $r > 1$, we have

$$\mathfrak{M}_\phi\left(\frac{a+b}{2}\right) \leqq \tfrac{1}{2}\{\mathfrak{M}_\phi(a) + \mathfrak{M}_\phi(b)\}, \tag{3.16.1}$$

$$\mathfrak{S}_\phi\left(\frac{a+b}{2}\right) \leqq \tfrac{1}{2}\{\mathfrak{S}_\phi(a) + \mathfrak{S}_\phi(b)\}, \tag{3.16.2}$$

$$\mathfrak{T}_\phi\left(\frac{a+b}{2}\right) \leqq \tfrac{1}{2}\{\mathfrak{T}_\phi(a) + \mathfrak{T}_\phi(b)\}, \tag{3.16.3}$$

all these inequalities being essentially equivalent and included in Theorem **24**.

The inequalities are not equivalent for general ϕ; all of them are of the form

$$2\phi^{-1}\left\{\Sigma p\,\phi\left(\frac{a+b}{2}\right)\right\} \leqq \phi^{-1}\{\Sigma p\,\phi(a)\} + \phi^{-1}\{\Sigma p\,\phi(b)\}, \tag{3.16.4}$$

but the differences between the weights p are now significant. In (3.16.1), $\Sigma p = 1$; in (3.16.2), $p = 1$; in (3.16.3), the p are any positive numbers. We call these three cases the cases (I), (II), and (III). In discussing them, we shall suppose that

$$\phi > 0, \quad \phi' > 0$$

for $x > 0$.

The inequality (3.16.4) asserts that $\phi^{-1}\{\Sigma p\phi(x)\}$ is, for given p, a convex function of the n variables $x_1, x_2, \ldots, x_n$; or, after § 3.12,[a] that

$$\chi(t) = \phi^{-1}\{\Sigma p\phi(x+ut)\},$$

where the x, p, u are fixed and the x and p positive, is convex in t for all t for which all $x+ut$ are positive. If ϕ is twice differentiable this condition is, by Theorem **94**, equivalent to $\chi''(0) \geqq 0$. A straightforward calculation shows that

(3.16.5)
$$\{\phi'(\chi)\}^3\chi'' = \{\phi'(\chi)\}^2\Sigma pu^2\phi''(x) - \phi''(\chi)\{\Sigma pu\phi'(x)\}^2,$$

where

$$\chi = \chi(0) = \phi^{-1}\{\Sigma p\phi(x)\} \tag{3.16.6}$$

and $\chi'' = \chi''(0)$. We have therefore to consider in what circumstances

$$\{\phi'(\chi)\}^2\Sigma pu^2\phi''(x) - \phi''(\chi)\{\Sigma pu\phi'(x)\}^2 \geqq 0. \tag{3.16.7}$$

It is easy to see that (3.16.7) cannot be true generally without restriction on the sign of ϕ''. Suppose for example that $\phi' > 0$ and that ϕ'' is continuous and sometimes negative. We can then choose x_1 and x_2 so that $\phi''(x_1) < 0$, $\phi''(x_2) < 0$, and u_1 and u_2 so that

$$p_1u_1\phi'(x_1) + p_2u_2\phi'(x_2) = 0.$$

In this case (3.16.7), for $n = 2$, reduces to

$$\{\phi'(\chi)\}^2\{p_1u_1^2\phi''(x_1) + p_2u_2^2\phi''(x_2)\} \geqq 0,$$

which is false. We shall therefore suppose in what follows that

$$\phi > 0, \ \phi' > 0, \ \phi'' > 0.$$

We can write (3.16.7) in the form

$$\psi(\chi) = \frac{\phi'^2(\chi)}{\phi''(\chi)} \geqq \frac{\{\Sigma pu\phi'(x)\}^2}{\Sigma pu^2\phi''(x)}. \tag{3.16.8}$$

[a] We take for granted the obvious extensions of § 3.12 from two to n variables.

Now, by Theorem 7,

$$(3.16.9)\quad (\Sigma pu\phi')^2 = \left\{\Sigma \sqrt{(p\phi'')}\, u \,.\, \sqrt{\left(\frac{p\phi'^2}{\phi''}\right)}\right\}^2 \leqq \Sigma pu^2\phi'' \,\Sigma p\frac{\phi'^2}{\phi''}.$$

Hence (3.16.8) is certainly true for all x, u if

$$(3.16.10)\qquad \psi(\chi) \geqq \Sigma p \frac{\{\phi'(x)\}^2}{\phi''(x)} = \Sigma p \psi(x),$$

for all x. Further, there is equality in (3.16.9) if

$$u = \frac{\phi'(x)}{\phi''(x)} \quad (\nu = 1, 2, \ldots, n),$$

so that (3.16.10) is both a sufficient and a necessary condition for the truth of (3.16.8). Finally, if we write $y = \phi(x)$ and

$$(3.16.11)\quad \Phi(y) = \psi(x) = \psi\{\phi^{-1}(y)\} = \frac{\{\phi'[\phi^{-1}(y)]\}^2}{\phi''\{\phi^{-1}(y)\}},$$

(3.16.10) assumes the form

$$(3.16.12)\qquad \Phi(\Sigma py) \geqq \Sigma p \Phi(y).$$

We now consider the three cases (I), (II), (III) separately.

(i) In case (I), (3.16.12) is true if and only if $\Phi(y)$ is a concave function of y.

(ii) In case (II), (3.16.12) is (3.14.2), reversed and with y, Φ for x, f. A sufficient (though not a necessary) condition is that Φ/y is an increasing function of y or, what is the same thing, that $\phi\phi''/\phi'^2$ is a decreasing function of x.

(iii) In case (III), (3.16.12) is (3.14.3), with the appropriate variations. It can be true generally only when $\Phi(y)$ is a multiple of y, or when $\phi\phi''/\phi'^2$ is constant, in which case ϕ is of one of the forms

$$(3.16.13)\qquad (ax+b)^c \ (a > 0,\ c > 1^{\text{a}}), \quad e^{ax+b}.$$

In these cases it is true.

There are alternative forms of the conditions (i) and (ii) which show better their relations to one another. We shall suppose ϕ''' continuous, as we may do without affecting seriously the interest

[a] Since $\phi'' > 0$.

of the results. Then from

$$\Phi(y)=\frac{\phi'^2(x)}{\phi''(x)}$$

we deduce
$$\Phi'(y)=\frac{d}{dx}\frac{\phi'(x)}{\phi''(x)}+1,$$

$$\Phi''(y)=\frac{1}{\phi'(x)}\frac{d^2}{dx^2}\frac{\phi'(x)}{\phi''(x)}.$$

Hence $\Phi(y)$ is concave if and only if $\phi'(x)/\phi''(x)$ is concave, or, what is the same thing, if $\phi'\phi'''/\phi''^2$ is increasing. These are alternative forms of (i), and an alternative form of (ii) is 'ϕ/ϕ' is convex'.

Summing up our conclusions, we have

106.[a] *Suppose that ϕ'''' is continuous and that $\phi>0$, $\phi'>0$, $\phi''>0$. Then*

(i) *it is necessary and sufficient for* (3.16.1) *that ϕ'/ϕ'' should be concave, or $\phi'\phi'''/\phi''^2$ increasing;*

(ii) *it is sufficient (but not necessary) for* (3.16.2) *that ϕ/ϕ' should be convex, or $\phi\phi''/\phi'^2$ decreasing;*

(iii) *it is necessary and sufficient for* (3.16.3) *that ϕ should be one of the functions* (3.16.13).

We leave it for the reader to formulate the variations of this theorem, when (for example) $\phi>0$, $\phi'<0$, $\phi''>0$, or when the inequalities are reversed. It is instructive to verify that (i) is satisfied (from a certain x onwards) when $\phi=x^p/\log x$, where $p>1$, but not when $\phi=x^p\log x$, while for (ii) the situation is reversed.

3.17. Comparison of sets.

Theorem **105** asserts that

$$\mathfrak{S}_\psi(a)\leqq\mathfrak{S}_\chi(a)$$

for a given pair of functions ψ and χ and all a. The theorems of this section are of a different type, involving given sets (a) and (a') and a variable function ϕ. We consider the conditions under which

$$\mathfrak{S}_\phi(a')\leqq\mathfrak{S}_\phi(a),$$

[a] The first results of this character are due to Bosanquet (**1**): Bosanquet considers case (II).

or, what is the same thing for increasing ϕ,

(3.17.1)
$$\phi(a_1')+\phi(a_2')+\ldots+\phi(a_n')\leqq\phi(a_1)+\phi(a_2)+\ldots+\phi(a_n)$$
for given a and a' and all ϕ of a certain class.

107. *Suppose that the sets* (a) *and* (a') *are arranged in descending order of magnitude. Then a necessary and sufficient condition that* (3.17.1) *should be true for all continuous and increasing* ϕ *is that*
$$a_\nu'\leqq a_\nu \quad (\nu=1,2,\ldots,n).$$

The sufficiency of the condition is obvious. To prove it necessary, suppose that $a_\mu'>a_\mu$ for some μ, that $a_\mu<b<a_\mu'$, and that $\phi^*(x)$ is defined by
$$\phi^*(x)=0 \ (x\leqq b), \quad \phi^*(x)=1 \ (x>b).$$
Then
$$\Sigma\phi^*(a')\geqq\mu>\mu-1\geqq\Sigma\phi^*(a).$$
Hence (3.17.1) is false for ϕ^*, and therefore also false for an appropriately chosen continuous increasing approximation to ϕ^*.

Our next theorem is connected with the theorems of §§ 2.18–2.20.

108. *In order that* (3.17.1) *should be true for all continuous convex* ϕ, *it is necessary and sufficient either that* (1) $(a')\prec(a)$, *i.e. that* (a') *is majorised by* (a) *in the sense of* § 2.18, *or that* (2) (a') *is an average of* (a) *in the sense of* § 2.20.

If these conditions are satisfied, and $\phi''(x)$ *exists for all* x, *and is positive, then equality can occur in* (3.17.1) *only when the sets* (a) *and* (a') *are identical*[a].

We have proved already (Theorem **46**) that the two conditions are equivalent. It is therefore enough to prove that the first is necessary and the second sufficient. We may suppose (a) and (a') arranged in descending order.

(i) *Condition* (1) *is necessary.* Condition (1) asserts that

(3.17.2) $$a_1'+a_2'+\ldots+a_\nu'\leqq a_1+a_2+\ldots+a_\nu \quad (\nu=1,2,\ldots,n),$$
with equality in the case $\nu=n$.

The functions x and $-x$ are both continuous and convex in any

[a] Schur (**2**) proves that (2) is a sufficient condition, and the remark concerning the case of equality is also due to him. For the complete theorem, see Hardy, Littlewood, and Pólya (**2**). Karamata (**1**) considers condition (1).

interval. Hence, if (3.17.1) is true, $\Sigma a' \leqq \Sigma a$ and $\Sigma(-a') \leqq \Sigma(-a)$, i.e. $\Sigma a' = \Sigma a$, which is (3.17.2) for $\nu = n$, with equality.

Next, let

$$\phi(x) = 0 \quad (x \leqq a_\nu), \qquad \phi(x) = x - a_\nu \quad (x > a_\nu).$$

Then $\phi(x)$ is continuous and convex in any interval, and $\phi(x) \geqq 0$, $\phi(x) \geqq x - a_\nu$. Hence

$$a_1' + a_2' + \ldots + a_\nu' - \nu a_\nu \leqq \Sigma\phi(a') \leqq \Sigma\phi(a) = a_1 + a_2 + \ldots + a_\nu - \nu a_\nu,$$

which is (3.17.2).

(ii) *Condition* (2) *is sufficient*. If (a') is an average of (a), we have

$$a_\mu' = p_{\mu 1} a_1 + p_{\mu 2} a_2 + \ldots + p_{\mu n} a_n,$$

where

$$p_{\mu\nu} \geqq 0, \qquad \sum_{\mu=1}^{n} p_{\mu\nu} = 1, \qquad \sum_{\nu=1}^{n} p_{\mu\nu} = 1$$

for all μ and ν. If ϕ is convex, then

$$\phi(a_\mu') \leqq p_{\mu 1}\,\phi(a_1) + \ldots + p_{\mu n}\,\phi(a_n), \tag{3.17.3}$$

and (3.17.1) follows by summation.

(iii) If there is equality in (3.17.1), there must be equality in each of (3.17.3).

If $\phi''(x) > 0$, *and every* $p_{\mu\nu}$ *is positive*, then it follows from Theorem 95 that all the a are equal, in which case all the a' are also equal and the common values are the same.

In general, however, some of the $p_{\mu\nu}$ will be zero. We shall say that a_μ' and a_ν are *immediately connected* if $p_{\mu\nu} > 0$, i.e. if a_ν occurs effectively in the formula for a_μ'; and that any two elements (whether a or a') are *connected* if they can be joined by a chain of elements in which each consecutive pair is immediately connected.

Consider now the complete set C of elements connected with a_1. We may write this set (changing the numeration of the elements if necessary) as

$$a_1, a_2, \ldots, a_r, \quad a_1', a_2', \ldots, a_s'; \tag{C}$$

the a' of C involve the a of C, and no other a, and no other a' involves an a of C. Hence, using the sum-properties of the p,

$$s = \sum_{\mu=1}^{s} \sum_{\nu=1}^{r} p_{\mu\nu} = \sum_{\nu=1}^{r} \sum_{\mu=1}^{s} p_{\mu\nu} = r;$$

so that C contains just as many a' as a. It follows from Theorem **95**, and from the equality in (3.17.3), that all a immediately connected with an a' are equal to that a'. Hence all connected a and a' are equal, and C contains r elements of each set, all equal to a_1.

We now repeat the argument, starting from a_{r+1}, and we conclude that both (a) and (a') consist of a certain number of blocks of equal elements, the values of the elements in corresponding blocks being the same.

Incidentally we have proved

109. *If* $\phi''(x) > 0$, $p_{\mu\nu} > 0$, $\sum_{\mu} p_{\mu\nu} = 1$, $\sum_{\nu} p_{\mu\nu} = 1$, *and* $a_\mu' = \sum_{\nu} p_{\mu\nu} a_\nu$, *then*

$$\Sigma\phi(a') < \Sigma\phi(a), \tag{3.17.4}$$

unless all the a and a' are equal.

If all the a' are equal, (3.17.4) is a special case of Theorem **95**. A special case of Theorem **108** which is often useful is

110. *If $\phi(x)$ is continuous and convex, and $|h'| \leqq |h|$, then*

$$\phi(x-h') + \phi(x+h') \leqq \phi(x-h) + \phi(x+h). \tag{3.17.5}$$

3.18. Further general properties of convex functions. We have assumed since § 3.6 that $\phi(x)$ is continuous. We shall now discard this hypothesis and consider the direct consequences of (3.5.1). The general lesson of the theorems which follow will be that *a convex function is either very regular or very irregular*, and in particular that a convex function which is not 'entirely irregular' is necessarily continuous (so that the hypothesis of continuity is a good deal less restrictive than might have been expected).

111. *Suppose that $\phi(x)$ is convex in the open interval (H, K), and bounded above in some interval i interior to (H, K). Then $\phi(x)$ is continuous in the open interval (H, K). Further, $\phi(x)$ has everywhere left-hand and right-hand derivatives; the right-hand derivative is not less than the left-hand derivative; and both derivatives increase with x.*

It follows that a discontinuous convex function is unbounded in every interval.

We prove first that $\phi(x)$ is bounded above in every interval interior to (H, K). The kernel of the proof is this. The argument of § **3.6** shows that

$$\phi(\Sigma qx) \leqq \Sigma q\phi(x)$$

for any *rational* q; it was only in the passage to irrational q that we used the hypothesis of continuity. Suppose now that i is (h, k) and that the upper bound of ϕ in i is G. It is enough to prove ϕ bounded above in (l, h) and (k, m), where l and m are any numbers such that $H<l<h<k<m<K$. If x is in (l, h), we can find a ξ in i so that x divides (l, ξ) rationally, and then $\phi(x)$ must lie below a bound depending on $\phi(l)$ and G, and so be bounded above in (l, h). Similarly, it must be bounded above in (k, m).

To state the argument precisely, let h be the left-hand end of i and G the upper bound of ϕ in i, and suppose that

$$H<l<x<h.$$

We can choose integers m and $n>m$ so that

$$\xi = l + \frac{n}{m}(x-l)$$

lies in i, and then

$$\phi(x) = \phi\left\{\frac{m\xi+(n-m)l}{n}\right\} \leqq \frac{m}{n}\phi(\xi) + \frac{n-m}{n}\phi(l)$$

$$\leqq \frac{m}{n}G + \frac{n-m}{n}\phi(l) \leqq \operatorname{Max}\{G, \phi(l)\}.$$

Hence $\phi(x)$ is bounded above in (l, h).

In proving the remainder of the theorem we may restrict ourselves to an interval (H', K') interior to (H, K), or, what is the same thing, we may suppose ϕ bounded above in the original interval. Suppose then that $\phi(x) \leqq G$ in (H, K), that $H<x<K$, that m and $n>m$ are positive integers, and that δ is a number (positive or negative) small enough to leave $x+n\delta$ inside (H, K). Then

$$\phi(x+m\delta) = \phi\left\{\frac{m(x+n\delta)+(n-m)x}{n}\right\} \leqq \frac{m}{n}\phi(x+n\delta) + \frac{n-m}{n}\phi(x),$$

or $$\frac{\phi(x+n\delta)-\phi(x)}{n} \geqq \frac{\phi(x+m\delta)-\phi(x)}{m} \quad (m<n).$$

Replacing δ by $-\delta$, and combining the two inequalities, we obtain

$$(3.18.1)\quad \frac{\phi(x+n\delta)-\phi(x)}{n} \geqq \frac{\phi(x+m\delta)-\phi(x)}{m} \geqq \frac{\phi(x)-\phi(x-m\delta)}{m} \geqq \frac{\phi(x)-\phi(x-n\delta)}{n}$$

(the central inequality following directly from the convexity of ϕ).

If in (3.18.1) we take $m=1$, and remember that $\phi \leqq G$, we find that

$$(3.18.2)\quad \frac{G-\phi(x)}{n} \geqq \phi(x+\delta)-\phi(x) \geqq \phi(x)-\phi(x-\delta) \geqq \frac{\phi(x)-G}{n}.$$

We now suppose that $\delta \to 0$ and $n \to \infty$, but so that $x \pm n\delta$ remains inside the interval. It then follows from (3.18.2) that $\phi(x+\delta)$ and $\phi(x-\delta)$ tend to $\phi(x)$, and so that ϕ is continuous.

We next suppose $\delta > 0$, and replace δ in (3.18.1) by δ/n. We have then

$$(3.18.3)\quad \frac{\phi(x+\delta)-\phi(x)}{\delta} \geqq \frac{\phi(x+\delta')-\phi(x)}{\delta'} \geqq \frac{\phi(x)-\phi(x-\delta')}{\delta'} \geqq \frac{\phi(x)-\phi(x-\delta)}{\delta},$$

where $\delta' = m\delta/n$ is any rational multiple of δ less than δ. Since ϕ is continuous, (3.18.3) is true for any $\delta' < \delta$. It follows that the quotients on the extreme left and right decrease and increase respectively when δ decreases to zero, and so that each tends to a limit. Hence ϕ possesses right-handed and left-handed derivatives ϕ_r' and ϕ_l', and $\phi_l' \leqq \phi_r'$.

Finally, we may write $x-\delta'=x_1$, $x=x_2$, $x+\delta=x_3$ (or $x-\delta=x_1$, $x=x_2$, $x+\delta'=x_3$), when (3.18.3) gives

$$\frac{\phi(x_3)-\phi(x_2)}{x_3-x_2} \geqq \frac{\phi(x_2)-\phi(x_1)}{x_2-x_1}.$$

A fortiori, if $x_1 < x_2 < x_3 < x_4$, we have

$$\frac{\phi(x_4)-\phi(x_3)}{x_4-x_3} \geqq \frac{\phi(x_2)-\phi(x_1)}{x_2-x_1}.$$

Making $x_3 \to x_4$, $x_2 \to x_1$, we obtain

$$\phi_r'(x_4) \geqq \phi_l'(x_4) \geqq \phi_r'(x_1) \geqq \phi_l'(x_1), \tag{3.18.4}$$

which completes the proof of the theorem.

It is plain from what precedes that

$$\phi_l'(x_4) \geqq \frac{\phi(x_3)-\phi(x_2)}{x_3-x_2} \geqq \phi_r'(x_1), \tag{3.18.5}$$

if $x_1 \leqq x_2 < x_3 \leqq x_4$.

Theorem **111** asserts nothing about the existence of an ordinary differential coefficient $\phi'(x)$. It is however easy to prove that $\phi'(x)$ exists except perhaps for an enumerable set of values of x. The function $\phi_l'(x)$, being monotonic, is continuous except perhaps in such a set. If it is continuous at x_1 then, by (3.18.4), $\phi_r'(x_1)$ lies between $\phi_l'(x_1)$ and $\phi_l'(x_4)$, which tends to $\phi_l'(x_1)$ when $x_4 \to x_1$. Hence $\phi_r'(x_1) = \phi_l'(x_1)$, and $\phi'(x)$ exists for $x = x_1$.

It is also plain from (3.18.5) that, if $\phi(x)$ is continuous and convex in an open interval (a, b), then

$$\left| \frac{\phi(x')-\phi(x)}{x'-x} \right|$$

is bounded for all x and x' of any closed sub-interval of (a, b).

3.19. Further properties of continuous convex functions. We now suppose $\phi(x)$ convex and continuous. It follows from (3.18.5) that if $H < \xi < K$ and

$$\phi_l'(\xi) \leqq \lambda \leqq \phi_r'(\xi)$$

then the line

$$y - \phi(\xi) = \lambda(x-\xi) \tag{3.19.1}$$

will lie wholly under (on or below) the curve. In other words

112. *If $\phi(x)$ is convex and continuous then there is at least one line through every point of the curve $y = \phi(x)$ which lies wholly under the curve.*

A line through a point of a curve which lies wholly on one side of the curve (under or over) is called a *Stützgerade* of the curve. If $\phi(x)$ is concave, then the graph of $\phi(x)$ has at every point a *Stützgerade* over the curve. If $\phi(x)$ is both convex and concave the two lines must coincide and $\phi(x)$ must be linear.

It is easy to see the truth of Theorem **112** directly. If $\xi<x<x'$, and P, Q, Q' are the points on the curve corresponding to ξ, x, x', then PQ lies under PQ', and the slope of PQ decreases as x approaches ξ, and so tends to a limit ν. Similarly, if $x<\xi$ and x tends to ξ, the slope of QP increases to a limit μ. If μ were greater than ν, and x_1, x_2 were respectively less than and greater than ξ, and sufficiently near to ξ, then P would lie above P_1P_2, in contradiction to the convexity of the curve. Hence $\mu \leqq \nu$, and (3.19.1) lies under the curve if λ has any value between μ and ν inclusive.

In this proof we do not appeal to Theorem **111**, but the proof depends on just those geometrical ideas which underlie the more formal and analytical argument of § 3.18.

Suppose now, conversely, that $\phi(x)$ is continuous and has the property asserted in Theorem **112**. If x_1 and x_2 are two values of x, P_1 and P_2 the corresponding points on the curve, and P the point corresponding to $\xi=\frac{1}{2}(x_1+x_2)$, then both P_1 and P_2 lie over a certain line through P, and the middle point of P_1P_2 lies over P. Hence $\phi(x)$ is convex.

We have thus proved that the property of Theorem **112** affords a necessary and sufficient condition for the convexity of a continuous function, and might be used as an alternative definition of convexity. That is to say, we might define convexity, for continuous functions, as follows: *a continuous function $\phi(x)$ is said to be convex in (H, K) if to any ξ of (H, K) corresponds a number $\lambda=\lambda(\xi)$ such that*

$$\phi(\xi)+\lambda(x-\xi)\leqq\phi(x)$$

for all x of (H, K).

This definition of a convex function is quite as 'natural' as that implied in (3.5.2), and it is interesting to deduce some of the characteristic properties of continuous convex functions directly from it. For example, the inequality (3.4.2) may be proved as follows[a].

Writing as usual

$$\mathfrak{A}(b)=\Sigma qb,$$

and taking $\xi=\mathfrak{A}(a)$, a value which lies in the interval of variation of the a, we have

$$\phi\{\mathfrak{A}(a)\}+\lambda(a-\xi)\leqq\phi(a)$$

[a] Jessen (**2**).

for a certain $\lambda = \lambda(\xi)$ and all a. Performing the operation $\mathfrak{A}$ on each side, we obtain

$$\phi\{\mathfrak{A}(a)\} + \lambda\{\mathfrak{A}(a) - \xi\} \leqq \mathfrak{A}\{\phi(a)\},$$

or

$$\phi\{\mathfrak{A}(a)\} \leqq \mathfrak{A}\{\phi(a)\},$$

which is (3.4.2). It is instructive to compare this argument with that of § 3.10 (ii).

3.20. Discontinuous convex functions. Discontinuous convex functions are, by Theorem **111**, unbounded in every interval, and their existence has not been proved except under the assumption of Zermelo's Axiom or (what is for our purpose equivalent) the assumption that the continuum can be well-ordered.

If

$$f(x+y) = f(x) + f(y), \tag{3.20.1}$$

then

$$f(2x) = 2f(x)$$

and

$$2f\left(\frac{x+y}{2}\right) = f(x+y) = f(x) + f(y).$$

Thus a solution of (3.20.1) is certainly convex.

It was proved by Hamel (**1**)[a] that, if Zermelo's Axiom is true, there exist *bases* $[\alpha, \beta, \gamma, \ldots]$ for the real numbers, that is to say, sets of real numbers $\alpha, \beta, \gamma, \ldots$ such that every real a is expressible uniquely in the form of a finite sum

$$x = a\alpha + b\beta + \ldots + l\lambda$$

with rational coefficients $a, b, \ldots, l$. If we assume this, we can at once write down discontinuous solutions of (3.20.1); we give $f(x)$ arbitrary values $f(\alpha), f(\beta), \ldots$ for $x = \alpha, \beta, \ldots$, and define $f(x)$ generally by

$$f(x) = af(\alpha) + bf(\beta) + \ldots + lf(\lambda).$$

Then, if $y = a'\alpha + \ldots$, we have

$$f(x+y) = f\{(a+a')\alpha + \ldots\} = (a+a')f(\alpha) + \ldots = f(x) + f(y).$$

For more detailed study of the properties of convex functions, of the solutions of the equation (3.20.1) and of inequalities associated with it, we may refer to Darboux (**1**), Fréchet (**1, 2**), F. Bernstein (**1**), Bernstein and Doetsch (**1**), Blumberg (**1**), Sierpiński (**1, 2**), Cooper (**3**), and Ostrowski (**1**). Blumberg and Sierpiński prove that *any convex measurable function is continuous*, and Ostrowski obtains a still more general result.

[a] See also Hahn (**1**, 581).

MISCELLANEOUS THEOREMS AND EXAMPLES

113. If α is a constant, $\alpha \neq 0$, and $\chi = \alpha\phi$, then

$$\mathfrak{S}_\chi(a) = \mathfrak{S}_\phi(a), \quad \mathfrak{T}_\chi(a) = \mathfrak{T}_\phi(a).$$

[The corresponding property of $\mathfrak{M}$ is included in Theorem **83**.]

114. An increasing convex function of a convex function is convex.

115. If every chord of a continuous curve contains a point which lies above the curve, then every point of every chord, except the end-points, lies above the curve.

116. If $\phi(x)$ is convex and continuous, $a<b<c$, and $\phi(a)=\phi(b)=\phi(c)$, then $\phi(x)$ is constant in (a, c).

117. If all the numbers are positive, then

$$x \log \frac{x}{a} + y \log \frac{y}{b} > (x+y) \log \frac{x+y}{a+b},$$

unless $x/a = y/b$.

[$(x \log x)'' > 0$.]

118. If $f(x)$ is positive and twice differentiable, then a necessary and sufficient condition that $\log f(x)$ should be convex is that $ff'' - f'^2 \geqq 0$.

119. If $\phi(x)$ is continuous for $x>0$, and one of the functions $x\phi(x)$ and $\phi(1/x)$ is convex, then so is the other.

120. If $\phi(x)$ is positive, twice differentiable and convex, then so are

$$x^{\frac{1}{2}(s+1)}\, \phi(x^{-s}) \quad (s \geqq 1), \qquad e^{\frac{1}{2}x} \phi(e^{-x})$$

(the first for positive x).

121. If ψ and χ are continuous and strictly monotonic, and χ increasing, then a necessary and sufficient condition that

$$\psi^{-1}\{\psi^{q_1}(a_1) \ldots \psi^{q_n}(a_n)\} \leqq \chi^{-1}\{\chi^{q_1}(a_1) \ldots \chi^{q_n}(a_n)\},$$

for all a and q, is that

$$\phi(y) = \log\{\chi\psi^{-1}(e^y)\}$$

should be convex.

[Compare Theorem **92**.]

122. Suppose that

(i) $$\phi(x_1)(x_3 - x_2) + \phi(x_2)(x_1 - x_3) + \phi(x_3)(x_2 - x_1) \geqq 0,$$

or (what is the same thing)

(ii) $$\begin{vmatrix} 1 & x_1 & \phi(x_1) \\ 1 & x_2 & \phi(x_2) \\ 1 & x_3 & \phi(x_3) \end{vmatrix} \geqq 0,$$

for all x_1, x_2, x_3 of an open interval I for which $x_1 < x_2 < x_3$. Then $\phi(x)$ is continuous in I, and has finite left-handed and right-handed derivatives at every point of I.

If $\phi(x)$ is twice differentiable, then (i) and (ii) are equivalent to the differential inequality

$$\phi''(x) \geqq 0.$$

[(i) and (ii) are alternative forms of (3.5.2), and $\phi(x)$ is convex, so that the theorem is a restatement of parts of Theorems **111** and **94**.]

123. Suppose that

(i) $$\phi(x_1)\sin(x_3-x_2)+\phi(x_2)\sin(x_1-x_3)+\phi(x_3)\sin(x_2-x_1)\geqq 0,$$

or (what is the same thing)

(ii) $$\begin{vmatrix} \cos x_1 & \sin x_1 & \phi(x_1) \\ \cos x_2 & \sin x_2 & \phi(x_2) \\ \cos x_3 & \sin x_3 & \phi(x_3) \end{vmatrix} \geqq 0,$$

for all x_1, x_2, x_3 of an open interval I for which $x_1<x_2<x_3<x_1+\pi$. Then $\phi(x)$ is continuous in I and has finite left-handed and right-handed derivatives at every point of I.

If $\phi(x)$ is twice differentiable, then (i) and (ii) are equivalent to the differential inequality

$$\phi''(x)+\phi(x)\geqq 0.$$

[The result is important in the study of convex curves and of the behaviour of analytic functions in angular domains. See Pólya (**1**, 320; **4**, 573–576).]

124. A necessary and sufficient condition that a continuous function $\phi(x)$ should be convex in an interval I is that, if α is any real number and i any closed interval included in I, then $\phi(x)+\alpha x$ should attain its maximum in i at one of the ends of i. If also x and $\phi(x)$ are positive, then a necessary and sufficient condition that $\log\phi(x)$ should be a convex function of $\log x$ is that $x^{\alpha}\phi(x)$ should have the same maximal property.

[For applications of this theorem, which results immediately from the definitions, see Saks (**1**).]

125. A necessary and sufficient condition that a continuous function $\phi(x)$ should be convex in (a, b) is that

(i) $$\phi(x)\leqq \frac{1}{2h}\int_{x-h}^{x+h}\phi(t)\,dt$$

for $$a\leqq x-h<x<x+h\leqq b.$$

[This is a corollary of Theorem **124**. If $\phi(x)$ satisfies (i), so does $\phi(x)+\alpha x$; and it is plain that any continuous function which satisfies (i) must possess the property of Theorem **124**.[a]

Theorem **125** may also be proved independently; and there are various generalisations. In particular we need only suppose (i) true for every x and arbitrarily small $h=h(x)$.]

[a] For a formal proof, use Theorem **183**.

126. If $\phi(x)$ is convex and continuous for all x, and not constant, then $\phi(x)$ tends to infinity, for one or other approach of x to infinity, in such a manner as to be ultimately greater than a constant multiple of $|x|$.

127. If $\phi''>0$ for $x>0$, and $\phi(0)\geqq 0$, then ϕ/x increases for $x>0$.

[This follows at once from the equations

$$x^2\frac{d}{dx}\left(\frac{\phi}{x}\right)=x\phi'-\phi,\quad \frac{d}{dx}(x\phi'-\phi)=x\phi''.]$$

128. If $\phi''>0$ for $x\geqq 0$ and

$$\lim_{x\to\infty}(x\phi'-\phi)\leqq 0,$$

then ϕ/x decreases (strictly) for $x>0$.

[The limit certainly exists, since $x\phi'-\phi$ increases. The result follows from the equations used in proving Theorem **127**. The cases considered in Theorems **127** and **128** are the extreme cases possible when $\phi''>0$; if neither condition is satisfied, ϕ/x has a minimum for some positive x.]

129. If $\phi''>0$ for all x, $\phi(0)=0$, and ϕ/x is interpreted as $\phi'(0)$ for $x=0$, then ϕ/x increases for all x.

130. If the set $a_1, a_2, \dots, a_{2n+1}$ is convex in the sense of § 3.5, i.e. if

$$\Delta^2 a_\nu = a_\nu - 2a_{\nu+1} + a_{\nu+2} \geqq 0 \quad (\nu=1,2,\dots,2n-1),$$

then
$$\frac{a_1+a_3+\dots+a_{2n+1}}{n+1}\geqq\frac{a_2+a_4+\dots+a_{2n}}{n},$$

with inequality except when the numbers are in arithmetical progression.

[Nanson (**1**). Add up the inequalities

$$r(n-r+1)\,\Delta^2 a_{2r-1}\geqq 0,\quad r(n-r)\,\Delta^2 a_{2r}\geqq 0.$$

Theorem **130** may also be proved as an example of Theorem **108**: the set formed by the numbers 2, 4, ..., $2n$, each taken $n+1$ times, is majorised by that formed by the numbers 1, 3, ..., $2n+1$, each taken n times.]

131. If $0<x<1$, then

$$\frac{1-x^{n+1}}{n+1}>\frac{1-x^n}{n}\sqrt{x}.$$

[Put $x=y^2$ and $a_\nu=y^\nu$ in Theorem **130**.]

132. C is the centre of a circle and $A_0A_1\dots A_nC$ a polygon, whose vertices, except C, lie on the circle in the order indicated. C, A_0, A_n, are fixed and $A_1, A_2, \dots, A_{n-1}$ vary. Then the area and perimeter of the polygon are greatest when

$$A_0A_1=A_1A_2=\dots=A_{n-1}A_n.$$

[Let the angle $A_{\nu-1}CA_\nu$ be α_ν. Since $(\sin x)''<0$ for $0<x<\pi$, we have, by Theorem **95**,

$$\frac{1}{n}\Sigma\sin\alpha_\nu<\sin\left(\frac{\Sigma\alpha_\nu}{n}\right),$$

unless all the α_ν are equal, and a similar inequality in which α_ν is replaced by $\frac{1}{2}\alpha_\nu$. These inequalities give the two parts of the theorem. When A_n coincides with A_0, they reduce to familiar maximal properties of regular polygons.]

133. Suppose that f and g are continuous increasing inverse functions which vanish at the origin, that $F=xf$, $G=xg$, and that g satisfies the inequality

$$g(xy) \leqq g(x)\,g(y).$$

Suppose further that $\Sigma ab \leqq AB$ for all positive a and b such that $\Sigma G(b) \leqq G(B)$. Then

$$\Sigma F(a) \leqq \frac{1}{F(1/A)}.$$

[Cooper (3). The result is included in Theorem **15** when f is a power of x.]

134. If $\phi(x)$ is convex and continuous for $x \geqq 0$, $\nu=1, 2, 3, \ldots$, and the a_ν are non-negative and decreasing, then

$$\phi(0)+\Sigma\{\phi(na_n)-\phi[(n-1)\,a_n]\} \leqq \phi(\Sigma a_n).$$

If also $\phi'(x)$ is a strictly increasing function, there is equality only when the a_ν are equal up to a certain point and then zero.

[Hardy, Littlewood, and Pólya (2). Write

$$s_0=0, \quad s_\nu=a_1+a_2+\ldots+a_\nu \quad (\nu \geqq 1),$$

and

$$s_\nu+(\nu-1)\,a_\nu=s_{\nu-1}+\nu a_\nu=2x,$$

$$s_\nu-(\nu-1)\,a_\nu=2h, \quad s_{\nu-1}-\nu a_\nu=2h'.$$

It is easily verified that $|h'| \leqq h$, with equality only if $a_\nu=0$ or

$$a_1=a_2=\ldots=a_\nu.$$

It follows from Theorem **110** that

$$\phi(\nu a_\nu)-\phi\{(\nu-1)\,a_\nu\} \leqq \phi(s_\nu)-\phi(s_{\nu-1}),$$

and the result follows by summation.]

135. If $q>1$ and a_ν decreases, then

$$\Sigma\{\nu^q-(\nu-1)^q\}\,a_\nu{}^q \leqq (\Sigma a_\nu)^q.$$

[Example of Theorem **134**.]

136. Suppose that a is a function of $\nu_1, \nu_2, \ldots, \nu_m$; that $i_1, i_2, \ldots, i_m$ is a permutation of the numbers 1, 2, ..., m; and that the ψ and χ are continuous and strictly monotonic and the χ increasing. Then, in order that

$$\mathfrak{M}_{\psi_m}^{\nu_m} \ldots \mathfrak{M}_{\psi_1}^{\nu_1}(a) \leqq \mathfrak{M}_{\chi_{i_m}}^{\nu_{i_m}} \ldots \mathfrak{M}_{\chi_{i_1}}^{\nu_{i_1}}(a)$$

for all a and q, it is necessary that

(1) χ_μ is convex with respect to ψ_μ, for $\mu=1, 2, \ldots, m$;

(2) χ_λ is convex with respect to ψ_μ, if $\lambda>\mu$ and λ and μ correspond to an inversion in the permutation $i_1, i_2, \ldots, i_m$ (i.e. if the order is $\ldots \mu, \ldots, \lambda, \ldots$ in the series 1, 2, ..., m but $\ldots \lambda, \ldots, \mu, \ldots$ in $i_1, i_2, \ldots, i_m$).

[Jessen (3).]

137. In order that

$$\mathfrak{M}_{r_m}^{\nu_m} \dots \mathfrak{M}_{r_1}^{\nu_1}(a) \leqq \mathfrak{M}_{s_{i_m}}^{\nu_{i_m}} \dots \mathfrak{M}_{s_{i_1}}^{\nu_{i_1}}(a)$$

for all a and q, it is necessary *and sufficient* that (1) $r_\mu \leqq s_\mu$ and (2) $r_\mu \leqq s_\lambda$ (the range of μ and λ being defined as in Theorem **136**).

[Jessen (**3**). The most important cases are:

(i) $m=1,\ r<s$.

(ii) $m=2,\ (i_1, i_2)=(2, 1),\ s_2=r_2 \geqq s_1=r_1$.

The two cases correspond to Theorems **16** and **26**. The kernel of the theorem is contained in the statement that, whenever the two sides of the inequality are comparable, the inequality may be proved by repeated application of the special cases corresponding to (i) and (ii).]

138. A continuous curve $y=\phi(x)$ defined in an open interval, say $0<x<1$, has the following property: through every point of the curve there is either (a) a line which lies under the curve, or (b) a line which lies over the curve. Then one and the same of (a) and (b) is true at all points of the curve, and the curve is convex or concave.

[It is easy to show that if S_a and S_b are the sets of values of x for which (a) or (b) is true, then S_a and S_b are closed (in the open interval). But a continuum cannot be the sum of two non-null closed and exclusive sets.]

139. Suppose that $\phi(x)$ is convex and bounded below in (H, K), and that $m(x)$ is the lower bound of $\phi(x)$ at x (the limit of the lower bound of $\phi(x)$ in an interval including x). Then $m(x)$ is a continuous convex function; and either (i) $\phi(x)$ is identical with $m(x)$, or (ii) the graph of $\phi(x)$ is dense in the region $H \leqq x \leqq K,\ y \geqq m(x)$.

If $\phi(x)$ is convex and not bounded below, its graph is dense in the strip $H \leqq x \leqq K$.

[Bernstein and Doetsch (**1**).]

CHAPTER IV

VARIOUS APPLICATIONS OF THE CALCULUS

4.1. Introduction. Particular inequalities arising in ordinary analysis are often proved more easily by some special device than by an appeal to any general theory. We therefore interrupt our systematic treatment of the subject at this point, and devote a short chapter to the illustration of the simplest and most useful of these devices. The subject-matter is arranged according to the methods and instruments used rather than the character of the results.

DIFFERENTIAL CALCULUS: FUNCTIONS OF ONE VARIABLE

4.2. Applications of the mean value theorem. Our first examples depend upon a straightforward use of the mean value theorem

$$f(x+h)-f(x)=hf'(x+\theta h) \quad (0<\theta<1), \tag{4.2.1}$$

or its generalisations with higher derivatives. It is a corollary of (4.2.1) that a function with a positive differential coefficient increases with x.

(1) We have

$$\log(x+1)-\log x=\frac{1}{\xi},$$

where $x<\xi<x+1$, when $x>0$. It follows that

$$\frac{d}{dx}[x\{\log(x+1)-\log x\}]=\log(x+1)-\log x-\frac{1}{x+1}>0,$$

$$\frac{d}{dx}[(x+1)\{\log(x+1)-\log x\}]=\log(x+1)-\log x-\frac{1}{x}<0.$$

Hence $\left(1+\frac{1}{x}\right)^x$ increases with x, while $\left(1+\frac{1}{x}\right)^{x+1}$ decreases.

Since the latter function is $\left(1-\frac{1}{y}\right)^{-y}$, where $y=x+1>1$, we obtain

140. $\left(1+\frac{1}{x}\right)^x$ *increases for* $x>0$; $\left(1-\frac{1}{x}\right)^{-x}$ *decreases for* $x>1$.

This is substantially the same as Theorem **35**.

(2) If $x>1$, $r>1$, we have

$$x^r=1+r(x-1)+\tfrac{1}{2}r(r-1)\xi^{r-2}(x-1)^2,$$

where $1<\xi<x$, and so

141. $x^r>1+r(x-1)+\tfrac{1}{2}r(r-1)\left(\frac{x-1}{x}\right)^2 \quad (x>1,\ r>1).$

This inequality was found, in a less precise form, in § 2.15.

(3) If $x \neq 0$ we have

$$e^x=1+x+\tfrac{1}{2}x^2e^{\theta x}, \tag{4.2.2}$$

where $0<\theta<1$, and so

142. $$e^x>1+x \quad (x\neq 0).$$

We can deduce another proof of Theorem **9**. If

$$\Sigma q=1, \quad \Sigma qa=\mathfrak{A},$$

and the a are not all equal, we can write $a=(1+x)\mathfrak{A}$, where $\Sigma qx=0$ and the x are not all zero. Then $1+x\leqq e^x$, with inequality for at least one x, and

$$\Pi a^q=\mathfrak{A}\,\Pi\,(1+x)^q<\mathfrak{A}\,e^{\Sigma qx}=\mathfrak{A}=\Sigma qa.$$

The argument is a special case of that used at the end of § 3.19.

(4) The function $\quad f(x)=e^x-1-x-\tfrac{1}{2}x^2,$

and its first two derivatives, vanish for $x=0$. There is no other zero of $f(x)$, since this would (by repeated application of Rolle's Theorem) involve the existence of a zero of $f'''(x)=e^x$. Hence

$$e^x>1+x+\tfrac{1}{2}x^2 \ (x>0), \quad e^x<1+x+\tfrac{1}{2}x^2 \ (x<0).$$

The same argument may be applied to any number of terms of the Taylor series of various functions. When the function is e^x, we obtain

143. *If n is odd then*

$$e^x > 1 + x + \frac{x^2}{2!} + \dots + \frac{x^n}{n!} \quad (x \neq 0). \tag{4.2.3}$$

If n is even then (4.2.3) is true for $x > 0$ and the reversed inequality for $x < 0$.

4.3. Further applications of elementary differential calculus. In this section we give some applications of a less immediate character.

(1) Repeated application of Rolle's Theorem leads easily to the following lemma[a]: *if*

$$f(x,y) = c_0 x^m + c_1 x^{m-1} y + \dots + c_m y^m = 0$$

has all its roots x/y real, then the same is true of all non-identical[b] *equations obtained from it by partial differentiations with respect to x and y. Further, if E is any one of these equations, and has a multiple root α, then α is also a root, of multiplicity one higher, of the equation from which E is derived by differentiation.*

We use this lemma to prove a theorem proved already in a less complete form in § 2.22.

144.[c] *If $a_1, a_2, \dots, a_n$ are n real, positive or negative, numbers, $p_0 = 1$, and p_μ is the arithmetic mean of the products of μ different a, then*

$$p_{\mu-1} p_{\mu+1} < p_\mu^2 \quad (\mu = 1, 2, \dots, n-1),$$

unless all the a are equal.

We suppose that no a is zero, since the specification of the cases of equality becomes more troublesome when zero a's are admitted.

Let
$$f(x,y) = (x + a_1 y)(x + a_2 y) \dots (x + a_n y)$$
$$= p_0 x^n + \binom{n}{1} p_1 x^{n-1} y + \binom{n}{2} p_2 x^{n-2} y^2 + \dots + p_n y^n.$$

Since no a vanishes, $p_n \neq 0$ and $x/y = 0$ is not a root of $f = 0$. Hence $x/y = 0$ cannot be a multiple root of any of the derived

[a] Maclaurin (**2**). See Pólya and Szegö (**1**, II, 45–47 and 230–232).

[b] That is to say, all equations whose coefficients are not all zero.

[c] Newton (**1**, 173). For further references, see § 2.22.

equations; and therefore no two consecutive p, such as p_μ and $p_{\mu+1}$, can vanish. Hence the equation

$$p_{\mu-1}x^2+2p_\mu xy+p_{\mu+1}y^2=0,$$

which is obtained from $f(x,y)=0$ by a series of differentiations, is not identical; and therefore its roots are real, so that

$$p_{\mu-1}p_{\mu+1}\leqq p_\mu^{\,2}.$$

Finally, the roots of the derived equation can be equal only if all the roots of the original equation are equal.

It will be observed that the a need not necessarily be positive, as they were in § 2.22.[a]

(2) Suppose that $\phi(x)=\log(\Sigma pa^x)$, and (what is no real limitation) that the a are all positive and unequal. Then

$$\phi'=\frac{\Sigma pa^x\log a}{\Sigma pa^x},\quad \phi''=\frac{\Sigma pa^x\,\Sigma pa^x(\log a)^2-(\Sigma pa^x\log a)^2}{(\Sigma pa^x)^2}>0$$

(Theorem **7**). An easy calculation shows that, if a_r is the greatest a, then

$$\phi(0)=\log\Sigma p,\quad \lim_{x\to\infty}(x\phi'-\phi)=-\log p_r.$$

It follows from Theorems **127** and **129** that ϕ/x increases for $x>0$ if $\Sigma p\leqq 1$, and for all x if $\Sigma p=1$. In the last case

$$\frac{\phi(x)}{x}=\log\mathfrak{M}_x(a),\quad \lim_{x\to 0}\frac{\phi(x)}{x}=\phi'(0)=\log\mathfrak{G}(a).$$

We thus obtain further proofs of Theorems **9** and **16**.

If, on the other hand, $p_\nu\geqq 1$ for every ν, ϕ/x decreases, by Theorem **128**. In particular $\mathfrak{S}_x(a)=(\Sigma a^x)^{1/x}$ decreases (Theorem **19**). The general case gives part of Theorem **23**.

(3) The following examples have applications in ballistics.

145. $\log\sec x<\frac{1}{2}\sin x\tan x\quad(0<x<\frac{1}{2}\pi)$.

146. *The function*

$$\rho(x)=\frac{8\log\sec x}{(g(x))^2},$$

where $g(x)=\int_0^x(1+\sec t)\,dt=x+\log(\sec x+\tan x)$,

decreases steadily from 1 *to* 0 *as* x *increases from* 0 *to* $\frac{1}{2}\pi$.

[Use Theorem **145** to show that $\frac{d}{dx}(g^3\rho'\cot x)<0$ and so $\rho'<0$.]

[a] Except in Theorem **55**. For positive a, see Theorems **51**, **54**, **77**, and § 3.5.

147. *The function*

$$\sigma(x)=\frac{\int_0^x (1+\sec t)\log\sec t\,dt}{\log\sec x\int_0^x (1+\sec t)\,dt}$$

increases steadily from $\frac{1}{3}$ *to* $\frac{1}{2}$ *as* x *increases from* 0 *to* $\frac{1}{2}\pi$.

There is a general theorem which will be found useful in the proof of Theorem **147.**

148. *If* f, g, *and* f'/g' *are positive increasing functions, then* f/g *either increases for all* x *in question, or decreases for all such* x, *or decreases to a minimum and then increases. In particular, if* $f(0)=g(0)=0$, *then* f/g *increases for* $x>0$.

To prove this, observe that

$$\frac{d}{dx}\left(\frac{f}{g}\right)=\left(\frac{f'}{g'}-\frac{f}{g}\right)\frac{g'}{g}$$

and consider the possible intersections of the curves $y=f/g$, $y=f'/g'$. At one of these intersections the first curve has a horizontal and the second a rising tangent, and therefore there can be at most one intersection.

If we take g as the independent variable, write $f(x)=\phi(g)$, and suppose, as in the last clause of the theorem, that

$$f(0)=g(0)=0,$$

or $\phi(0)=0$, then the theorem takes the form: *if* $\phi(0)=0$ *and* $\phi'(g)$ *increases for* $g>0$, *then* ϕ/g *increases for* $g>0$. This is a slight generalisation of part of Theorem **127.** Theorem **148** should also be compared with Theorems **128** and **129.**

4.4. Maxima and minima of functions of one variable.

A very common method for the proof of inequalities is that of finding the absolute maximum or minimum of a function $\phi(x)$ by a discussion of the sign of $\phi'(x)$.

(1) Since
$$\frac{d}{dx}\{(1-x)e^x\}=-xe^x,$$

the function $(1-x)e^x$ has just one maximum, for $x=0$. Hence

149.
$$e^x<\frac{1}{1-x}\quad (x<1,\ x\neq 0).$$

This is also a consequence of Theorem **142.**

(2) Since
$$\frac{d}{dx}(\log x-x+1)=\frac{1}{x}-1$$

the function $\log x-x+1$ has one maximum, for $x=1$. Hence

150.
$$\log x<x-1\quad (x>0,\ x\neq 1).$$

More generally $\log x < n(x^{1/n}-1)$, for any positive n, since we may write $x^{1/n}$ for x in Theorem **150**. This result is also a corollary of Theorem **36**.

(3) Let $\quad \phi(x) = 1 + xy - (1+x^k)^{1/k}(1+y^{k'})^{1/k'},$

where $k>1$, $x>0$, $y>0$. It is easily verified that $\phi(x)$ has a unique maximum 0 for $x^k = y^{k'}$.

This gives another proof of H_0 (Theorem **38**), and so of H (Theorem **11**).

(4) If x and y are positive, and $k>1$, then the function

$$\phi(x) = xy - \frac{x^k}{k} - \frac{y^{k'}}{k'}$$

has the derivative $y - x^{k/k'}$, and attains its maximum 0 for $x^k = y^{k'}$. We deduce Theorem **61** (and so Theorems **37** and **9**).

(5) The function

$$\phi(x) = xy - x\log x - e^{y-1},$$

where x is positive, attains its maximum 0 for $x = e^{y-1}$. We deduce Theorem **63**.

4.5. Use of Taylor's series. If $f(x) = \Sigma a_n x^n$ and $g(x) = \Sigma b_n x^n$ are two series with positive coefficients, and $a_n \leqq b_n$ for every n, we say that $f(x)$ is *majorised* by $g(x)$, and write $f \prec g$. It is plain that $f \prec g$ and $f_1 \prec g_1$ imply $ff_1 \prec gg_1$, and so on.

To illustrate the use of this idea in the proof of inequalities, we prove

151. *If* $s_n = a_1 + a_2 + \dots + a_n$, *where* $n>1$ *and the* a *are positive, then*

$$(1+a_1)(1+a_2)\dots(1+a_n) < 1 + \frac{s_n}{1!} + \frac{s_n^{\,2}}{2!} + \dots + \frac{s_n^{\,n}}{n!}.$$

In fact $1 + a_\nu x \prec e^{a_\nu x}$, so that

$$\Pi(1+ax) \prec e^{s_n x}.$$

The result follows by adding up the coefficients of $1, x, x^2, \dots, x^n$, and observing that there is strict inequality between the coefficients of x^2. It may also be proved by writing the left-hand side in the form

$$1 + np_1 + \frac{n(n-1)}{1.2}p_2 + \dots + p_n$$

(so that $np_1 = s_n$) and using Theorem **52**.

DIFFERENTIAL CALCULUS: FUNCTIONS OF SEVERAL VARIABLES

4.6. Applications of the theory of maxima and minima of functions of several variables. The most 'universal' weapon for the discovery and proof of inequalities is the general theory of maxima and minima of functions of any number of variables. Suppose that we wish to prove that two functions ϕ and ψ of the continuous variables $x_1, x_2, \ldots, x_n$ are comparable; let us say, to fix our ideas, that $\phi - \psi \geqq 0$. This will be so, if $\phi - \psi$ has a minimum, if and only if this minimum is non-negative; and this is a question which can always be attacked (at any rate when the functions are differentiable) by the standard arguments of the theory of maxima and minima.

The method is attractive theoretically, and always opens a first line of attack on the problem; but it is apt to lead to serious complications in detail (usually connected with the 'boundary values' of the variables), and it will be found that, however suggestive it may be, it rarely leads to the simplest solution. We illustrate these remarks by considering its application to the fundamental inequalities G and H.

(1) To prove G, consider

$$f(x_1, x_2, \ldots, x_{n-1}) = x_1^{q_1} x_2^{q_2} \ldots x_n^{q_n},$$

where
$$x_n = \frac{1}{q_n}(\mathfrak{A} - q_1 x_1 - \ldots - q_{n-1} x_{n-1}),$$

in the closed and bounded domain $x_1 \geqq 0, \ldots, x_n \geqq 0$. It is continuous, and therefore attains a maximum, which is not on the boundary (where f vanishes). At the maximum

$$0 = \frac{1}{f}\frac{\partial f}{\partial x_\nu} = \frac{q_\nu}{x_\nu} - \frac{q_n}{x_n}\frac{q_\nu}{q_n} \quad (\nu = 1, 2, \ldots, n-1),$$

and so the x are all equal to $\mathfrak{A}$. In this case no serious complication is introduced by the boundary values[a].

(2) We may use H (for two sets of variables) as an illustration of the 'method of Lagrange'. Consider

$$f(x_1, x_2, \ldots, x_n) = b_1 x_1 + b_2 x_2 + \ldots + b_n x_n,$$

[a] Compare § 2.6 (i).

where $b_\nu > 0$, subject to the condition that

$$\phi(x_1, x_2, \ldots, x_n) = x_1^k + x_2^k + \ldots + x_n^k \quad (k > 1)$$

is a positive constant X. The $(n-1)$-dimensional domain defined by $x \geqq 0$, $\phi = X$ is closed and bounded, and some x vanishes at every point of its boundary.

If the maximum is attained at an internal point, then, at that point,

$$\frac{\phi_{x_\nu}}{f_{x_\nu}} = \frac{k x_\nu^{k-1}}{b_\nu} = \lambda$$

is independent of ν; and an elementary calculation shows that

$$f = X^{1/k} (\Sigma b^{k'})^{1/k'} = (\Sigma x^k)^{1/k} (\Sigma b^{k'})^{1/k'}.$$

There remains the possibility that the maximum should be attained at a boundary point, where some x, say x_n, is zero. This possibility may be excluded by an inductive argument, since, if we assume that the inequality has been proved already for $n-1$ variables, and that $x_n = 0$, we have

$$f = \sum_1^{n-1} b_\nu x_\nu \leqq \left(\sum_1^{n-1} x_\nu^k \right)^{1/k} \left(\sum_1^{n-1} b_\nu^{k'} \right)^{1/k'} < (\Sigma x_\nu^k)^{1/k} (\Sigma b_\nu^{k'})^{1/k'}.$$

The weakness of the method is that, if we are to argue by induction at all, it is better to prove the whole theorem inductively, and then we come back to one of the proofs of H given already.

(3) It is quite usual that the method should, as in case (2), prove troublesome when developed in detail; but even in such cases it is very useful as indicating a possible solution of the problem.

A great many of our theorems assert inequalities between two symmetric functions $f(x_1, x_2, \ldots, x_n)$ and $g(x_1, x_2, \ldots, x_n)$, homogeneous of the same degree and positive for all positive x. This is true, for example, of Theorems **9**, **16** and **17** (for unit weights, the crucial case), Theorem **45** (in the case of comparability) and Theorems **51** and **52**.

When we use Lagrange's method, we must consider the maximum of f for constant g, say for $g = 1$. Lagrange's equations are

$$\frac{\partial f}{\partial x_\nu} = \lambda \frac{\partial g}{\partial x_\nu} \quad (\nu = 1, 2, \ldots, n).$$

These equations have always a solution with $x_1 = x_2 = \ldots = x_n$, and λ is the value of f for this system of values of x. If λ is a strict absolute maximum of f, then $f \leqq \lambda g$, and there is inequality except when all the x are equal.

All this is in fact true in the cases mentioned, but there are other cases in which the solution does not give the maximum of f. This happens for example in Theorem **45**, in the case of non-comparability.

INTEGRAL CALCULUS

4.7. Comparison of series and integrals. There are many inequalities which are proved most easily by arguments based on the integral calculus; and often, by consideration of areas or volumes, in an 'intuitive' way. We give here a few of the most useful general theorems, in which the integrals considered are of the elementary Riemann or Riemann-Stieltjes type. In Ch. VI we shall consider inequalities between integrals systematically, and there we shall use the general Lebesgue and Lebesgue-Stieltjes integrals.

The theorems which follow immediately are due in principle to Maclaurin and Cauchy[a].

152. *If $f(x)$ decreases for $x \geqq 0$, then*

$$f(1)+f(2)+\ldots+f(n) \leqq \int_0^n f(x)\,dx \leqq f(0)+f(1)+\ldots+f(n-1).$$

There is inequality if $f(x)$ decreases strictly.

In fact
$$f(\nu+1) \leqq \int_\nu^{\nu+1} f(x)\,dx \leqq f(\nu)$$

(with inequality if $f(x)$ decreases strictly).

Further theorems of the same type, which we state without proof, are:

153. *If $a_0 < a_1 < a_2 < \ldots$, and $f(x)$ decreases for $x \geqq a_0$, then*

$$\sum_{\nu=1}^{n} (a_\nu - a_{\nu-1}) f(a_\nu) \leqq \int_{a_0}^{a_n} f(x)\,dx \leqq \sum_{\nu=1}^{n} (a_\nu - a_{\nu-1}) f(a_{\nu-1}).$$

[a] Maclaurin (**1**, I, 289); Cauchy (**2**, 222).

154. *If $f(x) \geqq 0$, and $f(x)$ decreases in $(0, \xi)$ and increases in $(\xi, 1)$, where $0 \leqq \xi \leqq 1$, then*

$$\frac{1}{n}\left\{f\left(\frac{1}{n}\right)+f\left(\frac{2}{n}\right)+\dots+f\left(\frac{n-1}{n}\right)\right\} \leqq \int_0^1 f(x)\,dx$$

$$\leqq \frac{1}{n}\left\{f(0)+f\left(\frac{1}{n}\right)+\dots+f(1)\right\}.$$

155. *If $f(x,y)$ is a decreasing function of x for fixed y and a decreasing function of y for fixed x, then*

$$\sum_{\mu=1}^{m}\sum_{\nu=1}^{n} f(\mu,\nu) \leqq \int_0^m \int_0^n f(x,y)\,dx\,dy \leqq \sum_{\mu=0}^{n-1}\sum_{\nu=0}^{n-1} f(\mu,\nu).$$

Applications of these theorems, particularly to the theory of the convergence of series, may be found in any textbook of analysis.

4.8. An inequality of Young. The simple but useful theorem which follows is due to W. H. Young[a] and is of a different type.

156. *Suppose that $\phi(0)=0$, and that $\phi(x)$ is continuous and strictly increasing for $x \geqq 0$; that $\psi(x)$ is the inverse function, so that $\psi(x)$ satisfies the same conditions; and that $a \geqq 0$, $b \geqq 0$. Then*

$$ab \leqq \int_0^a \phi(x)\,dx + \int_0^b \psi(x)\,dx.$$

There is equality only if $b=\phi(a)$.

The theorem becomes intuitive if we draw the curve $y=\phi(x)$ or $x=\psi(y)$, and the lines $x=0$, $x=a$, $y=0$, $y=b$, and consider the various areas bounded by them. A formal proof is included in that of the more general theorems which follow.

A corollary of Theorem **156** is

157. *If the conditions of Theorem* **156** *are satisfied, then*

$$ab \leqq a\phi(a)+b\psi(b).$$

Theorem **157** is weaker than Theorem **156**, but often as effective in applications.

We pass to more general theorems which include Theorem **156**.

[a] W. H. Young (2).

158. *Suppose that* $\nu=1, 2, \ldots, n$; *that* $a_\nu \geqq 0$; *that* $f_\nu(x)$ *is continuous, non-negative, and strictly increasing; and that one of the* $f_\nu(0)$ *is zero. Then*

$$\Pi f_\nu(a_\nu) \leqq \Sigma \int_0^{a_\nu} \prod_{\mu \neq \nu} f_\mu(x) \, . \, df_\nu(x);$$

and there is equality only if $a_1 = a_2 = \ldots = a_n$.[a]

The inequality may be made intuitive by considering the curve $x_\nu = f_\nu(t)$ in n-dimensional space, and the volumes bounded by the coordinate planes and the cylinders which project the curves on these planes.

To obtain a formal proof, put

$$F_\nu(x) = f_\nu(x) \quad (0 \leqq x \leqq a_\nu), \qquad F_\nu(x) = f_\nu(a_\nu) \quad (x \geqq a_\nu),$$

so that $F_\nu(x) \leqq f_\nu(x)$. If we suppose, as we may, that a_n is the largest of the a_ν, then, since $\Pi F_\nu(0) = 0$, we have

$$\Pi f_\nu(a_\nu) = \Pi F_\nu(a_n) = \int_0^{a_n} d\{\Pi F_\nu(x)\} = \sum_\nu \int_0^{a_n} \prod_{\mu \neq \nu} F_\mu(x) \, . \, dF_\nu(x)$$

$$= \sum_\nu \int_0^{a_\nu} \prod_{\mu \neq \nu} F_\mu(x) \, . \, df_\nu(x) \leqq \sum_\nu \int_0^{a_\nu} \prod_{\mu \neq \nu} f_\mu(x) \, . \, df_\nu(x);$$

and there is inequality unless every a_ν is equal to a_n.

159.[b] *Suppose that* $g_\nu(x)$ *is a system of* n *continuous and strictly increasing functions each of which vanishes for* $x=0$; *that*

$$\Pi g_\nu^{-1}(x) = x; \tag{4.8.1}$$

and that $a_\nu \geqq 0$. *Then*

$$\Pi a_\nu \leqq \Sigma \int_0^{a_\nu} \frac{g_\nu(x)}{x} \, dx.$$

There is inequality unless $g_1(a_1) = g_2(a_2) = \ldots = g_n(a_n)$.

If we put

$$g_\nu^{-1}(x) = \chi_\nu(x), \quad b_\nu = g_\nu(a_\nu), \quad a_\nu = g_\nu^{-1}(b_\nu) = \chi_\nu(b_\nu)$$

and apply Theorem **158** to the system $f_\nu(x) = \chi_\nu(x)$, we obtain

$$\Pi a_\nu = \Pi \chi_\nu(b_\nu) \leqq \Sigma \int_0^{b_\nu} \frac{x}{\chi_\nu(x)} \, d\chi_\nu(x) = \Sigma \int_0^{a_\nu} \frac{g_\nu(y)}{y} \, dy.$$

[a] Oppenheim (**1**). The proof is by T. G. Cowling.
[b] Cooper (**1**).

A system of n functions

$$\phi_\nu(x)=\frac{g_\nu(x)}{x}$$

connected by (4.8.1) is a generalisation of a pair of inverse functions. For suppose that $n=2$, $g_1(x)=x\phi(x)$, $g_2(x)=x\psi(x)$, and write (4.8.1) in the two forms

$$\frac{x}{g_1^{-1}(x)}=g_2^{-1}(x), \quad \frac{x}{g_2^{-1}(x)}=g_1^{-1}(x).$$

Then

$$\phi(x)=\frac{g_1(x)}{x}=g_2^{-1}\{g_1(x)\}, \quad \psi(x)=\frac{g_2(x)}{x}=g_1^{-1}\{g_2(x)\},$$

and $g_2^{-1}g_1$, $g_1^{-1}g_2$ are inverse; and the functions ϕ and ψ of Theorem **156** can always be represented in this form. Hence Theorem **159** includes Theorem **156**.

If in Theorem **159** we take $g_\nu(x)=x^{1/q_\nu}$, where $\Sigma q_\nu=1$, then (4.8.1) is satisfied, and we obtain

$$\Pi a_\nu \leqq \Sigma q_\nu a_\nu^{1/q_\nu},$$

which is Theorem **9**. If in Theorem **156** we take $\phi(x)=x^{k-1}$, where $k>1$, we have $\psi(x)=x^{k'-1}$, and we obtain Theorem **61**. If we take

$$\phi(x)=\log(x+1), \quad \psi(y)=e^y-1,$$

and write u, v for $a+1$ and $b+1$, we obtain Theorem **63**, for $u\geqq 1$, $v\geqq 1$.[a]

We pointed out above that Theorem **156** is intuitive from simple graphical considerations. If instead of reckoning areas we count up the number of lattice-points (points with integral coordinates) inside them, we obtain

160. *If $\phi(x)$ increases strictly with x, $\phi(0)=0$, and $\psi(x)$ is the function inverse to $\phi(x)$, then*

$$mn\leqq \sum_0^m [\phi(\mu)]+\sum_0^n [\psi(\nu)],$$

where $[y]$ is the integral part of y.

This theorem is less interesting in itself, but illustrates a type of argument often effective in the Theory of Numbers.

[a] Actually the result is true for $u>0$ and all v. See § 4.4 (5).

CHAPTER V

INFINITE SERIES

5.1. Introduction. Our theorems so far have related to finite sums, and we have now to consider their extensions to infinite series. The general conclusion will be that our theorems remain valid for infinite series in so far as they retain significance.

Two preliminary remarks are necessary.

(1) The first concerns the *interpretation* of our formulae. An inequality $X < Y$ (or $X \leqq Y$), where X and Y are infinite series, is always to be interpreted as meaning 'if Y is convergent, then X is convergent, and $X < Y$ (or $X \leqq Y$)'. More generally, an inequality of the type

$$X < \Sigma A Y^b \dots Z^c \tag{5.1.1}$$

(or $X \leqq \Sigma A Y^b \dots Z^c$), where $Y, \dots, Z$ are any finite number of infinite series, Σ is a finite sum, and $A, b, \dots, c$ are positive, is to be interpreted as meaning 'if $Y, \dots, Z$ are convergent then X is convergent, and X satisfies the inequality'. Neglect of this understanding would lead to confusion when it is '$<$' which stands in the inequality. We could read 'Y' as '∞' in the case of divergence; then '$X \leqq \infty$' would convey no information, but '$X < \infty$' would imply the convergence of X, and this implication would usually be false.

Some inequalities will occur which are not of the form (5.1.1). These are usually secondary, and should be reduced to the form (5.1.1) if there is any doubt about their interpretation. Thus

$$X^a < A Y^b$$

should be interpreted as

$$X < A^{1/a} Y^{b/a},$$

which is of the form (5.1.1); and $X > Y$ should be interpreted as $Y < X$.

There is one important inequality, viz.

$$\Sigma ab > (\Sigma a^k)^{1/k} (\Sigma b^{k'})^{1/k'}, \tag{5.1.2}$$

where $k<1$, $k \neq 0$,[a] which we have written deliberately in a form unlike (5.1.1). We might have written it as

$$\Sigma a^k < (\Sigma ab)^k (\Sigma b^{k'})^{-k/k'} \tag{5.1.3}$$

when $0<k<1$, or as

$$\Sigma b^{k'} < (\Sigma ab)^{k'} (\Sigma a^k)^{-k'/k} \tag{5.1.4}$$

when $k<0$. These are of the form (5.1.1), and are the forms which arise primarily in the proof of Theorem **13**. We prefer the form (5.1.2) for formal reasons, and because it shows clearly the contrast between the two cases of the theorem, but we must use the other forms if we wish to show explicitly and exactly the implications about convergence.

There are a few cases where the inequality asserted is not one between infinite series but involves the results of other limit operations. Thus, when we extend the inequality $\mathfrak{G}(a) < \operatorname{Max} a$ (Theorem **2**) we obtain an inequality between an infinite product and the upper bound of an infinite set. Such an inequality '$X<Y$' is to be interpreted in the same way, viz. as 'if Y is finite then X is finite and $X<Y$'.

(2) The second remark concerns *method*, and should be read in conjunction with § 1.7. Suppose, for example, that we wish to prove the inequality

$$(\Sigma ab)^2 \leqq \Sigma a^2 \Sigma b^2$$

for infinite series. We know the inequality for finite sums (Theorem **7**) and, everything being positive, our conclusion follows by a passage to the limit.

We cannot extend Theorem **7**, in its complete form, to infinite series in so simple a manner, since in the limiting process '$<$' degenerates into '$\leqq$', and we are unable to pick out the possible cases of equality. Here and elsewhere, we must avoid limiting processes; instead of deducing the infinite theorem from the finite one, we must verify that the proof given for the finite theorem remains valid, with that minimum of change required by the new context, in the infinite case. For example, either proof of Theorem **7** given in § 2.4 may be extended to the infinite

[a] (2.8.4) of Theorem **13**.

case by the addition of a few obvious comments concerning convergence.

It will not be necessary to retrace the path followed in Ch. II systematically. The few new points which arise are neither difficult nor particularly interesting, and, in so far as they are important, recur in a more interesting form in Ch. VI. We shall therefore arrange the substance of this chapter informally, illustrating and commenting upon the new possibilities, and ending with an enumeration of some of the more important theorems of Ch. II which remain valid with the new interpretation.

5.2. The means $\mathfrak{M}_r$. We begin by some comments on a new point which arises in the definition of the means $\mathfrak{M}_r$. We have now an infinity of terms a and weights p, and there are two cases to consider, according as Σp is convergent or divergent.

(i) If Σp is convergent we may suppose that the sum is 1 and write q for p. In this case $\mathfrak{M}_r$ is defined, for $r > 0$, by

$$\mathfrak{M}_r(a) = (\Sigma q a^r)^{1/r}, \tag{5.2.1}$$

and may be regarded as a 'mean' in the sense of § 2.2 or a 'weighted sum' in the sense of § 2.10 (iv). It is finite or infinite according as $\Sigma q a^r$ is convergent or divergent.

(ii) If Σp is divergent, we can still define $\mathfrak{M}_r$ as a limit, e.g. by

$$\mathfrak{M}_r(a) = \lim_{n \to \infty} \left(\sum_1^n p_\nu a_\nu{}^r \Big/ \sum_1^n p_\nu \right)^{1/r}, \tag{5.2.2}$$

or as the corresponding upper limit $\overline{\lim}$. The latter definition is not particularly interesting, though it would preserve most of our theorems. If we define $\mathfrak{M}_r$ by (5.2.2), we are met by a difficulty: *the existence of $\mathfrak{M}_r$ for a given r does not ensure its existence for any other r.* In fact we can determine the a so that $\mathfrak{M}_r$ shall exist for a given set $r_1, r_2, \ldots, r_m$ of values of r and for no others. We shall therefore confine our attention to case (i).

For the general question of the existence of $\mathfrak{M}_r$, see, for example, Besicovitch (**1**). We may illustrate the point by showing briefly how to find a so that either of the limits

$$\lim \frac{1}{n}(a_1 + a_2 + \ldots + a_n), \quad \lim \frac{1}{n}(a_1{}^2 + a_2{}^2 + \ldots + a_n{}^2)$$

may exist without the other: here $p = 1$.

Take first two sequences

$$\alpha_1, \alpha_2, \ldots, \alpha_\varpi, \alpha_1, \alpha_2, \ldots; \quad \beta_1, \beta_2, \ldots, \beta_\varpi, \beta_1, \beta_2, \ldots$$

with period ϖ. When $a = \alpha$, both limits exist and have the values

$$A_1 = \frac{\alpha_1 + \alpha_2 + \ldots + \alpha_\varpi}{\varpi}, \quad A_2 = \frac{\alpha_1^2 + \alpha_2^2 + \ldots + \alpha_\varpi^2}{\varpi};$$

and when $a = \beta$ they have the corresponding values B_1, B_2.

Now take the a as follows:

$$\begin{array}{llll} \alpha_1, & \alpha_2, & \ldots, & \alpha_\varpi \quad \text{(repeated } N_1 \text{ times)}, \\ \beta_1, & \beta_2, & \ldots, & \beta_\varpi \quad \text{(repeated } N_2 \text{ times)}, \\ \alpha_1, & \alpha_2, & \ldots, & \alpha_\varpi \quad \text{(repeated } N_3 \text{ times)}, \\ \ldots, & \ldots, & \ldots, & \ldots \quad \ldots\ldots\ldots\ldots\ldots\ldots \end{array}$$

It is easy to see that, by supposing the sequence $N_1, N_2/N_1, N_3/N_2, \ldots$ to tend to infinity with great rapidity, we can make $(a_1 + \ldots + a_n)/n$ and $(a_1^2 + \ldots + a_n^2)/n$ oscillate between A_1, B_1 and A_2, B_2 respectively. The conditions for convergence will then be $A_1 = B_1$ and $A_2 = B_2$ respectively, and we can obviously choose the α and β so that either of these conditions shall be satisfied without the other.

We therefore restrict ourselves to case (i). We define $\mathfrak{M}_r$, for positive or negative r, by (5.2.1), with the convention that $\mathfrak{M}_r = 0$ if r is negative and an a zero or $\Sigma q a^r$ divergent. We define $\mathfrak{G}$ (and $\mathfrak{M}_0$) by

$$(5.2.3) \qquad \mathfrak{G}(a) = \mathfrak{M}_0(a) = \Pi a^q = \exp(\Sigma q \log a),$$

with the conventions that $\mathfrak{G} = \infty$ if Πa^q diverges to ∞ (i.e. if $\Sigma q \log a$ diverges to $+\infty$), and $\mathfrak{G} = 0$ if Πa^q diverges to 0 (i.e. if $\Sigma q \log a$ diverges to $-\infty$). It is to be observed that $\log a$ may have either sign, and that the definition of $\mathfrak{G}$ fails if $\Sigma q \log a$ is oscillatory. In this case $\mathfrak{G}$ is meaningless.

It follows from Theorems **36** and **150** that

$$(5.2.4) \qquad \log t < \frac{t^r - 1}{r} < \frac{t^s - 1}{s} \quad (0 < r < s,\ t > 0),$$

unless $t = 1$, when there is equality. We define $\log^+ t$ and $\log^- t$ by

$$\log^+ t = \log t \ (t > 1), \qquad \log^+ t = 0 \ (0 < t \leqq 1),$$
$$\log^- t = \log t \ (0 < t \leqq 1), \quad \log^- t = 0 \ (t > 1),$$

so that

$$\log^+ t \geqq 0, \ \log^- t \leqq 0, \ \log t = \log^+ t + \log^- t, \ \log^- t = -\log^+ \frac{1}{t}.$$

It then follows from (5.2.4) that

$$0 \leqq \Sigma q \log^+ a \leqq \frac{1}{r} \Sigma' q (a^r - 1) \leqq \frac{1}{s} \Sigma' q (a^s - 1),$$

where Σ' denotes a summation extended over the a which exceed 1. Hence, if $\mathfrak{M}_s(a)$ is finite for some positive s, then $\mathfrak{M}_r(a)$ is finite for $0<r<s$, and $\Sigma q \log^+ a$ is convergent. We can prove similarly that, if $\mathfrak{M}_{-s}(a)$ is positive for some positive s, then $\mathfrak{M}_{-r}(a)$ is positive for $-s<-r<0$, and $\Sigma q \log^- a$ is convergent. In the first case $\mathfrak{G}$ is positive and finite or zero, in the second it is positive and finite or infinite. If $\Sigma q \log a$ oscillates then $\Sigma q \log^+ a$ and $\Sigma q \log^- a$ are both divergent, and this is only possible when $\mathfrak{M}_r(a)=\infty$ for all positive r and $\mathfrak{M}_r(a)=0$ for all negative r. It is only in this case that the definition of $\mathfrak{G}(a)$ can fail.

There is cne new point which, as we shall see in § 5.9, affects the specification of the cases of equality in some of our theorems. This point arises from the possibility that, when $r \leqq 0$, $\mathfrak{M}_r(a)$ may be zero although no a is zero. If $r>0$ then, as in Ch. II, $\mathfrak{M}_r(a)$ can be zero only if (a) is null, in which case $\mathfrak{M}_r(a)=0$ for all r. But when $r \leqq 0$ there is a difference. The $\mathfrak{M}_r(a)$ of Ch. II was zero, for such an r, if and only if some a was zero, and then $\mathfrak{M}_r(a)$ was zero for all $r \leqq 0$. It is now possible, when $r \leqq 0$, that $\mathfrak{M}_s(a)$ should be zero for $s<r$ and positive for $s \geqq r$, or zero for $s \leqq r$ and positive for $s>r$.

Thus in Theorem **1** there were two exceptional cases;

$$\operatorname{Min} a < \mathfrak{M}_r(a) < \operatorname{Max} a,$$

unless either all the a are equal, or else $r<0$ and an a is zero. All that we can say now is 'unless either all the a are equal (in which case both inequalities reduce to equalities) or else $r<0$ and $\mathfrak{M}_r(a)=0$ (in which case $\operatorname{Min} a=0$ and the first inequality reduces to an equality)'. Substantially the same point arises in connection with Theorems **2**, **5**, **10**, **16**, **24** and **25** (to quote only cases referred to in our summary in § 5.9).

5.3. The generalisation of Theorems 3 and 9. We use the inequalities (5.2.4), and the equation

$$\lim_{r \to 0} \frac{t^r - 1}{r} = \log t.$$

Taking $t=a/\Sigma qa=a/\mathfrak{A}$, $r=1$ in (5.2.4), we have

$$\log a-\log \mathfrak{A} \leqq \frac{a}{\mathfrak{A}}-1,$$

$$\log \mathfrak{G}-\log \mathfrak{A}=\Sigma q(\log a-\log \mathfrak{A}) \leqq 1-1=0,$$

with equality only if every a is $\mathfrak{A}$. This proves the analogue of Theorem **9**.

Suppose now that $\mathfrak{M}_s$ is finite for some positive s. Then $\mathfrak{G}$ is positive and finite or zero: the proof below applies to either case[a]. Given $\epsilon>0$, we can choose N so that

$$\sum_{n\leqq N} q\log a<\log(\mathfrak{G}+\epsilon), \tag{5.3.1}$$

$$\sum_{n>N} q\frac{a^s-1}{s}<\epsilon, \tag{5.3.2}$$

and then r_0 so that $0<r_0<s$ and

$$\sum_{n\leqq N} q\frac{a^r-1}{r}<\sum_{n\leqq N} q\log a+\epsilon \tag{5.3.3}$$

for $0<r<r_0$. We have then

$$\log \mathfrak{G}(a)=\frac{1}{r}\log \mathfrak{G}(a^r) \leqq \frac{1}{r}\log \mathfrak{A}(a^r) \leqq \frac{\mathfrak{A}(a^r)-1}{r}$$

$$=\sum_{n\leqq N} q\frac{a^r-1}{r}+\sum_{n>N} q\frac{a^r-1}{r}<\sum_{n\leqq N} q\log a+\epsilon+\sum_{n>N} q\frac{a^s-1}{s}$$

$$<\log(\mathfrak{G}+\epsilon)+2\epsilon.$$

Hence $$\log \mathfrak{M}_r(a)=\frac{1}{r}\log \mathfrak{A}(a^r)\to\log \mathfrak{G}(a)$$

when $r\to+0$. We leave it to the reader to prove that, if $\mathfrak{M}_r$ is positive for some negative r, then $\mathfrak{M}_r\to\mathfrak{G}$ when $r\to-0$.

5.4. Hölder's inequality and its extensions. The proofs of Hölder's inequality, and other theorems of the same type, given in Ch. II apply equally to infinite series. We may observe in passing that the series may be *multiple* series. Thus

$$(\Sigma\Sigma a_{\mu\nu}b_{\mu\nu})^2 \leqq \Sigma\Sigma a_{\mu\nu}^2\,\Sigma\Sigma b_{\mu\nu}^2.$$

[a] It is modelled on the proof by F. Riesz (7) of the corresponding integral theorem: see § 6.8.

For example, suppose that Σu_μ^2 and Σv_ν^2 are convergent, and take

$$a_{\mu\nu}=u_\mu v_\nu, \quad b_{\mu\nu}=\frac{1}{(\mu+\nu)^{1+\delta}} \quad (\delta>0).$$

Since $\Sigma\Sigma\,(\mu+\nu)^{-2-2\delta}$ is convergent, it follows that

$$\Sigma\Sigma\frac{u_\mu v_\nu}{(\mu+\nu)^{1+\delta}}$$

is convergent. This is an imperfect form of a theorem to be proved later (Theorem **315**).

The theorems concerning $\mathfrak{M}_r$ deduced from Hölder's inequality (Theorems **16** and **17**) are unchanged, except that the statement of the second exceptional case of Theorem **16** must be modified in accordance with our remarks at the end of § 5.2. Here we must say 'unless $s \leqq 0$ and $\mathfrak{M}_s(a)=0$'.

One new point of greater interest arises in connection with this group of theorems. There is a theorem, suggested by Theorem **15** but not a corollary of it (even when it has been extended to infinite series), which has no analogue for finite sums.

161.[a] *If $k>1$ and Σab is convergent for all b for which $\Sigma b^{k'}$ is convergent, then Σa^k is convergent.*

We deduce this from another theorem due to Abel[b], which is of great interest in itself.

162. *If Σa_n is divergent and*

$$A_n=a_1+a_2+\ldots+a_n,$$

[a] Landau (**1**).

[b] Abel (**1**). There are theorems of the same type involving an arbitrary function $f(x)$. Thus, if Σa_n is divergent, $f(x)$ is positive and decreasing, and

$$I=\int_1^\infty f(x)\,dx,$$

then the convergence of I involves that of $\Sigma a_n f(A_n)$, and the divergence of I involves that of $\Sigma a_n f(A_{n-1})$: see, for example, de la Vallée Poussin (**1**, 398–399), Littlewood (**1**). This theorem, though of a more general character, does not actually include Theorem **162**: it is not true that the divergence of I necessarily involves that of $\Sigma a_n f(A_n)$. For an example to the contrary take

$$a_n=2^{2^n}, \quad f(x)=\frac{1}{x\log x}.$$

so that $A_n \to \infty$, *then*

$$\text{(i)}\ \Sigma \frac{a_n}{A_n} \textit{ is divergent,}$$

$$\text{(ii)}\ \Sigma \frac{a_n}{A_n^{1+\delta}} \textit{ is convergent for every positive } \delta.$$

(i) We have

$$\frac{a_{n+1}}{A_{n+1}} + \frac{a_{n+2}}{A_{n+2}} + \dots + \frac{a_{n+r}}{A_{n+r}} \geq \frac{A_{n+r} - A_n}{A_{n+r}} = 1 - \frac{A_n}{A_{n+r}},$$

which tends to 1 when n is fixed and $r \to \infty$, and is therefore greater than $\frac{1}{2}$ for any n and some corresponding r. This proves (i).

(ii) We may obviously suppose $0 < \delta < 1$. Then

$$\Sigma \frac{A_n^{\delta} - A_{n-1}^{\delta}}{A_{n-1}^{\delta} A_n^{\delta}} = \Sigma \left(\frac{1}{A_{n-1}^{\delta}} - \frac{1}{A_n^{\delta}} \right)$$

is convergent. By Theorem **41**, the numerator on the left is not less than $\delta A_n^{\delta-1}(A_n - A_{n-1}) = \delta a_n A_n^{\delta-1}$. It follows that

$$\Sigma \frac{a_n}{A_n A_{n-1}^{\delta}}$$

is convergent. We prove in fact a little more than (ii).

To deduce Theorem **161**, write

$$a^k = u, \quad ab = uv, \quad b^{k'} = uv^{k'}.$$

We then have to prove that, if Σu_n is divergent, there is a v_n such that $\Sigma u_n v_n$ is divergent and $\Sigma u_n v_n^{k'}$ convergent. We take $v_n = 1/U_n$, where $U_n = u_1 + u_2 + \dots + u_n$, and the conclusion follows from Theorem **162**.

5.5. The means $\mathfrak{M}_r$ (*continued*). There is little to be added about the means $\mathfrak{M}_r$, but one or two further remarks are required. We begin with a remark concerning the generalisation of Theorem **4**. This theorem, in so far as it concerns positive r, must be interpreted as follows: *if the a are bounded, and* $a^* = \operatorname{Max} a$ *is their upper bound, then*

$$\mathfrak{M}_r \to \mathfrak{M}_\infty = a^*,$$

when $r \to +\infty$; *if the a are unbounded, but $\mathfrak{M}_r$ is finite for all positive r, then* $\mathfrak{M}_r \to \infty$.

The question of the continuity of $\mathfrak{M}_r$ for a finite positive or negative r is no longer quite trivial. We state a comprehensive theorem, but give no proof, since all the points involved arise in a more interesting form in Ch. VI (§§ 6.10–6.11).

If $a_n = C$, $\mathfrak{M}_r = C$ for all r (whether C be positive or zero). We exclude this case, and also the case in which $\mathfrak{G}$ is meaningless, when $\mathfrak{M}_r = \infty$ for $r > 0$ and $\mathfrak{M}_r = 0$ for $r < 0$. We write

$$\mathfrak{L}_r(a) = \log \mathfrak{M}_r(a)$$

(with the conventions $\log \infty = +\infty$, $\log 0 = -\infty$).

163. *Apart from the cases just mentioned, the set of values I for which $\mathfrak{L}_r(a)$ is finite is either the null set or a closed, half-closed, or open interval (u, v), where $-\infty \leqq u \leqq v \leqq +\infty$, which has $r = 0$ as an internal or end-point, so that $u \leqq 0 \leqq v$, but is otherwise arbitrary* (and in particular may include all real values or none). *$\mathfrak{L}_r$ is $+\infty$ to the right, and $-\infty$ to the left, of I; is a continuous and strictly increasing function of r in the interior* (if it exists) *of I; and tends, when r approaches an end-point of I from inside I, to a limit equal to its value at the end-point.*

5.6. The sums $\mathfrak{S}_r$. The definition of $\mathfrak{S}_r$ given in § 2.10 is unchanged, and there is little to be said about the theorems concerning it, though those which involve continuity in r are naturally less immediate. Theorem **20** must be interpreted as meaning 'if $\mathfrak{S}_r$ is convergent for some (sufficiently large) r, then it is convergent for all greater r and ...', and Theorem **21** as meaning 'if $\mathfrak{S}_r$ is convergent for all positive r (however small) then...'. The extension of Theorem **20** may be proved as follows. If $\mathfrak{S}_R$ is convergent for a positive R, then $a_n \to 0$, and $\mathfrak{S}_r$ is convergent for $r > R$. There is a largest a, which we may suppose on grounds of homogeneity to be 1, and we may suppose the a arranged in descending order. If then

$$a_1 = a_2 = \dots = a_N = 1 > a_{N+1},$$

we have

$$\mathfrak{S}_r = (N + a_{N+1}^r + a_{N+2}^r + \dots)^{1/r},$$

for $r > R$. The series here lies between 1 and

$$N + a_{N+1}^R + a_{N+2}^R + \dots,$$

from which the theorem follows.

There is one new theorem (trivial in the finite case).

164. *If $\mathfrak{S}_R$ is convergent, then $\mathfrak{S}_r$ is continuous for $r > R$ and continuous on the right for $r = R$. If $\mathfrak{S}_R$ is divergent, but $\mathfrak{S}_r$ convergent for $r > R$, then $\mathfrak{S}_r \to \infty$ when $r \to R$.*

The proof may be left to the reader.

5.7. Minkowski's inequality. The main arguments of §§ 2.11–2.12 require no alteration. Theorems **24–26** suggest a further generalisation, with both summations infinite.

165.[a] *If $r > 1$ and a_{mn} is not of the form $b_m c_n$, then*

$$\{\sum_n q_n (\sum_m p_m a_{mn})^r\}^{1/r} < \sum_m p_m (\sum_n q_n a_{mn}^r)^{1/r}.$$

The inequality is reversed when $0 < r < 1$.

There is no real loss of generality in supposing $p = 1$, $q = 1$, and the proof goes as before. Similarly, corresponding to Theorem **27**, we have

166.
$$\sum_n (\sum_m a_{mn})^r > \sum_m \sum_n a_{mn}^r$$

if $r > 1$ (with a reversed inequality if $0 < r < 1$), unless, for every n, $a_{mn} = 0$ for every m save one.

5.8. Tchebychef's inequality. As one further illustration we take Tchebychef's inequality (Theorem **43**).

We may suppose $\Sigma p = 1$. The identity

$$\sum_1^n p_\mu \sum_1^n p_\nu a_\nu b_\nu - \sum_1^n p_\mu a_\mu \sum_1^n p_\nu b_\nu = \tfrac{1}{2} \sum_1^n \sum_1^n p_\mu p_\nu (a_\mu - a_\nu)(b_\mu - b_\nu)$$

shows, provided of course that neither (a) nor (b) is null, (i) that if (a) and (b) are similarly ordered then the convergence of Σpab implies that of Σpa and Σpb, and (ii) that if (a) and (b) are oppositely ordered then the convergence of Σpa and Σpb implies that of Σpab. In either case we may put $n = \infty$ in the identity, and our conclusions follow as before.

5.9. A summary. The theorem which follows is substantially an enumeration of the principal theorems of Ch. II which remain valid, with the glosses which we have explained in the preceding sections, for infinite series.

[a] Here, as in Theorem **26**, we abandon our usual convention about q.

167. *Theorems* **1, 2, 3, 4, 5, 7, 9, 10, 11, 12, 13, 14, 15, 16, 17, 18, 19, 20, 21, 22, 24, 25, 27, 28** *and* **43** *remain valid when the series concerned are infinite, provided that the inequalities asserted are interpreted in accordance with the conventions laid down in* § 5.1, *and that the statement of the exceptional cases in Theorems* **1, 2, 5, 10, 16, 24** *and* **25** *is modified in the manner explained in* § 5.2.

It may be worth while to supplement the last clause of the theorem by a more explicit statement. The last words of the theorems must be replaced by

(**1**) '*or else* $r<0$ *and* $\mathfrak{M}_r(a)=0$',

(**2**) '*or* $\mathfrak{G}(a)=0$',

(**5**) '*or* $r \leqq 0$ *and* $\mathfrak{M}_r(a)=0$',

(**10**) '*or* (2) $\mathfrak{G}(a+b+\ldots+l)=0$',

(**16**) '*or* $s \leqq 0$ *and* $\mathfrak{M}_s(a)=0$',

(**24**) '*or* $r \leqq 0$ *and* $\mathfrak{M}_r(a+b+\ldots+l)=0$',

(**25**) '*or* $r<0$ *and* $(\Sigma(a+b+\ldots+l)^r)^{1/r}=0$'.

We may add also that (as is explained in § 6.4) most of the theorems referred to in Theorem **167** (especially those concerning $\mathfrak{M}_r$) may be derived by specialisation from the corresponding theorems for integrals. In Ch. VI, however, we often ignore negative values of r.

MISCELLANEOUS THEOREMS AND EXAMPLES

The theorems which follow are for the most part connected with Theorems **156** and **157**. We suppose in Theorems **168–175** that $f(x)$ and $g(x)$ are inverse functions which vanish for $x=0$ and increase steadily as x increases, and that

$$F(x)=\int_0^x f(u)\,du, \quad G(x)=\int_0^x g(t)\,dt.$$

168. If $\Sigma F(a_n)$ and $\Sigma G(b_n)$ are convergent, then $\Sigma a_n b_n$ is convergent, and

$$\Sigma a_n b_n \leqq \Sigma F(a_n)+\Sigma G(b_n).$$

[Corollary of Theorem **156**.]

169. If $\Sigma a_n f(a_n)$ and $\Sigma b_n g(b_n)$ are convergent, then $\Sigma a_n b_n$ is convergent, and

$$\Sigma a_n b_n \leqq \Sigma a_n f(a_n)+\Sigma b_n g(b_n).$$

[Corollary of Theorem **157**.]

170. It is possible to choose f (and so g, F, G) and a_n in such a manner that $\Sigma F(a_n)$ is divergent, but $\Sigma a_n b_n$ is convergent for all b_n for which $\Sigma G(b_n)$ is convergent.

171. It is also possible to make $\Sigma a_n f(a_n)$ divergent, but $\Sigma a_n b_n$ convergent whenever $\Sigma b_n g(b_n)$ is convergent.

[The point of the last two theorems is to show that Theorems **168** and **169** have no converses in the sense in which Theorem **161** gives a converse of Hölder's inequality and the convergence test deduced from it. Theorem **171** is proved by Cooper (**3**), and Theorem **170**, which includes Theorem **171** and is a little stronger, may be proved in the same manner.]

172. If $\Sigma \dfrac{b_n}{\log(1/b_n)}$ is convergent, then $\Sigma \dfrac{b_n}{\log n}$ is convergent.

[Cooper (**3**): Theorem **172** is used in Cooper's proof of Theorem **171.**]

173. If $g(x)$ satisfies the inequality

$$g(xy) \leqq g(x)\, g(y),$$

and if $\Sigma a_n b_n$ is convergent whenever $\Sigma b_n g(b_n)$ is convergent, then $\Sigma a_n f(a_n)$ is convergent. Similarly, if $\Sigma a_n b_n$ is convergent whenever $\Sigma G(b_n)$ is convergent, then $\Sigma F(a_n)$ is convergent.

[See Cooper (**3**) for the first form which in this case is stronger, the second form being a corollary.]

174. If $\Sigma a_n b_n$ is convergent whenever $\Sigma G(b_n)$ is convergent, then there is a number $\lambda = \lambda(a)$, depending upon the sequence (a), for which $\Sigma F(\lambda a_n)$ is convergent.

175. If the conditions of Theorem **174** are satisfied, and $F(cx) \leqq kF(x)$ for small x, a $c > 1$, and some k, then $\Sigma F(a_n)$ is convergent.

[For the last two theorems see Birnbaum and Orlicz (**1**).]

176. If a_n and b_n tend to zero, k is positive, and

$$\Sigma \frac{a_n}{\log(1/a_n)}, \quad \Sigma e^{-k/b_n}$$

are convergent, then $\Sigma a_n b_n$ is convergent.

[Use Theorem **169.**]

177. If $x > 0$, $a_n \geqq 0$, and $f(x) = \Sigma a_n x^n$, then $f(x)$ is a convex function of x and $\log f(x)$ of $\log x$.

[Plainly $f''(x) \geqq 0$. To prove the second result, let $x = e^{-y}$, $f(x) = g(y)$. Then

$$gg'' - g'^2 = \Sigma a_n e^{-ny}\, \Sigma n^2 a_n e^{-ny} - (\Sigma n a_n e^{-ny})^2 \geqq 0,$$

by Theorem **7**. The result follows from Theorem **118.**]

178. If $a_n \geqq 0$, $\lambda_n > \lambda_{n-1} \geqq 0$, and $f(x) = \Sigma a_n e^{-\lambda_n x}$, then $\log f(x)$ is a convex function of x.

179. If $a_n > 0$ and $\lambda_n, \mu_n, \ldots, \nu_n, x, y, \ldots, z$ are real, then the domain D of convergence of the series

$$\Sigma a e^{\lambda x + \mu y + \ldots + \nu z} = f(x, y, \ldots, z)$$

is convex, and $\log f$ is a convex function of $x, y, \ldots, z$ in D.

[Because (by the extension of Theorem **11** to infinite series)

$$f\{x_1 t + x_2(1-t), \ldots, z_1 t + z_2(1-t)\} \leqq \{f(x_1, \ldots, z_1)\}^t \{f(x_2, \ldots, z_2)\}^{1-t}.$$

Here our conventions concerning convergence are important.]

180. $$\Sigma a_n^2 < 2\,(\Sigma n^2 a_n^2)^{\frac{1}{2}}\, (\Sigma (a_n - a_{n+1})^2)^{\frac{1}{2}},$$

unless $a_n = 0$ for all n.

[See Theorem **226.**]

CHAPTER VI

INTEGRALS

6.1. Preliminary remarks on Lebesgue integrals. The integrals considered in this chapter are Lebesgue integrals, except in §§ 6.15–6.22, where we are concerned with Stieltjes integrals. It may be convenient that we should state here how much knowledge of the theory we assume. This is for the most part very little, and all that the reader usually needs to know is that there is *some* definition of an integral which possesses the properties specified below. There are naturally many of our theorems which remain significant and true with the older definitions, but the subject becomes *easier*, as well as more comprehensive, if the definitions presupposed have the proper degree of generality.

We take for granted the idea of a *measurable set*, usually in one but occasionally in more dimensions. The sets which we consider may be bounded or unbounded. The definition of measure applies in the first instance to bounded sets: an unbounded set is said to be measurable if any bounded part of it is measurable, and its measure is the upper bound of the measures of its bounded components.

We shall generally assume, without special remark, that any set E with which we are concerned is measurable. We denote the measure of E by mE or sometimes, where there is no risk of ambiguity, simply by E. If E is unbounded, mE may be ∞.

We also take for granted the idea of a *measurable function*. Sums, products, and limits of measurable functions are measurable. All functions definable by the ordinary processes of analysis are measurable, and we shall confine our attention to measurable functions; we shall not usually repeat explicitly that a function which occurs in our work is assumed to be measurable.

Next, we take for granted the definition of the integral, of a bounded or unbounded function, over any (bounded or unbounded) interval or measurable set of points. A bounded measurable function is integrable over any bounded measurable set. We call the class of (bounded or unbounded) functions integrable over the interval or set E in question the class L or, if it is desirable to emphasise the set in question, the class $L(E)$. If f belongs to L, we say that 'f is L', and write $f \in L$ (and similarly for other classes).

If $f=1$ then
$$\int_E f\,dx = mE.$$

If $f \in L$, then $|f| \in L$. If f^+ and f^- are the functions equal to f when f is positive and negative respectively, and to zero otherwise, so that
$$f^+ = \operatorname{Max}(f, 0), \quad f^- = \operatorname{Min}(f, 0), \quad f = f^+ + f^-, \quad |f| = f^+ - f^-,$$
then $f^+ \in L$ and $f^- \in L$, and[a]
$$\int f\,dx = \int f^+ dx + \int f^- dx, \quad \int |f|\,dx = \int f^+ dx - \int f^- dx.$$

If $f \geqq 0$, and $(f)_n = \operatorname{Min}(f, n)$, then
$$\int f\,dx = \lim_{n\to\infty} \int (f)_n\,dx$$
(substantially by definition).

If $f \in L$, and (g is measurable and) $|g| < C|f|$, then $g \in L$.

If $f_1, f_2, \ldots, f_n \in L$ then
$$\int (a_1 f_1 + a_2 f_2 + \ldots + a_n f_n)\,dx = a_1 \int f_1\,dx + a_2 \int f_2\,dx + \ldots + a_n \int f_n\,dx.$$

If $p > 0$ and (f is measurable and) $|f|^p \in L$, we say that f belongs to the Lebesgue class L^p, or $f \in L^p$. These classes are most important when $p \geqq 1$. L^1 is identical with L.

If the integration is over a finite interval (or bounded set), then L^p includes every L^q for which $q > p$; $f \in L^q$ implies $f \in L^p$. A bounded function belongs to every L^q. These propositions are not true for an infinite interval; f may belong to L^p, in $(0, \infty)$, for one value of p only.

If the interval is finite and $f \in L^q$, $p < q$, then $|f|^p < 1 + |f|^q$, so that $f \in L^p$.

If the interval is $(0, a)$, where $a < 1$, then (a) $x^{-1/p}$ belongs to $L^{p-\delta}$ for every $\delta > 0$, but not to L^p; (b) $x^{-1/p}\left(\log \frac{1}{x}\right)^{-2/p}$ belongs to L^p, but not to

[a] We state the results for one variable and omit explicit reference to the range or set of integration.

$L^{p+\delta}$; (c) $\log(1/x)$ belongs to every L^p; and (d) $e^{1/x}$ belongs to no L^p. If the interval is $(0, \infty)$, then $x^{-\frac{1}{2}}(1+|\log x|)^{-1}$ belongs to L^2 but to no other class L^p.

6.2. Remarks on null sets and null functions. A set of measure zero is called a *null set*. Null sets are negligible in the theory of integration. If $f=g$ except in a null set, we say that f and g are *equivalent*, and write $f \equiv g$. Equivalent functions have the same integral (if any).

If $f \equiv 0$, we call f a *null function*, and say that *f is null*.

Similarly, we define 'equivalent in E', 'null in E': f is null in E if $f=0$ at all points of E except the points of a null set. In such cases we shall not repeat the reference to E when the context makes it obvious, as for example when we are considering integrals over E.

If a property $P(x)$ is possessed by all x except the x of a null set, we shall say that it is possessed by *almost all x*, or that $P(x)$ is true for almost all x, or *almost always*. Thus a null function is almost always zero.

We shall generally assume that our functions $f, g, \ldots$ are almost always finite; but there will be occasions when we have to consider functions infinite in a set of positive measure. Thus if f is generally positive, but zero in a set E of positive measure, and $r < 0$, then we must regard f^r as infinite in E, and $\int f^r dx$ as having the value ∞.

If E is null then

$$\int_E f\,dx = 0$$

for all f. We shall assume without special remark that a set E over which an integral is extended is not null.

If $f \geqq 0$, then a necessary and sufficient condition that $\int f\,dx = 0$ is that f should be null.

It may be worth while to call attention explicitly to the theorem which replaces this in the theory of Riemann integration. We denote the class of Riemann integrable functions by R (it is contained in L). A necessary and sufficient condition that f should be R is that f should be bounded and that its set of discontinuities should be null.

If f is R, and $f \geqq 0$, then a necessary and sufficient condition that $\int f\,dx = 0$ is that $f=0$ at all points of continuity of f.

For (1) if the condition is satisfied, then $f \equiv 0$ and so $\int f\,dx = 0$. And (2) if it is not satisfied, then there is a point of continuity ξ at which $f(\xi) > 0$, and an interval including ξ throughout which $f(x) > \frac{1}{2}f(\xi)$; so that $\int f dx > 0$.

This theorem enables us to specify the cases of equality in our inequalities when they are restricted to functions of R. In fact most of our theorems have a dual interpretation. The primary interpretation is that in which the integrals are Lebesgue integrals and 'null' and 'equivalent' are interpreted as in the theory of Lebesgue. In the secondary interpretation integrals are 'Riemann integrals', a 'null function' is a function which is zero at all its points of continuity, and 'equivalent functions' are functions whose difference is null in this sense.

6.3. Further remarks concerning integration. What has been set out in §§ 6.1 and 6.2 is a sufficient foundation for most of the subsequent theorems: for example, for the most complete forms of Hölder's and Minkowski's inequalities (Theorems **188** and **198**). There will be a few occasions on which we shall have to appeal to more difficult theorems, and we enumerate these here.

(a) *Integration by parts.* The theorem required is: *if f and g are integrals (absolutely continuous functions), then*

$$\int_a^b fg'\,dx = \Big[fg\Big]_a^b - \int_a^b f'g\,dx.$$

(b) *Passage to the limit under the integral sign.* The two main theorems are

(i) *If $|s_n(x)| < \phi(x)$, where $\phi \in L$, and $s_n(x)$ tends to a limit $s(x)$ for all or almost all x, then*

$$\int s_n(x)\,dx \to \int s(x)\,dx.$$

(ii) *If $s_n(x) \in L$ for every n, $s_n(x)$ increases with n for all or almost all x, and*

$$\lim s_n(x) = s(x),$$

then

$$\int s_n(x)\,dx \to \int s(x)\,dx.$$

In (ii) the integral on the right may be infinite, when the result is to be interpreted as $\int s_n(x)\,dx \to \infty$; in particular this happens if $s(x) = \infty$ in a set of positive measure. In each of these theorems n may be an integer which tends to infinity or a continuous parameter which tends to a limit.

It follows from (i), as is shown in books on the theory of functions of a real variable, that a function $f(x)$ whose incrementary ratio

$$\frac{f(x+h)-f(x)}{h}$$

is bounded (and which therefore has a derivative almost everywhere) is the integral of its derivative. Combining this remark with that at the end of § 3.18 we see that a continuous convex function $f(x)$ is the integral of its derivative $f'(x)$, or of its one-sided derivatives $f_l'(x), f_r'(x)$. It is therefore the integral of an increasing function. On the other hand, if $f(x)$ is the integral of an increasing function $g(x)$, and $h>0$, then

$$f(x+h)-f(x)=\int_x^{x+h} g(u)\,du \geqq \int_{x-h}^{x} g(u)\,du=f(x)-f(x-h),$$

so that $f(x)$ is convex. Hence *the class of continuous convex functions is identical with that of integrals of increasing functions.*

An increasing function belongs to R, so that the integrals in question exist as Riemann integrals, and the theorem could be proved without any use of the theory of Lebesgue.

(c) *Substitution.* The standard theorem is: *if f and g are integrable, $g \geqq 0$, G is an integral of g, and $a=G(\alpha)$, $b=G(\beta)$, then*

$$\int_a^b f(x)\,dx=\int_\alpha^\beta f\{G(y)\}g(y)\,dy. \tag{6.3.1}$$

Here any of a, b, α, β may be infinite.

This theorem covers all cases in which we shall require the rule for transformation of an integral by change of the independent variable[a]. But we shall generally use only trivial substitutions such as $x=y+a$ or $x=ay$, when the validity of the rule follows at once from the definitions.

(d) *Multiple and repeated integrals.* The only theorem appealed to is 'Fubini's Theorem'. *If $f(x,y)$ is (areally measurable and) non-negative, and any one of the integrals*

$$\int_a^A\int_b^B f\,dx\,dy,\quad \int_a^A dx\int_b^B f\,dy,\quad \int_b^B dy\int_a^A f\,dx$$

exists, then the other integrals exist, and all are equal. Here the limits are finite or infinite, and the case of divergence is included; if one integral diverges the others diverge.

[a] We may add two additional remarks concerning the formula (6.3.1).

(1) If we suppose, as in the text, that g is non-negative and integrable, but assume only the measurability (and not the integrability) of f, then the existence of the right-hand side of (6.3.1) is a sufficient, as well as a necessary, condition for the existence of the left-hand side, i.e. for the integrability of f.

(2) The integrability of $f(x)$, though it implies that of $f\{G(y)\}g(y)$, does not imply even the measurability of $f\{G(y)\}$.

Suppose then that $f(x,y)$ is measurable and non-negative. The double integral is zero if and only if $f(x,y)$ is null, i.e. if the set in which $f(x,y)>0$ has measure zero. The first repeated integral is zero if and only if $f(x,y)$ is, for almost all x, null in y; and the second if $f(x,y)$ is, for almost all y, null in x. Hence these three senses of 'a null non-negative function of two variables' are equivalent.

6.4. Remarks on methods of proof. Inequalities proved for finite sums may often be extended to integrals by the use of limiting processes, but something is usually lost in the argument. We may illustrate this by considering the analogue for integrals of Theorem 7.

Suppose first that $f(x)$ and $g(x)$ are non-negative and Riemann integrable in (0, 1); and take

$$a_\nu = f\left(\frac{\nu}{n}\right), \quad b_\nu = g\left(\frac{\nu}{n}\right)$$

in Theorem 7. Dividing by n^2, we obtain[a]

$$\left\{\frac{1}{n}\Sigma f\left(\frac{\nu}{n}\right) g\left(\frac{\nu}{n}\right)\right\}^2 \leqq \frac{1}{n}\Sigma f^2\left(\frac{\nu}{n}\right) . \frac{1}{n}\Sigma g^2\left(\frac{\nu}{n}\right);$$

and, making $n\to\infty$,

$$\left(\int_0^1 fg\,dx\right)^2 \leqq \int_0^1 f^2 dx \int_0^1 g^2 dx. \tag{6.4.1}$$

If we use the Lebesgue integral we must argue differently[b]. Suppose that f and g are non-negative and L^2 in (0, 1), and that e_{rs} is the set in which

$$\frac{r-1}{n} \leqq f^2 < \frac{r}{n}, \quad \frac{s-1}{n} \leqq g^2 < \frac{s}{n} \quad (r, s = 1, 2, 3, \ldots).$$

Then

$$\left(\int_0^1 fg\,dx\right)^2 \leqq \left[\Sigma\Sigma\left(\frac{r}{n}\right)^{\frac{1}{2}}\left(\frac{s}{n}\right)^{\frac{1}{2}} e_{rs}\right]^2 \leqq \Sigma\Sigma\frac{r}{n}e_{rs}\,\Sigma\Sigma\frac{s}{n}e_{rs},$$

by Theorem 7. Now

$$\Sigma\Sigma\frac{r}{n}e_{rs} = \Sigma\Sigma\frac{1}{n}e_{rs} + \Sigma\Sigma\frac{r-1}{n}e_{rs} \leqq \frac{1}{n} + \int_0^1 f^2 dx,$$

[a] It is here that 'homogeneity in Σ' (§ 1.4) is essential.

[b] The precise form of argument used here was suggested to us by Mr H. D. Ursell.

and there is a similar inequality involving g. Hence, making $n \to \infty$, we obtain (6.4.1).

In either case our final result is imperfect. Even if we can use Theorem **7** with '$<$', this will degenerate into '$\leqq$' when we pass to the limit, and we shall lose touch with the cases of equality.

The passage in the opposite direction, from an integral inequality to an inequality for sums, is much simpler, and can be effected by suitable specialisation. Consider, for example, the inequality

$$\exp\left\{\int_0^1 \log f(x)\,dx\right\} < \int_0^1 f(x)\,dx \tag{6.4.2}$$

(§ 6.7, Theorem **184**). If

$$q_\nu > 0, \quad q_1 + q_2 + \dots + q_n = 1,$$

and we define $f(x)$ by

$$f(x) = a_\nu \quad (q_1 + \dots + q_{\nu-1} \leqq x < q_1 + \dots + q_{\nu-1} + q_\nu),$$

it being understood that $q_1 + \dots + q_{\nu-1}$ means 0 when $\nu = 1$, we obtain Theorem **9**. The conditions under which inequality degenerates into equality in Theorem **9** also follow immediately from the corresponding conditions for (6.4.2).

This method of proof is often useful, since integrals are often more manageable than series. We shall meet with examples in Ch. IX.

6.5. Further remarks on method: the inequality of Schwarz. We meet the difficulty of § 6.4, as in our treatment of infinite series, by going back to the proofs of the theorems of Ch. II, and observing that, with the obvious changes, they can be applied to integrals of the most general type. We may illustrate the point here by considering 'Schwarz's[a] inequality', the analogue of Theorem **7**.

181. $(\int fg\,dx)^2 < \int f^2\,dx \int g^2\,dx$, *unless* $Af \equiv Bg$, *where* A *and* B *are constants, not both zero.*

Here, and later, we suppress the limits of integration when there is nothing to be gained by showing them explicitly; they

[a] Or Buniakowsky's (see p. 16).

may be finite or infinite, or the integrals may be over any measurable set E, in which case of course $Af \equiv Bg$ means $Af \equiv Bg$ in E. We also adopt conventions corresponding to those of § 5.1: '$X < Y$' means 'if Y is finite then X is finite and $X < Y$'; and inequalities of other forms like those mentioned in § 5.1 are to be interpreted similarly. Thus every inequality contains implicitly an assertion about 'convergence', which we shall only make explicit occasionally. For example, Theorem **181** asserts implicitly that 'if $\int f^2 dx$ and $\int g^2 dx$ are finite, then $\int fg\,dx$ is finite; if f and g are L^2, then fg is L'.

The proofs corresponding to those of § 2.4 run as follows.

(i) We have

$$\begin{aligned}\int f^2 dx \int g^2 dx - \left(\int fg\,dx\right)^2 &= \tfrac{1}{2}\int f^2(x)\,dx \int g^2(y)\,dy + \tfrac{1}{2}\int g^2(x)\,dx \int f^2(y)\,dy \\ &\qquad - \int f(x)g(x)\,dx \int f(y)g(y)\,dy \\ &= \tfrac{1}{2}\int dy \int \{f(x)g(y) - g(x)f(y)\}^2 dx \geqq 0.\end{aligned}$$

It remains to discuss the possibility of equality. In the first place, there is certainly equality if $Af \equiv Bg$. Next, if there is equality, and g is null, then $Af \equiv Bg$ with $A = 0$, $B = 1$. We may therefore assume that g is not null, so that the set E in which $g(y) \neq 0$ has positive measure. If

$$\int dy \int \{f(x)g(y) - g(x)f(y)\}^2 dx = 0$$

then

$$\int \{f(x)g(y) - g(x)f(y)\}^2 dx = 0 \tag{6.5.1}$$

for almost all y, and therefore for some y belonging to E. We may therefore suppose that $g(y_0) \neq 0$ and that (6.5.1) is true for $y = y_0$. But then

$$f(x)g(y_0) - g(x)f(y_0) = 0$$

for almost all x, and this completes the proof.

(ii) The quadratic form

$$\int (\lambda f + \mu g)^2 dx = \lambda^2 \int f^2 dx + 2\lambda\mu \int fg\,dx + \mu^2 \int g^2 dx$$

is positive. We can now complete the proof as in § 2.4.

The analogue of Theorem **181** for multiple integrals may be proved similarly. We shall not usually mention such extensions explicitly, but we shall occasionally take them for granted. It is to be understood that, when we do this, the extension may be proved in the same manner as the original theorem.

We can translate the proof of Theorem 8 in the same manner, and so obtain

182.[a]
$$\begin{vmatrix} \int f^2 dx & \int fg\,dx & \dots & \int fh\,dx \\ \dots\dots & \dots\dots & \dots & \dots\dots \\ \int hf\,dx & \int hg\,dx & \dots & \int h^2 dx \end{vmatrix} > 0,$$

unless the functions f, g, ..., h are linearly dependent, i.e. unless there are constants A, B, ..., C, not all zero, such that

$$Af + Bg + \dots + Ch \equiv 0.$$

MEANS $\mathfrak{M}_r(f)$

6.6. Definition of the means $\mathfrak{M}_r(f)$ when $r \neq 0$. In what follows the sign of integration, used without specification of the range, refers to a finite or infinite interval (a, b) or to a measurable set E.[b] $f(x)$ is finite almost everywhere in E and non-negative; $p(x)$, the 'weight function', is finite and positive[c] everywhere in E, and integrable over E. The parameter r is real and not zero.

Our hypotheses involve $0 < \int p\,dx < \infty$. It is often convenient to suppose

$$\int p\,dx = 1:$$

in this case (cf. § 2.2) we write q for p.

We write

$$\mathfrak{M}_r(f) = \mathfrak{M}_r(f, p) = \left(\frac{\int pf^r dx}{\int p\,dx}\right)^{1/r} \quad (r \neq 0), \tag{6.6.1}$$

$$\mathfrak{A}(f) = \mathfrak{M}_1(f), \tag{6.6.2}$$

so that

$$\mathfrak{M}_r(f) = \{\mathfrak{A}(f^r)\}^{1/r}; \tag{6.6.3}$$

with the following conventions. If $\int pf^r dx$ is infinite, we write

$$\int pf^r dx = \infty, \quad \mathfrak{M}_r(f) = \infty \ (r > 0), \quad \mathfrak{M}_r(f) = 0 \ (r < 0).$$

[a] Gram (1).

[b] When $r > 0$ we can reduce every case to that of the interval $(-\infty, \infty)$, by supposing $f = 0$ in the set complementary to E.

[c] The hypothesis $p \geq 0$, instead of $p > 0$, would lead to slightly different results concerning the cases of equality (e.g. $pf \equiv pC$ instead of $f \equiv C$). This case could be reduced to the apparently more special case by replacing E by the sub-set of E in which $p > 0$.

In particular $\mathfrak{M}_r(f)=0$ if $r<0$ and $f=0$ in a set of positive measure. If we agree further to regard 0 and ∞ as reciprocals of one another, we have

$$\mathfrak{M}_r(f)=\frac{1}{\mathfrak{M}_{-r}(1/f)}. \tag{6.6.4}$$

This formula enables us to pass from positive to negative r, and we shall simplify the following theorems by restricting ourselves, for the most part, to positive r.

If $f\equiv 0$, $\mathfrak{M}_r(f)=0$ for all r. If $f\equiv C$, where C is positive and finite, then $\mathfrak{M}_r(f)=C$ for all r. If $f\equiv\infty$,[a] then $\mathfrak{M}_r(f)=\infty$ for all r. Apart from these cases, $\mathfrak{M}_r(f)=\infty$ is possible only when $r>0$, and $\mathfrak{M}_r(f)=0$ when $r<0$.

We define Max f, the 'effective upper bound' of f, as the largest ξ which has the property:

'if $\epsilon>0$, there is a set $e(\epsilon)$ of positive measure in which $f>\xi-\epsilon$'.

If there is no such ξ, we write Max $f=\infty$. For functions continuous in a closed interval, Max f is the ordinary maximum. Min f is defined similarly; Min $f\geqq 0$ and

$$\operatorname{Min} f=\frac{1}{\operatorname{Max}(1/f)}. \tag{6.6.5}$$

Equivalent functions have the same Max and Min.

Suppose for example that the range of integration is $(0,\infty)$, and that $f(x)$ and $q(x)$ are the step functions defined by

$$f(x)=a_n,\quad q(x)=q_n\quad (n-1\leqq x<n,\quad n=1,2,3,\dots).$$

Then
$$\mathfrak{M}_r(f)=(\Sigma q a^r)^{1/r}=\mathfrak{M}_r(a),$$
according to the definition of (5.2.1). Similarly

$$\operatorname{Max} f=\operatorname{Max} a,\quad \operatorname{Min} f=\operatorname{Min} a,$$

and (if we anticipate the definition of § 6.7) $\mathfrak{G}(f)=\mathfrak{G}(a)$. This specialisation enables us to include many theorems of Ch. II and Ch. V in the corresponding theorems of this chapter.

Alternatively we might (as in § 6.4) suppose that the range of integration is (0, 1), and define $f(x)$ and $q(x)$ by

$$f(x)=a_n\quad (q_1+\dots+q_{n-1}\leqq x<q_1+\dots+q_n),\quad q(x)=1.$$

In this case also $\mathfrak{M}_r(f)$ reduces to $\mathfrak{M}_r(a)$.

[a] To admit this case is to abandon momentarily the understanding of § 6.2, that f is assumed to be finite almost everywhere.

183. *If $r \neq 0$ and $\mathfrak{M}_r(f)$ is finite and positive, then*

$$\operatorname{Min} f < \mathfrak{M}_r(f) < \operatorname{Max} f,$$

unless $f \equiv C$.

Here r is of either sign. The proof is like that of Theorem **1**. Suppose first that $r = 1$. Then, using a weight function $q(x)$, we have

$$\int q(f - \mathfrak{A})\,dx = 0.$$

Hence either $f \equiv \mathfrak{A}$ or $f - \mathfrak{A}$ is positive and negative, each in a set of positive measure. This proves the result for $r = 1$, and we extend it to the general case by use of (6.6.3).

If we wish to state Theorem **183** in a form corresponding more exactly to that of Theorem **1** and its extension in Ch. V[a], we must say '$\operatorname{Min} f < \mathfrak{M}_r(f) < \operatorname{Max} f$ *unless* $f \equiv C$ *or else* $r < 0$ *and* $\mathfrak{M}_r(f) = 0$'. We have then two cases of equality corresponding exactly to those distinguished in § 5.2, the 'primary' case in which $f \equiv C$, in which both inequalities reduce to equalities, and the 'secondary' case, occurring only for $r < 0$, in which one inequality only reduces to an equality. This distinction recurs in many of our theorems, when $r < 0$, as it recurred in Chs. II and V; but it is less conspicuous here because we often ignore negative values of r.

6.7. The geometric mean of a function.

We define the geometric mean $\mathfrak{G}(f)$ by

(6.7.1) $$\mathfrak{G}(f) = \mathfrak{G}(f, p) = \exp\left(\frac{\int p \log f\,dx}{\int p\,dx}\right),$$

or

(6.7.2) $$\log \mathfrak{G}(f) = \mathfrak{A}(\log f),$$

so that, in particular, if $p = q$, $\int q\,dx = 1$, we have

(6.7.3) $$\mathfrak{J}(f) = \log \mathfrak{G}(f) = \int q \log f\,dx.$$

Certain preliminary explanations are necessary.

Since $\log f$ is not necessarily positive, the possibilities concerning the convergence of $\mathfrak{J}$ are more complex than those which we have considered hitherto.

[a] See § 5.2.

If we denote by $\mathfrak{J}^+$ and $\mathfrak{J}^-$ the integrals formed with $\log^+ f$ and $\log^- f$, as $\mathfrak{J}$ is formed with $\log f$,[a] then there are four possibilities: (a) $\mathfrak{J}^+$ and $\mathfrak{J}^-$ both finite, (b) $\mathfrak{J}^+$ finite, $\mathfrak{J}^- = -\infty$, (c) $\mathfrak{J}^+ = \infty$, $\mathfrak{J}^-$ finite, (d) $\mathfrak{J}^+ = \infty$, $\mathfrak{J}^- = -\infty$. The four cases are exemplified by the functions

$$x, \quad e^{-1/x}, \quad e^{1/x}, \quad \exp\left(\frac{1}{x^2}\sin\frac{1}{x}\right)$$

in (0, 1), with $q(x) = 1$.

If $\mathfrak{M}_r(f)$ is finite for some $r > 0$, then, since

$$\log^+ f \leqq \operatorname{Max}\left(\frac{f^r - 1}{r}, 0\right),$$

$\mathfrak{J}^+(f)$ will be finite, and we shall be concerned only with cases (a) and (b). In case (a), $\mathfrak{J}(f)$ exists as a Lebesgue integral, and $\mathfrak{G}(f)$ is positive and finite. In case (b) we write

$$\mathfrak{J}(f) = -\infty, \quad \mathfrak{G}(f) = 0.$$

Similarly, if $\mathfrak{M}_r(1/f)$ is finite for some $r > 0$, we are in case (a) or case (c); in the latter we write

$$\mathfrak{J}(f) = \infty, \quad \mathfrak{G}(f) = \infty.$$

In case (d) the symbol $\mathfrak{G}(f)$ is meaningless. In this case $\mathfrak{M}_r(f)$ and $\mathfrak{M}_r(1/f)$ are infinite for every positive r, and $\mathfrak{M}_r(f) = 0$ for every negative r.

In case (a) we have

$$\mathfrak{G}\left(\frac{1}{f}\right) = \frac{1}{\mathfrak{G}(f)}, \tag{6.7.4}$$

both sides being positive and finite; and a moment's consideration shows that this equation holds in all cases, if we adopt the same convention about 0 and ∞ as in (6.6.4), and the additional convention that one side of the equation is meaningless if the other is meaningless.

We now prove the analogue of Theorem **9**.

184. *If $\mathfrak{A}(f)$ is finite then*

$$\mathfrak{G}(f) < \mathfrak{A}(f), \tag{6.7.5}$$

[a] See § 5.2 for the definitions of $\log^+$ and $\log^-$.

unless $f \equiv C$, *where* C *is constant. More generally, if* $\mathfrak{M}_r(f)$, *where* $r > 0$, *is finite, then*

$$(6.7.6) \qquad \mathfrak{G}(f) < \mathfrak{M}_r(f),$$

with the same reservation[a].

Suppose first that $r = 1$, $\mathfrak{M}_r = \mathfrak{A}$. If $\mathfrak{A}(f) = 0$, $f \equiv 0$, and so $\mathfrak{J}(f) = -\infty$, $\mathfrak{G}(f) = 0 = \mathfrak{A}(f)$. We may therefore suppose $\mathfrak{A}(f) > 0$.

Since, by Theorem **150**,

$$(6.7.7) \qquad \log t < t - 1,$$

if $t > 0$, $t \neq 1$, we have

$$\log f - \log \mathfrak{A}(f) \leqq \frac{f}{\mathfrak{A}(f)} - 1,$$

$$\mathfrak{A}(\log f) - \log \mathfrak{A}(f) \leqq \frac{\mathfrak{A}(f)}{\mathfrak{A}(f)} - 1 = 0,$$

$$\log \mathfrak{G}(f) = \mathfrak{A}(\log f) \leqq \log \mathfrak{A}(f).$$

Equality can occur only if $f \equiv \mathfrak{A}(f)$.

The result for general r now follows from (6.6.3).

In Theorem **184** we have stated the hypotheses 'if $\mathfrak{A}(f)$ is finite', 'if $\mathfrak{M}_r(f)$ is finite' explicitly. As we have explained in §§ 5.1 and 6.5, we shall often save space by omitting such hypotheses in accordance with our conventions. We shall also denote constants by $C, A, B, a, b, \ldots$ without explanation, when there is no danger of ambiguity. Two C's occurring in the same connection will not necessarily be the same.

We add two corollaries (extensions of Theorem **10**).

185. $\mathfrak{G}(f) + \mathfrak{G}(g) < \mathfrak{G}(f+g)$, *unless* $Af \equiv Bg$, *where* A, B *are not both zero, or* $\mathfrak{G}(f+g) = 0$.

We may suppose that $\mathfrak{G}(f+g) > 0$. Then, by Theorem **184**,

$$\frac{\mathfrak{G}(f)}{\mathfrak{G}(f+g)} = \mathfrak{G}\left(\frac{f}{f+g}\right) \leqq \mathfrak{A}\left(\frac{f}{f+g}\right).$$

The addition of the two inequalities of this type gives the result.

More generally

186. $$\mathfrak{G}(f_1) + \mathfrak{G}(f_2) + \mathfrak{G}(f_3) + \ldots < \mathfrak{G}(f_1 + f_2 + f_3 + \ldots)$$

(*the series being finite or infinite*), *unless* $f_n \equiv C_n \Sigma f_n$ *or* $\mathfrak{G}(\Sigma f_n) = 0$.

[a] For the proof which follows see F. Riesz (7).

6.8. Further properties of the geometric mean. Our next theorem corresponds to Theorem **3** (for positive r).

187. *If $\mathfrak{M}_r(f)$ is finite for some positive r, then*

$$\mathfrak{M}_r(f) \to \mathfrak{G}(f) \tag{6.8.1}$$

when $r \to +0$.

It should be observed that $\mathfrak{G}(f)$ may be finite even when $\mathfrak{M}_r(f) = \infty$ for all $r > 0$. This is so, for example, if $f(x) = \exp(x^{-\frac{1}{2}})$, $q(x) = 1$, and the range is $(0, 1)$.

When E is a closed interval or set, and f is continuous and positive, the proof is immediate. In this case $f \geqq \delta > 0$, $\log f$ is bounded, and

$$\mathfrak{M}_r{}^r = \int e^{r \log f} q\,dx = \int [1 + r \log f + O\{r^2 (\log f)^2\}]\, q\,dx$$
$$= 1 + r\mathfrak{J} + O(r^2),$$
$$\lim \log \mathfrak{M}_r = \lim \frac{1}{r} \log\{1 + r\mathfrak{J} + O(r^2)\} = \mathfrak{J}.$$

There is some difficulty in extending this argument to the general case. The difficulty can however be avoided as follows[a]. By (6.7.6) and (6.7.7), we have

$$\log \mathfrak{G}(f) \leqq \log \mathfrak{M}_r(f) = \frac{1}{r} \log \mathfrak{A}(f^r) \tag{6.8.2}$$
$$\leqq \frac{1}{r}\{\mathfrak{A}(f^r) - 1\} = \int \frac{f^r - 1}{r}\, q\,dx.$$

When r decreases to zero, $(t^r - 1)/r$ decreases (by Theorem **36**) and tends to the limit $\log t$. Hence[b]

$$\lim \frac{1}{r}\{\mathfrak{A}(f^r) - 1\} = \mathfrak{A}(\log f) = \log \mathfrak{G}(f), \tag{6.8.3}$$

the right-hand side being finite or $-\infty$. Combining (6.8.2) and (6.8.3) we see that

$$\log \mathfrak{G}(f) \leqq \underline{\lim} \log \mathfrak{M}_r(f) \leqq \overline{\lim} \log \mathfrak{M}_r(f) \leqq \log \mathfrak{G}(f),$$

which proves the theorem.

6.9. Hölder's inequality for integrals. We consider next the

[a] F. Riesz (7). Other, less simple, proofs have been given by Besicovitch, Hardy, and Littlewood: see Hardy (7).

[b] See § 6.3 (*b*) (ii).

theorems for integrals which correspond to Theorems **11–15**. It is convenient to introduce another definition which enables us to shorten our statements of cases of equality. Two functions f, g will be said to be *effectively proportional* if there are constants A, B, not both zero, such that $Af \equiv Bg$. The idea has occurred already in Theorems **181** and **185**. A null function is effectively proportional to any function. We shall say that f, g, h, ... are effectively proportional if every pair are so.

188. *If $\alpha, \beta, \ldots, \lambda$ are positive and $\alpha+\beta+\ldots+\lambda=1$, then*

$$\int f^{\alpha} g^{\beta} \ldots l^{\lambda} dx < \left(\int f dx\right)^{\alpha} \left(\int g dx\right)^{\beta} \ldots \left(\int l dx\right)^{\lambda}, \tag{6.9.1}$$

unless one of the functions is null or all are effectively proportional.

Assuming no function null, we have, by Theorem **9**,

$$\frac{\int f^{\alpha} g^{\beta} \ldots l^{\lambda} dx}{\left(\int f dx\right)^{\alpha} \left(\int g dx\right)^{\beta} \ldots \left(\int l dx\right)^{\lambda}} = \int \left(\frac{f}{\int f dx}\right)^{\alpha} \left(\frac{g}{\int g dx}\right)^{\beta} \ldots \left(\frac{l}{\int l dx}\right)^{\lambda} dx$$
$$\leq \int \left(\frac{\alpha f}{\int f dx} + \frac{\beta g}{\int g dx} + \ldots + \frac{\lambda l}{\int l dx}\right) dx = 1,$$

with inequality unless

$$\frac{f}{\int f dx} \equiv \frac{g}{\int g dx} \equiv \ldots \equiv \frac{l}{\int l dx}.$$

As a corollary[a] we have

189. *If $k > 1$ then*

$$\int fg\, dx < \left(\int f^{k} dx\right)^{1/k} \left(\int g^{k'} dx\right)^{1/k'} \tag{6.9.2}$$

unless f^k and $g^{k'}$ are effectively proportional.

If $0 < k < 1$ or $k < 0$ then

$$\int fg\, dx > \left(\int f^{k} dx\right)^{1/k} \left(\int g^{k'} dx\right)^{1/k'} \tag{6.9.3}$$

unless either (a) f^k and $g^{k'}$ are effectively proportional or (b) fg is null.

The second half of the theorem requires a little explanation. Suppose first that $0 < k < 1$ and that $\int g^{k'} dx$ is finite, so that g is almost always positive. If then we write $l = 1/k$, so that $l > 1$, and

$$f = (uv)^{l}, \quad g = v^{-l},$$

so that

$$fg = u^{l}, \quad f^{k} = uv, \quad g^{k'} = v^{l'},$$

[a] Compare § 2.8.

then u and v are defined for almost all x, and

$$\int uv\,dx < \left(\int u^l dx\right)^{1/l} \left(\int v^{l'} dx\right)^{1/l'}$$

or

$$\int f^k dx < \left(\int fg\,dx\right)^k \left(\int g^{k'} dx\right)^{1-k},$$

unless u^l, $v^{l'}$ are effectively proportional or, what is the same thing, unless f^k, $g^{k'}$ are effectively proportional. Since $\int g^{k'} dx$ is finite and not zero[a], this is (6.9.3).

If $\int g^{k'} dx = \infty$, then

$$\left(\int g^{k'} dx\right)^{1/k'} = 0$$

(since $k' < 0$). Hence the right-hand side of (6.9.3) is zero, and there is inequality unless $\int fg\,dx = 0$, or fg is null.

When $k < 0$, $0 < k' < 1$, and the argument is substantially the same.

As we have explained in §§ 5.1 and 6.5, the theorem contains implicitly an assertion about convergence or finitude; if two of the integrals involved are finite, then so is the third. The integral which is finite if the other two are finite is $\int fg\,dx$ when $k > 1$, $\int f^k dx$ if $0 < k < 1$, and $\int g^{k'} dx$ if $k < 0$.

The theorem corresponding to Theorem **161** is very important, and, like Theorem **161**, is not a direct corollary of preceding theorems.

190.[b] *If $k > 1$ and fg belongs to L for every g which belongs to $L^{k'}$, then f belongs to L^k.*

We consider first the case in which (a, b) is finite (or mE finite), and suppose that $\int f^k dx = \infty$. We can find a function f^* which (1) has only an enumerable infinity of values a_i, and (2) satisfies $f^* \leqq f < f^* + \epsilon$. Since f^k does not exceed a constant multiple of $f^{*k} + (f - f^*)^k$, by Theorem **13**, we have $\int f^{*k} dx = \infty$. Hence, if e_i is the set in which $f^* = a_i$,

$$\Sigma a_i^k e_i = \infty.$$

It follows from Theorem **161**, taking

$$a_i^k e_i = u_i^k, \quad b_i^{k'} e_i = v_i^{k'}, \quad a_i b_i e_i = u_i v_i,$$

[a] $\int g^{k'} dx = 0$ would involve $g^{k'} \equiv 0$ and so $g \equiv \infty$, and this possibility is excluded by the understanding of § 6.2.

[b] F. Riesz (2).

that there is a b_i such that $\Sigma b_i^{k'} e_i$ is convergent and $\Sigma a_i b_i e_i = \infty$. We take $g(x) = b_i$ in e_i (for all i). Then

$$\int g^{k'} dx = \Sigma b_i^{k'} e_i$$

is convergent, but

$$\int f^* g\, dx = \Sigma a_i b_i e_i = \infty,$$

and hence $\int fg\, dx = \infty$, contrary to the hypothesis.

If the integrals are over an infinite range, say $(0, \infty)$, we can write

$$x = \frac{t}{1-t},$$

when

$$\int_0^\infty fg\, dx = \int_0^1 FG\, dt, \quad \int_0^\infty f^k dx = \int_0^1 F^k dt, \quad \int_0^\infty g^{k'} dx = \int_0^1 G^{k'} dt,$$

where

$$F(t) = (1-t)^{-2/k} f\left(\frac{t}{1-t}\right), \quad G(t) = (1-t)^{-2/k'} g\left(\frac{t}{1-t}\right).$$

The theorem is thus reduced to the finite case.

191. *If* $k > 1$, *then a necessary and sufficient condition that* $\int f^k dx \leqq F$ *is that* $\int fg\, dx \leqq F^{1/k} G^{1/k'}$ *for all* g *such that* $\int g^{k'} dx \leqq G$.

The condition is necessary, by Theorem **189**. If it is satisfied, then $\int f^k dx$ is finite, by Theorem **190**. If $\int f^k dx > F$, we choose g so that $g^{k'}$ is effectively proportional to f^k, and then, by Theorem **189**,

$$\int fg\, dx = \left(\int f^k dx\right)^{1/k} \left(\int g^{k'} dx\right)^{1/k'} > F^{1/k} G^{1/k'}.$$

The theorem may also be stated with '$<$' for '$\leqq$' in the first two inequalities: in order that $\int f^k dx < F$, it is necessary and sufficient that $\int fg\, dx < F^{1/k} G^{1/k'}$ whenever $\int g^{k'} dx \leqq G$.

We can prove Theorem **191** without appealing to the more difficult Theorem **190**. If $\int f^k dx > F$ then $\int (f)_n{}^k dx > F$ for sufficiently large n. Then, choosing g effectively proportional to $(f_n)^{k-1}$, we have

$$\int fg\, dx \geqq \int (f)_n g\, dx = \left(\int (f)_n{}^k dx\right)^{1/k} G^{1/k'} > F^{1/k} G^{1/k'},$$

in contradiction to the hypothesis of the theorem.

Another proof of Theorem **190** (and of the associated Theorem **161**) has been given by Banach (**1**, 85–86).

An example of the use of Theorem **191** appears in § 6.13, in the proof of Theorem **202**, and others in Ch. IX.[a] In § 6.13 Theorem **202** is proved in

[a] See in particular §§ 9.3 and 9.7 (2).

two different ways, of which one depends explicitly on Theorem **191** while the other does not use it, and the logical status of Theorem **191** in proofs of this character is explained in detail.

6.10. General properties of the means $\mathfrak{M}_r(f)$. We shall now prove a number of theorems which include the analogues of those of § 2.9. The properties to be investigated are a little more complex than they were there, and we shall require some additional conventions before we can state them comprehensively. We suppose first that $r > 0$; the theorems which we prove in this case, with those which we have proved concerning $\mathfrak{G}(f)$, will give us the substance of what is required, and we shall be able to state the results for unrestricted r more summarily, leaving most of the details of verification to the reader.

192. *If $0 < r < s$ and $\mathfrak{M}_s$ is finite, then*

$$\mathfrak{M}_r < \mathfrak{M}_s,$$

unless $f \equiv C$.

If $r = s\alpha$, so that $0 < \alpha < 1$, we have, by Theorem **188**,

$$\int qf^r\,dx = \int (qf^s)^\alpha q^{1-\alpha}\,dx < \left(\int qf^s\,dx\right)^\alpha \left(\int q\,dx\right)^{1-\alpha} = \left(\int qf^s\,dx\right)^\alpha,$$

unless $qf^s \equiv Cq$. Since $q > 0$, this is the result required.

193. *If $\mathfrak{M}_r$ is finite for every r, then $\mathfrak{M}_r \to \operatorname{Max} f$ when $r \to +\infty$.*

(i) Suppose $\mu = \operatorname{Max} f$ finite. Then (a) $\mathfrak{M}_r \leqq \mu$, and (b) $f > \mu - \epsilon$ in a set e of positive measure ζ, so that

$$\int_e q\,dx = \zeta > 0, \quad \mathfrak{M}_r \geqq (\mu - \epsilon)\,\zeta^{1/r}, \quad \underline{\lim}\,\mathfrak{M}_r \geqq \mu - \epsilon.$$

(ii) Suppose $\operatorname{Max} f = \infty$. Then, for any $G > 0$, $f > G$ in a set e of positive measure, and, as above, $\underline{\lim}\,\mathfrak{M}_r \geqq G$.

From (6.6.4), (6.6.5) and Theorem **193** it follows that

$$\mathfrak{M}_r \to \operatorname{Min} f$$

when $r \to -\infty$.

194. *If $0 < s < \infty$ and $\mathfrak{M}_s$ is finite, then $\mathfrak{M}_r$ is continuous for $0 < r < s$ and continuous on the left for $r = s$. If $\mathfrak{M}_s = \infty$, but $\mathfrak{M}_r$ is finite for $0 < r < s$, then $\mathfrak{M}_r \to \infty$ when $r \to s$.*

(i) Suppose $\mathfrak{M}_s$ finite. Then

$$qf^r \leqq q \operatorname{Max}(1, f^s),$$

a majorant of class L independent of r; and the results follow from § 6.3 (b) (i).[a]

(ii) Suppose $\mathfrak{M}_s = \infty$. We can choose n so that

$$\int (qf^s)_n \, dx > G.$$

But $(qf^r)_n$ is a continuous function of r, and so[b]

$$\int (qf^r)_n \, dx > \tfrac{1}{2} G$$

for $r > s - \epsilon$. Hence $\int qf^r \, dx > \frac{1}{2} G$, which proves the theorem.

6.11. General properties of the means $\mathfrak{M}_r(f)$ (*continued*). In the preceding sections we have, in the main, confined our attention to means for which $r \geqq 0$, leaving it to the reader to deduce the corresponding results for means of negative order from the formulae (6.6.4) and (6.6.5). In this section we consider the means more comprehensively. We write, as is natural after Theorems **187** and **193**,

$$(6.11.1) \quad \mathfrak{G}(f) = \mathfrak{M}_0(f), \ \operatorname{Max} f = \mathfrak{M}_{+\infty}(f), \ \operatorname{Min} f = \mathfrak{M}_{-\infty}(f).$$

$\mathfrak{M}_0(f)$ may be meaningless, but only if $\mathfrak{M}_r(f) = \infty$ for all $r > 0$ and $\mathfrak{M}_r(f) = 0$ for all $r < 0$.

We begin by disposing of two exceptional cases.

(A) If $f \equiv C$ then $\mathfrak{M}_r = C$ for all r, and this is true even in the extreme cases $C = 0$ and $C = \infty$.[c]

(B) We may have

$$\mathfrak{M}_r = 0 \ (r < 0), \quad \mathfrak{M}_0 \text{ meaningless}, \quad \mathfrak{M}_r = \infty \ (r > 0).$$

These cases we dismiss. We then leave it to the reader to verify the truth of the assertions in (1) and (2) below, which cover all cases other than the exceptional cases (A) and (B).

(1) $\mathfrak{M}_r < \mathfrak{M}_s$ for $-\infty \leqq r < s \leqq \infty$, unless ($a$) $\mathfrak{M}_r = \mathfrak{M}_s = \infty$ (which can happen only if $r \geqq 0$), or (b) $\mathfrak{M}_r = \mathfrak{M}_s = 0$ (which can happen only if $s \leqq 0$).

[a] Continuity for $r < s$ can also be deduced from Theorems **111** and **197** (see § 6.12).
[b] By § 6.3 (b) (i).
[c] Strictly, the second case is excluded by the understanding of § 6.2.

(2) We denote by $\mathfrak{M}_{r-0}$ and $\mathfrak{M}_{r+0}$ the limits (which always exist) of $\mathfrak{M}_t$ when $t \to r$ from below and from above respectively.

If $r > 0$ then $\mathfrak{M}_{r-0} = \mathfrak{M}_r$, and $\mathfrak{M}_{r+0} = \mathfrak{M}_r$ except when $\mathfrak{M}_r$ is positive and finite but $\mathfrak{M}_t = \infty$ for $t > r$, in which case

$$\mathfrak{M}_{r+0} = \infty > \mathfrak{M}_r.$$

If $r < 0$ then $\mathfrak{M}_{r+0} = \mathfrak{M}_r$, and $\mathfrak{M}_{r-0} = \mathfrak{M}_r$ except when $\mathfrak{M}_r$ is positive and finite but $\mathfrak{M}_t = 0$ for $t < r$, in which case

$$\mathfrak{M}_{r-0} = 0 < \mathfrak{M}_r.$$

If $r = 0$ there are exceptional cases corresponding to each of those indicated above. If $\mathfrak{M}_0$ is 0 or ∞ then either (*a*) $\mathfrak{M}_{-0}$ and $\mathfrak{M}_{+0}$ are each equal to $\mathfrak{M}_0$, or else (*b*)

$$\mathfrak{M}_{-0} = \mathfrak{M}_0 = 0, \quad \mathfrak{M}_{+0} = \infty$$

or

$$\mathfrak{M}_{-0} = 0, \quad \mathfrak{M}_0 = \mathfrak{M}_{+0} = \infty.$$

If $\mathfrak{M}_0$ is positive and finite, then each of $\mathfrak{M}_{-0}$ and $\mathfrak{M}_{+0}$, if also positive and finite, is equal to $\mathfrak{M}_0$; but $\mathfrak{M}_{-0}$ may also be 0 or $\mathfrak{M}_{+0}$ may be ∞.

Finally, all possibilities not explicitly excluded may actually occur[a].

The results may be stated more symmetrically and concisely in terms of

$$\mathfrak{L}_r = \log \mathfrak{M}_r:$$

we agree that $\log \infty = +\infty$ and $\log 0 = -\infty$. We put aside the cases corresponding to cases (A) and (B) above, viz.

(*a*) $f \equiv C$ (where C may be 0 or ∞), when $\mathfrak{L}_r = \log C$ for all r;

(*b*) $\mathfrak{L}_0$ meaningless, when $\mathfrak{L}_r = +\infty$ for $r > 0$ and $\mathfrak{L}_r = -\infty$ for $r < 0$.

195. *Apart from the cases just mentioned, the set of values of r for which $\mathfrak{L}_r = \log \mathfrak{M}_r$ is finite is either the null set or a closed, half-closed, or open interval I or (u, v), where $-\infty \leqq u \leqq v \leqq \infty$, which includes the point $r = 0$* (so that $u \leqq 0 \leqq v$), *but is otherwise arbitrary* (so that, for example, u may be $-\infty$ and v be $+\infty$, or u and v may both be 0). *$\mathfrak{L}_r$ is $+\infty$ for values of r to the right of I and $-\infty$ for values to the left.*

Inside I, $\mathfrak{L}_r$ is continuous and strictly increasing. If r tends to an end-point of I through values of r interior to I, then $\mathfrak{L}_r$ tends to a limit (finite or infinite) equal to its value at the end-point in question.

6.12. Convexity of $\log \mathfrak{M}_r^r$. In this section (as in Theorem **17**) we suppose $r > 0$.

[a] See Theorem **231**.

196. *If* $0<r<s<t$, *and* $\mathfrak{M}_t$ *is finite, then*

$$\mathfrak{M}_s{}^s < (\mathfrak{M}_r{}^r)^{\frac{t-s}{t-r}} (\mathfrak{M}_t{}^t)^{\frac{s-r}{t-r}},$$

unless $f\equiv 0$ *in a part of* E *and* $f\equiv C$ *in the complementary part.*

The proof is based on Theorem **188**, and is the analogue of that of Theorem **17**. For equality,

$$qf^r \equiv Cqf^t.$$

As a corollary, we have

197. $\log \mathfrak{M}_r{}^r(f) = r\log \mathfrak{M}_r(f)$ *is a convex function of* r.

Compare Theorem **87**. The reader will find it instructive to deduce the continuity of $\mathfrak{M}_r$ (Theorem **194**) from Theorem **197**.

6.13. Minkowski's inequality for integrals. The inequalities of the Minkowski type are derived in substantially the same way as in § 2.11. The ordinary form of Minkowski's inequality for integrals is

198. *If* $k>1$ *then*

(6.13.1) $$\{\int (f+g+\ldots+l)^k dx\}^{1/k} < (\int f^k dx)^{1/k} + \ldots + (\int l^k dx)^{1/k},$$

and if $0<k<1$ *then*

(6.13.2) $$\{\int (f+g+\ldots+l)^k dx\}^{1/k} > (\int f^k dx)^{1/k} + \ldots + (\int l^k dx)^{1/k},$$

unless $f, g, \ldots, l$ *are effectively proportional.*

The inequality (6.13.2) *is still true generally when* $k<0$, *but there is a second case of exception, when both sides of the inequality vanish.*

We deduce this from Theorem **189** much as we deduced Theorem **24** from Theorem **13**. Since the cases of equality are a little puzzling, we write out the proof of (6.13.2) in detail.

If $S=f+g+\ldots+l$ then

(6.13.3) $$\int S^k dx = \int f S^{k-1} dx + \int g S^{k-1} dx + \ldots + \int l S^{k-1} dx.$$

Suppose first that $0<k<1$. By Theorem **19**,

$$S^k \leqq f^k + g^k + \ldots + l^k.$$

Hence, if $\int f^k dx, \ldots$ are finite, $\int S^k dx$ is finite. Also $\int S^k dx > 0$

unless $S \equiv 0$ and so $f, g, \ldots$ are all null. We may therefore suppose $\int S^k dx$ positive and finite.

By Theorem **189**,

$$\int fS^{k-1} dx > (\int f^k dx)^{1/k} (\int S^k dx)^{1/k'},$$

unless (*a*) f^k and S^k are effectively proportional or (*b*) $fS^{k-1} \equiv 0$. Since $k-1<0$, and S is finite almost everywhere, the second alternative can occur only if f is null, and so reduces to a case of the first. Hence (6.13.3) gives

$$(6.13.4) \quad \int S^k dx > \{(\int f^k dx)^{1/k} + \ldots + (\int l^k dx)^{1/k}\} (\int S^k dx)^{1/k'},$$

unless $f, g, \ldots, l$ are effectively proportional; and the conclusion follows.

The argument goes similarly when $k<0$, provided $\int S^k dx$ is positive and finite. If $\int S^k dx = 0$ then, since $k<0$, S is infinite almost everywhere, which is impossible since every f is finite almost everywhere. If $\int S^k dx$ is infinite then (again since $k<0$) $\int f^k dx, \ldots$ are all infinite, and both sides of (6.13.2) are zero. This is the second exceptional case mentioned in the enunciation, and occurs, for example, when

$$f = g = \ldots = l = 0$$

in a set E of positive measure.

We have excluded the cases $k=1$ and $k=0$ from the statement of Theorem **198**. The first is trivial and the second is included in Theorem **186**. We leave it to the reader to state Theorem **198** in a form corresponding to that of Theorem **24**.

Corresponding to Theorem **27**, we have

199. *If* $k>1$ *then*

$$(6.13.5) \quad \int (f+g+\ldots+l)^k dx > \int f^k dx + \ldots + \int l^k dx,$$

and if $0<k<1$ *then*

$$(6.13.6) \quad \int (f+g+\ldots+l)^k dx < \int f^k dx + \ldots + \int l^k dx,$$

unless, for almost all x, *all but one of* $f, g, \ldots, l$ *are zero. If all of* $f, g, \ldots, l$ *are almost always positive, then* (6.13.6) *is true also for* $k<0$.

Theorem **198**, with $k>1$, is a special case of the first of the following three more general theorems, in which the series are finite or infinite and the ranges of integration arbitrary. We confine ourselves to the case $k > 1$; in general, the sign of inequality is reversed when $k < 1$.

200. *If* $k > 1$ *then*

$$[\int\{\Sigma f_m(x)\}^k dx]^{1/k} < \Sigma\{\int f_m{}^k(x)\,dx\}^{1/k}, \tag{6.13.7}$$

unless $f_m(x) \equiv C_m \phi(x)$.

201. *If* $k > 1$ *then*

$$[\Sigma\{\int f_n(x)\,dx\}^k]^{1/k} < \int\{\Sigma f_n{}^k(x)\}^{1/k} dx, \tag{6.13.8}$$

unless $f_n(x) \equiv C_n \phi(x)$.

202. *If* $k > 1$ *then*

$$[\int\{\int f(x,y)\,dy\}^k dx]^{1/k} < \int\{\int f^k(x,y)\,dx\}^{1/k} dy, \tag{6.13.9}$$

unless $f(x,y) \equiv \phi(x)\psi(y)$.

In each theorem there is equality in the exceptional case.

Consider for example Theorem **202** (the least elementary of the theorems). We begin by proving the theorem with '$\leqq$'. We give two proofs, in the first of which we appeal to Theorem **191**. In each proof the chain of equalities and inequalities which arises is to be interpreted in the sense 'if the right-hand side of any equality or inequality is finite, then so is the left-hand side, and the two are related as stated'. The inversions of the order of integration are justified by Fubini's Theorem.

We write

$$J = J(x) = \int f(x,y)\,dy.$$

(i) In order that

$$\int J^k dx \leqq M^k \tag{6.13.10}$$

it is, by Theorem **191**, necessary and sufficient that

$$\int Jg\,dx \leqq M \tag{6.13.11}$$

for all g for which

$$\int g^{k'} dx \leqq 1. \tag{6.13.12}$$

Now

$$\begin{aligned} \int Jg\,dx &= \int g(x)\,dx \int f(x,y)\,dy \\ &= \int dy\,(\int g(x) f(x,y)\,dx) \leqq \int dy\,(\int f^k(x,y)\,dx)^{1/k}, \end{aligned} \tag{6.13.13}$$

by Theorem **189** and (6.13.12). Hence we may take

$$M = \int dy\,(\int f^k dx)^{1/k}$$

in (6.13.10), which proves the theorem (with '$\leqq$').

(ii) If $\int J^k dx = 0$, then $J=0$ for almost all x, and so (for almost all x) $f=0$ for almost all y. Hence, after § 6.3 (d), $f(x,y)\equiv 0$.

We may therefore suppose that $\int J^k dx > 0$. Let us assume for a moment that $\int J^k dx$ is finite. Then

$$\int J^k dx = \int J^{k-1} dx \int f dy = \int dy \int J^{k-1} f dx \leqq \int dy \{(\int f^k dx)^{1/k} (\int J^k dx)^{1/k'}\}$$
$$= (\int J^k dx)^{1/k'} \int (\int f^k dx)^{1/k} dy,$$

and so

$$(6.13.14) \qquad (\int J^k dx)^{1/k} \leqq \int (\int f^k dx)^{1/k} dy,$$

which is (6.13.9), with '$\leqq$' for '$<$'.

In this proof we have assumed $\int J^k dx$ *finite*, an assumption which was not required in proof (i). In order to get rid of the assumption, we must approximate to f by some function for which the assumption is certainly justified. Suppose for example that the integrations are over finite intervals or sets of finite measure, that $(f)_n$ is defined as in § 6.1, and that

$$J_n = \int (f)_n dy.$$

Then $\int J_n{}^k dx$ is certainly finite, and so

$$(\int J_n{}^k dx)^{1/k} \leqq \int \{\int (f)_n{}^k dx\}^{1/k} dy \leqq \int (\int f^k dx)^{1/k} dy.$$

From this (6.13.9) follows, with '$\leqq$', by making $n\to\infty$.

The arguments under (i) and (ii) are essentially of the same character, the part of the *arbitrary* g in (i) being played in (ii) by the *definite* function

$$g = \frac{J^{k-1}}{(\int J^k dx)^{1/k'}},$$

which satisfies (6.13.12) *if* $\int J^k dx$ *is finite*. By using this particular g, we avoid an appeal to a rather sophisticated general theorem, but at the cost of some additional complications. A similar alternative presents itself whenever we make use of Theorem **191**.

It remains to discuss the possibility of equality in (6.13.9). There will be inequality if[a]

$$\int Jg dx < M$$

for all g subject to (6.13.12). There is inequality in

$$\int dy (\int gf dx) \leqq \int dy \{(\int f^k dx)^{1/k} (\int g^{k'} dx)^{1/k'}\},$$

[a] See the last remark of § 6.9.

unless, for almost all y, f^k and $g^{k'}$ are effectively proportional, i.e. unless (for almost all y)

$$(6.13.15) \qquad \rho(y)\, f^k(x,y) = \sigma(y)\, g^{k'}(x),$$

where $\rho^2+\sigma^2>0$, for almost all x. If $\rho(y)$ were zero, for a y for which (6.13.15) holds, $g(x)$ would be null, which is false. Hence, in (6.13.15), $\rho(y)>0$, and so

$$f(x,y)=\phi(x)\,\psi(y),$$

where $\phi=g^{k'/k}$, $\psi=(\sigma/\rho)^{1/k}$. This equation holds, for almost all y, for almost all x, and therefore, by § 6.3 (d), for almost all x, y.

The proofs of Theorems **200** and **201** follow similar lines. Thus, in proving Theorem **201**, we write

$$J_n=\int f_n\,dx$$

and argue as follows. In order that $\Sigma J_n{}^k<M^k$, it is necessary and sufficient, by Theorem **15**,[a] that $\Sigma b_n J_n<M$ whenever $\Sigma b_n{}^{k'}\leqq 1$. Also

$$\Sigma b_n J_n=\Sigma b_n \int f_n\,dx=\int(\Sigma b_n f_n)\,dx\leqq\int dx\,(\Sigma f_n{}^k)^{1/k}(\Sigma b_n{}^{k'})^{1/k'}\leqq\int(\Sigma f_n{}^k)^{1/k}\,dx;$$

and so on. The summation under the integral sign is justified by (ii) of § 6.3 (b).

The analogue of Theorem **26** is

203. *If* $0<r<s$ *then*

$$\mathfrak{M}_s{}^{(y)}\,\mathfrak{M}_r{}^{(x)} f(x,y)<\mathfrak{M}_r{}^{(x)}\,\mathfrak{M}_s{}^{(y)} f(x,y),$$

unless $f(x,y)\equiv\phi(x)\,\psi(y)$.

For an explicit proof see Jessen (**1**).

6.14. Mean values depending on an arbitrary function.

There is a theory of integral mean values involving an arbitrary function similar to that developed in Ch. III. We do not set it out in detail here because it would be so largely a repetition, in a slightly different form, of what we have said already. We confine ourselves to proving the analogue of Theorem **95**.[b]

204. *Suppose that* $\alpha\leqq f(x)\leqq\beta$, *where* α *and* β *may be finite or infinite, and that* $f(x)$ *is almost always different from* α *and* β; *that the range of integration and the weight function* $p(x)$ *satisfy the*

[a] Extended to infinite series.

[b] A number of other analogues of theorems of Ch. III are stated among the miscellaneous theorems at the end of this chapter. A fuller treatment of some of them will be found in Jessen's papers **2** and **3**. A good deal of the content of these papers has been incorporated, with the appropriate modifications, into Ch. III.

conditions of § 6.6; *and that* $\phi''(t)$ *is positive and finite for* $\alpha < t < \beta$. *Then*

$$\phi\left(\frac{\int fp\,dx}{\int p\,dx}\right) \leqq \frac{\int \phi(f)\,p\,dx}{\int p\,dx}, \tag{6.14.1}$$

whenever the right-hand side exists and is finite; and there is equality only when $f \equiv C$.

It is possible that $\int fp\,dx = \infty$ or $\int fp\,dx = -\infty$; (6.14.1) is then still true if interpreted in the obvious manner. It is not possible (when the right-hand side is finite) that $\int fp\,dx$ should not exist, i.e. $\int f^+ p\,dx = \infty$ and $\int f^- p\,dx = -\infty$. For in this case $\alpha = -\infty$, $\beta = \infty$, and $\phi(f)$, being convex and not constant, must tend to infinity, with rapidity at least that of a multiple of $|f|$, either for large positive or for large negative values of f,[a] so that $\int \phi(f)\,p\,dx$ cannot exist and be finite.

We take $p = q$, $\int q\,dx = 1$, and suppose first that $\mathfrak{M} = \int fq\,dx$ is finite. If f is not effectively constant, $\alpha < \mathfrak{M} < \beta$. Also f is finite, and $\alpha < f < \beta$, for almost all x; so that, for almost all x,

$$\phi(f) = \phi(\mathfrak{M}) + (f - \mathfrak{M})\,\phi'(\mathfrak{M}) + \tfrac{1}{2}(f - \mathfrak{M})^2\,\phi''(\mu),$$

where μ lies between f and $\mathfrak{M}$, so that $\alpha < \mu < \beta$. Hence

$$\int \phi(f)\,q\,dx \geqq \phi(\mathfrak{M}),$$

which is (6.14.1). There is equality only if $(f - \mathfrak{M})^2\,\phi''(\mu) \equiv 0$; but $\alpha < \mu < \beta$, and so $\phi''(\mu) > 0$, for almost all x, so that then $f \equiv \mathfrak{M}$.

Next suppose (say) $\int fq\,dx = \infty$, so that $\beta = \infty$. Then

$$\phi\{\int (f)_n\,q\,dx\} \leqq \int \phi\{(f)_n\}\,q\,dx,$$

by what has been proved already. Since $\phi(f)$ is continuous and monotonic for large f, the integral on the right tends to $\int \phi(f)\,q\,dx$, while that on the left tends to $\phi(\infty)$. Hence $\phi(\infty)$ is finite, in which case ϕ is decreasing and

$$\phi(\infty) < \phi(f).$$

It follows that

$$\phi(\infty) = \phi(\infty)\int q\,dx \leqq \int \phi(f)\,q\,dx,$$

with equality only if $\phi(f) \equiv \phi(\infty)$, a possibility which we excluded. The case in which $\int fq\,dx = -\infty$ may be discussed similarly.

[a] See Theorem **126**.

It is possible that the left-hand side of (6.14.1) should be $-\infty$. The reader will find it instructive to verify that the various cases which we have contemplated can all occur.

If we take $\phi(t) = -\log t$, we obtain

$$\exp\left(\int q \log f\,dx\right) \leqq \int qf\,dx,$$

i.e. $\mathfrak{G}(f) \leqq \mathfrak{A}(f)$ (Theorem **184**). If we take $\phi = t^r$, we are led again to Hölder's inequality, and other examples may be constructed analogous to those of § 3.11. If we take $\phi(t) = t \log t$, we find

205. $$\frac{\int pf\,dx}{\int p\,dx} < \exp\left(\frac{\int pf \log f\,dx}{\int pf\,dx}\right),$$

unless $f \equiv C$.

We can extend the result of Theorem **204** (except for the specification of the cases of equality) to any convex and continuous ϕ.

206. *The inequality* (6.14.1) *is true whenever* $\phi(t)$ *is convex and continuous in* $\alpha < t < \beta$.

After § 3.19, we have

$$\phi(f) \geqq \phi(\mathfrak{M}) + \lambda(f - \mathfrak{M}),$$

where λ is any number between the left and right hand derivatives of $\phi(t)$ for $t = \mathfrak{M}$. Hence

$$\int \phi(f)\,q\,dx \geqq \phi(\mathfrak{M}),$$

which is (6.14.1).

STIELTJES INTEGRALS

6.15. The definition of the Stieltjes integral. We have so far considered series and integrals separately, and all the fundamental theorems have appeared in dual form; thus Hölder's inequality is contained in Theorems **13** and **189**. It is natural to look for an extension of these theorems which combines them into one, and we can find such an extension by using Stieltjes integrals.

Suppose that $\phi(x)$ increases (in the wide sense) in $a \leqq x \leqq b$, and that $\phi(a)=\alpha$, $\phi(b)=\beta$. We suppose α and β (but not necessarily a and b) finite[a]. The curve

$$y=y(x)=\phi(x)$$

is a rising curve which may have an enumerable set of ordinary discontinuities or of stretches of invariability. The inverse function

$$x=x(y)=x(\phi)$$

is defined uniquely except (*a*) in intervals (y_1, y_2) of y corresponding to discontinuities $x=\xi$ of ϕ and (*b*) for values of y which correspond to stretches of invariability of ϕ. If we agree that (y_1, y_2) is a stretch of invariability of $x(y)$, in which it has the value ξ, then $x(y)$ is defined except for the values (*b*), and is an increasing function of y for the values of y for which it is defined. Finally we complete the definition of $x(y)$, as an increasing function of y, by assigning to it, for a value (*b*) of y, any one of the values of x in the stretch of invariability. These values of y are enumerable, and our choice of $x(y)$ for any of them has no effect on the definitions which follow.

We now define the Stieltjes, or Lebesgue-Stieltjes, integral

$$\int_{x=a}^{x=b} f(x)\,d\phi(x)=\int_a^b f(x)\,d\phi,$$

of $f(x)$ with respect to $\phi(x)$, by

$$\int_a^b f(x)\,d\phi=\int_\alpha^\beta f\{x(\phi)\}\,d\phi, \tag{6.15.1}$$

whenever the integral on the right-hand side exists as a Lebesgue integral[b].

The definition (6.15.1), due to Radon (**1**), reduces the theory of Stieltjes integrals to that of Lebesgue integrals, and we may therefore expect that no new difficulties will arise. For full discussions of this and older definitions of the Stieltjes integral, we may refer to Hobson (**1**), Lebesgue (**1**), Pollard (**1**), Young (**7**).

[a] If, e.g., $b=\infty$, then $\beta=\lim_{x\to\infty}\phi(x)$.

[b] If g is any function of bounded variation, then $g=\phi-\psi$, where ϕ and ψ are increasing functions, and we may define the Stieltjes integral of f with respect to g by

$$\int f\,dg=\int f\,d\phi-\int f\,d\psi.$$

We shall not require this more general definition here.

We can define $$\int_E f(x)\,d\phi,$$
where ϕ is an increasing function, and E a set of values of x, similarly, that is to say by the equation

$$\int_E f(x)\,d\phi = \int_{\mathfrak{E}} f\{x(\phi)\}\,d\phi,$$

where $\mathfrak{E}$ is the set of values of ϕ corresponding to E. We must assume $\mathfrak{E}$ measurable. The integral

$$\int_E d\phi$$

is *the variation of ϕ in E.*

6.16. Special cases of the Stieltjes integral. The simplest cases are the following:

(*a*) $\phi = x$. In this case the Stieltjes integral reduces to the ordinary Lebesgue integral.

(*b*) *ϕ is an integral.* In this case

$$\int_a^b f(x)\,d\phi = \int_a^b f(x)\,\phi'(x)\,dx.$$

(*c*) *ϕ is a finite increasing step-function.*

Suppose that $a = a_1 < a_2 < \dots < a_n = b$, that $\phi(x) = \alpha_k$, where $\alpha_k < \alpha_{k+1}$, in $a_k < x < a_{k+1}$, and that $\phi(a_k)$, when $1 < k < n$, has any value consistent with the fact that ϕ increases. Then $x(y)$ is a step-function with values $a_1, a_2, \dots, a_n$, and

$$(6.16.1)\quad \int_a^b f\,d\phi = \int_\alpha^\beta f\{x(\phi)\}\,d\phi$$
$$= (\alpha_1 - \alpha)f(a_1) + (\alpha_2 - \alpha_1)f(a_2) + \dots + (\alpha_{n-1} - \alpha_{n-2})f(a_{n-1}) + (\beta - \alpha_{n-1})f(a_n)$$
$$= \Sigma \rho_k f(a_k),$$

where ρ_k is the saltus of ϕ at $x = a_k$. It is plain that any finite sum can be expressed as a Stieltjes integral; thus

$$\sum_1^n u_k = \int f(x)\,d\phi,$$

where ϕ is a step-function with unit jumps at $a_1, a_2, \dots, a_n$, and $u_k = f(a_k)$.

(*d*) These considerations extend at once to step-functions with infinitely many discontinuities, when the Stieltjes integral is $\Sigma \rho_k f(a_k)$, summed over all the discontinuities. Any convergent infinite series may be expressed in this way as a Stieltjes integral.

6.17. Extensions of earlier theorems. It will now be clear that all our fundamental theorems may be extended at once to Stieltjes integrals, and that the theorems thus obtained include those for Lebesgue integrals and also those for sums. We state the most representative of these theorems in the next section. Two preliminary remarks will be useful.

(1) When the Stieltjes integral is written as a Lebesgue integral, the variable of integration is ϕ. Our conditions for equality were always of the type $f \equiv g$, $f = g$ except in a set of measure zero. The exceptional set in our new theorems will be *of measure zero in* ϕ, and when we state this concept again in terms of x it becomes 'a set of values of x in which the variation of ϕ is zero', i.e. a set E such that the corresponding values of ϕ form a null set. Our conditions for equality must therefore all be interpreted in this sense. Thus 'f is effectively proportional to g' means that

$$Af = Bg,$$

where A and B are constants, not both zero, except at the points of a set over which the variation of ϕ is zero. It will be observed that such an exceptional set cannot include any point at which $\phi(x)$ is discontinuous.

A similar point occurs in the definition of Max$f$ and Min$f$. Thus Max$f$ is *the greatest number* ξ *such that, for every positive* ϵ, $f > \xi - \epsilon$ *in a set in which the variation of* ϕ *is positive.*

(2) Many inequalities '$X < Y$' are true for Lebesgue integrals when their analogues for Stieltjes integrals are true only with '$\leqq$'. Suppose, for example, that the integrations are over $(0, \infty)$ and that $\int f\,dx = 1$. Then, by Theorem **181**,

$$(6.17.1) \qquad (\int xf\,dx)^2 < \int f\,dx \int x^2 f\,dx = \int x^2 f\,dx,$$

unless $x^2 f \equiv Cf$ or $x^2 \equiv C$, which is untrue, so that (6.17.1) is true in any case. In the corresponding theorem for Stieltjes integrals we have $\int d\phi = 1$ and

$$(\int x\,d\phi)^2 \leqq \int d\phi \int x^2 d\phi = \int x^2 d\phi. \tag{6.17.2}$$

There is equality in (6.17.2) if $x^2 \equiv C$, i.e. if x is constant except in a set over which the variation of ϕ is zero, or, what is the same thing, if ϕ varies at one point only. Thus if $\phi = 0$ for $0 \leqq x < 1$, and $\phi(x) = 1$ for $x \geqq 1$, then

$$(\int x\,d\phi)^2 = 1 = \int x^2 d\phi.$$

6.18. The means $\mathfrak{M}_r(f;\phi)$. We write

$$\mathfrak{M}_r(f) = \mathfrak{M}_r(f;\phi) = \left(\frac{1}{\beta-\alpha}\int_a^b f^r d\phi\right)^{1/r} = \left(\frac{\int f^r d\phi}{\int d\phi}\right)^{1/r} \quad (r \neq 0),$$

$$\mathfrak{A}(f;\phi) = \mathfrak{M}_1(f;\phi),$$

$$\mathfrak{G}(f;\phi) = \exp\left(\frac{\int \log f\,d\phi}{\int d\phi}\right) = \mathfrak{M}_0(f;\phi).$$

These definitions presuppose that the integrals involved are finite. If $\int f^r d\phi = \infty$, we agree (following the conventions of § 6.6) that $\mathfrak{M}_r = \infty$ when $r > 0$ and $\mathfrak{M}_r = 0$ when $r < 0$. The points discussed in §§ 5.2 and 6.7 naturally recur here in connection with the definition of $\mathfrak{G}$.

The theorems corresponding to Theorems **183**, **184**, **187**, **189**, **192**, **193**, **197**, and **198** are as follows: we suppose for simplicity of statement that $r > 0$.

207. Min $f < \mathfrak{M}_r(f) <$ Max f *unless* $f \equiv C$.

208. $\mathfrak{G}(f) < \mathfrak{M}_r(f)$, *and in particular* $\mathfrak{G}(f) < \mathfrak{A}(f)$, *unless* $f \equiv C$.

209. *If* $\mathfrak{M}_r(f)$ *is finite for some* r, *then* $\mathfrak{M}_r(f) \to \mathfrak{G}(f)$ *when* $r \to +0$.

210. *If* $k > 1$, *then*

$$\int uv\,d\phi < (\int u^k d\phi)^{1/k} (\int v^{k'} d\phi)^{1/k'}$$

unless u^k *and* $v^{k'}$ *are effectively proportional. The inequality is reversed when* $0 < k < 1$ *or* $k < 0$, *except when* u^k *and* $v^{k'}$ *are effectively proportional, or the left-hand side is zero (in which case the right-hand side is also zero).*

This is Hölder's inequality; there are naturally corresponding generalisations of Theorems **11** (or **10**) and **188**.

211. *If* $r<s$, *then* $\mathfrak{M}_r(f)<\mathfrak{M}_s(f)$, *unless* $f\equiv C$.

212. *If* $\mathfrak{M}_r(f)$ *is finite for every positive* r, *then* $\mathfrak{M}_r(f)\rightarrow \operatorname{Max} f$ *when* $r\rightarrow +\infty$.

213. $\log \mathfrak{M}_r^r(f)$ *is a convex function of* r.

214. *If* $k>1$, *then*

$$\{\int (u+v)^k d\phi\}^{1/k} < (\int u^k d\phi)^{1/k} + (\int v^k d\phi)^{1/k},$$

unless u *and* v *are effectively proportional. The inequality is in general reversed if* $0<k<1$ *or* $k<0$.[a]

AXIOMATIC TREATMENT OF MEAN VALUES

6.19. Distribution functions. In Ch. III we defined the mean value

$$\mathfrak{M}_\phi = \mathfrak{M}_\phi(a,q) = \phi^{-1}\{\Sigma q\phi(a)\}$$

directly, and developed its characteristic properties from the definition. Here we reverse the process and give the 'axiomatic' treatment promised on p. 66. It is convenient to use the notation of Stieltjes integration, and it is for this reason that we have reserved the discussion until now; but the Stieltjes integrals which we use are actually all finite sums.

In what follows we consider a special class of step-functions, defined for all real x, which we call *finite distribution functions*. We call $F(x)$ a finite distribution function if

(i) it is constant in stretches and has only a finite number of discontinuities,

(ii) it increases (in the wide sense) from 0 to 1, so that

$$F(-\infty)=0, \quad F(\infty)=1,$$

(iii) $F(x)=\frac{1}{2}\{F(x-0)+F(x+0)\}$ for all x.

The distribution function which has jumps q at the points a provides a representation of both the values a and the weights q involved in $\mathfrak{M}_\phi(a)$. The simplest such function is

$$E(x)=\tfrac{1}{2}(1+\operatorname{sgn} x),$$

[a] We leave the specification of the exceptional cases to the reader.

which has the single jump 1 at $x=0$. If we write

$$E_\xi(x)=E(x-\xi),$$

then

$$F(x)=\Sigma q E_a(x), \tag{6.19.1}$$

where

$$a=a_\nu,\quad q=q_\nu\ (\nu=1,2,\ldots,n),\quad \Sigma q_\nu=1,\quad a_1<a_2<\ldots<a_n,$$

is the general finite distribution function with jumps q at the points a. Also

$$\int_{-\infty}^{\infty}\phi(x)\,dF(x)=\Sigma q\phi(a), \tag{6.19.2}$$

and the mean value (3.1.3) may be written as

$$\mathfrak{M}_\phi[F]=\phi^{-1}\left(\int_{-\infty}^{\infty}\phi(x)\,dF(x)\right). \tag{6.19.3}$$

Any finite distribution function is 0 for $x<A$ and 1 for $x>B$, A and B being finite numbers depending on F. In what follows we confine our attention to a sub-class of these functions, viz. those which satisfy

$$F(x)=0\ (x<A),\quad F(x)=1\ (x>B) \tag{6.19.4}$$

for a fixed A and B. In these circumstances we say that F belongs to $\mathfrak{D}(A,B)$.

If $\phi(x)$ is continuous and strictly monotonic in the closed interval (A,B), then $\mathfrak{M}_\phi[F]$ is defined, by (6.19.3), for all F of $\mathfrak{D}(A,B)$. The values of $\phi(x)$ outside (A,B) are not really involved in (6.19.3), and we may choose them as we please; it is natural to choose them so that $\phi(x)$ is continuous and strictly monotonic for $-\infty\leqq x\leqq\infty$.

6.20. Characterisation of mean values. Our object is to prove the following theorem.

215. *Suppose that there is a unique real number $\mathfrak{M}[F]$, corresponding to each F of $\mathfrak{D}(A,B)$, with the following properties:*

[1] $$\mathfrak{M}[E_\xi(x)]=\xi\quad(A\leqq\xi\leqq B);$$

[2] *if F_1 and F_2 belong to $\mathfrak{D}(A,B)$, $F_1\geqq F_2$ for all x, and $F_1>F_2$ for some x, then*

$$\mathfrak{M}[F_1]<\mathfrak{M}[F_2];$$

[3] *if* F, F^*, G *belong to* $\mathfrak{D}(A, B)$, *and*

$$\mathfrak{M}[F] = \mathfrak{M}[F^*],$$

then $$\mathfrak{M}[tF + (1-t)G] = \mathfrak{M}[tF^* + (1-t)G]$$

for $0 < t < 1$.

Then there is a function $\phi(x)$, *continuous and strictly increasing in the closed interval* (A, B), *for which*

$$\mathfrak{M}[F] = \mathfrak{M}_\phi[F] = \phi^{-1}\left(\int_{-\infty}^{\infty} \phi(x)\, dF(x)\right). \tag{6.20.1}$$

Conversely, if $\mathfrak{M}[F]$ *is defined by* (6.20.1), *for a* $\phi(x)$ *with the properties stated, then it satisfies* [1], [2], *and* [3], *so that these conditions are necessary and sufficient for the representation of* $\mathfrak{M}[F]$ *in the form* (6.20.1).[a]

We begin by proving the converse half of the theorem. If $\mathfrak{M}[F]$ is defined by (6.20.1), then it is obvious that it possesses property [1], and all but obvious that it possesses property [3], since

$$\begin{aligned} \phi(\mathfrak{M}[tF + (1-t)G]) &= t\int \phi\, dF + (1-t)\int \phi\, dG \\ &= t\int \phi\, dF^* + (1-t)\int \phi\, dG = \phi(\mathfrak{M}[tF^* + (1-t)G]). \end{aligned}$$

It remains to prove [2].

Suppose that F_1 and F_2 satisfy the conditions stated. Then there is a positive number μ and an interval (α, β) such that

$$F_1(x) > F_2(x) + \mu > F_2(x)$$

in (α, β).[b] Hence

$$\begin{aligned} \phi(\mathfrak{M}[F_2]) - \phi(\mathfrak{M}[F_1]) &= \int_{-\infty}^{\infty} \phi\, dF_2 - \int_{-\infty}^{\infty} \phi\, dF_1 \\ &= \int_{-\infty}^{\infty} (F_1 - F_2)\, d\phi\text{[c]} \\ &\geqq \int_{\alpha}^{\beta} (F_1 - F_2)\, d\phi \geqq \mu\{\phi(\beta) - \phi(\alpha)\} > 0. \end{aligned}$$

[a] See Nagumo (**1**), Kolmogoroff (**1**), de Finetti (**1**). We follow the lines of de Finetti's proof.

[b] There is an x_0 for which $F_1(x_0) > F_2(x_0)$ or

$$\tfrac{1}{2}\{F_1(x_0 - 0) + F_1(x_0 + 0)\} > \tfrac{1}{2}\{F_2(x_0 - 0) + F_2(x_0 + 0)\}.$$

Hence either $F_1(x_0 - 0) > F_2(x_0 - 0)$ or $F_1(x_0 + 0) > F_2(x_0 + 0)$. In the first case there is an interval satisfying the conditions to the left of x_0, in the second case one to the right.

[c] If we remember our understanding, at the end of § 6.19, about the definition of $\phi(x)$ outside (A, B).

6.21. Remarks on the characteristic properties. We have still to show that the properties [1]–[3] are sufficient to characterise the means $\mathfrak{M}_\phi$. We insert first some general remarks concerning the 'significance' of the properties.

(i) [1] asserts that 'if all the elements of a set have the same value, then their mean has that value'.

(ii) [2] asserts that '$\mathfrak{M}[F]$ is a strictly monotonic functional of F'. It would not be sufficient to assert that (under the conditions stated) $\mathfrak{M}[F_1] \leqq \mathfrak{M}[F_2]$, i.e. that '$\mathfrak{M}[F]$ is a monotonic functional'.

Let us consider some examples.

(*a*) The arithmetic mean

$$\mathfrak{A}(a, q) = \Sigma qa = \int x\,dF = \mathfrak{A}[F]$$

is a strictly monotonic functional of F. In this case $\phi(x) = x$.

(*b*) We may define 'Max a' as 'the lower bound of the values of x for which $F(x) = 1$' (F being any finite distribution function with jumps at the points a). Then Max $a = \mu[F]$ is a functional of F which is plainly monotonic: if $F_1 \geqq F_2$ for all x, then $\mu[F_1] \leqq \mu[F_2]$. But $\mu[F]$ is not strictly monotonic: if F_1 and F_2 are defined by

$$F_1 = F_2 = 0 \ (x<0); \quad F_1 = \tfrac{1}{2},\ F_2 = 0 \ (0<x<1); \quad F_1 = F_2 = 1 \ (x>1),$$

then

$$\mu[F_1] = \text{Max}\,(0, 1) = \text{Max}\,(1, 1) = \mu[F_2].$$

That $\mu[F]$ is not representable in the form (6.20.1) follows from the theorem itself; if it were, it would be strictly monotonic.

(*c*) The geometric mean $\mathfrak{G} = \mathfrak{G}(a, q)$ is a functional of F which is not strictly monotonic, since, for example, the sets $(0, a_2, \ldots)$ and $(0, b_2, \ldots)$ have the same $\mathfrak{G}$. It is representable by the formula

$$\mathfrak{G} = \exp\left(\int_0^\infty \log x\,dF(x)\right).$$

This is of the form (6.20.1), with $\phi(x) = \log x$ for $x > 0$; but $\mathfrak{G}$ is not represented in the manner prescribed by the theorem, since $\log x \to -\infty$ when $x \to 0$.

(iii) If we use [3] twice, the second time with F^*, G, G^*, $1-t$ in place of G, F, F^*, t, we see that

$$\mathfrak{M}[tF + (1-t)G] = \mathfrak{M}[tF^* + (1-t)G^*] \tag{6.21.1}$$

whenever $\mathfrak{M}[F] = \mathfrak{M}[F^*]$ and $\mathfrak{M}[G] = \mathfrak{M}[G^*]$. In other words (*a*) $\mathfrak{M}[tF + (1-t)G]$ *is determined uniquely by* $\mathfrak{M}[F]$, $\mathfrak{M}[G]$ *and* t.

More generally

$$(6.21.2) \qquad \mathfrak{M}[\Sigma q_\nu F_\nu] = \mathfrak{M}[\Sigma q_\nu F_\nu^*]$$

if $\mathfrak{M}[F_\nu] = \mathfrak{M}[F_\nu^*]$ and $\Sigma q_\nu = 1$.

A functional $\mathfrak{F}[F]$ is said to be *linear* if

$$\mathfrak{F}[tF + uG] = t\mathfrak{F}[F] + u\mathfrak{F}[G]:$$

in this case it has certainly property (*a*). If $\mathfrak{M}[F]$ satisfies (*a*) or [3], of which (*a*) is a consequence, we may call $\mathfrak{M}[F]$ *quasi-linear*. If also we agree to describe the property [1], as is natural, by *consistency*, we may state Theorem **215** shortly as follows: *the most general consistent, strictly increasing, and quasi-linear functional of F is that defined by* (6.20.1).

6.22. Completion of the proof of Theorem 215. The functions $E_A(x)$, $E_B(x)$, and $(1-t)E_A(x) + tE_B(x)$, where $0 < t < 1$, belong to $\mathfrak{D}(A, B)$.[a] We write

$$\psi(t) = \mathfrak{M}[(1-t)E_A + tE_B],$$

so that $\quad \psi(0) = \mathfrak{M}[E_A] = A, \quad \psi(1) = \mathfrak{M}[E_B] = B.$

Let us assume provisionally that $\psi(t)$ is strictly increasing and continuous. Then $\psi(t)$ has an inverse

$$\phi(u) = \psi^{-1}(u)$$

which is also continuous and increases strictly from 0 to 1 when u increases from A to B. If

$$u = \psi(t), \quad t = \phi(u),$$

then $\quad \mathfrak{M}[E_u] = u = \psi(t) = \mathfrak{M}[(1-\phi(u))E_A + \phi(u)E_B].$

Hence, using [3] in the extended form (6.21.2), and the expression (6.19.1) for any finite distribution function F, we obtain

$$\begin{aligned} \mathfrak{M}[F] &= \mathfrak{M}[\Sigma q E_a] \\ &= \mathfrak{M}[\Sigma q\{(1-\phi(a))E_A + \phi(a)E_B\}] \\ &= \mathfrak{M}[(1 - \Sigma q\phi(a))E_A + (\Sigma q\phi(a))E_B] \\ &= \psi(\Sigma q\phi(a)) = \phi^{-1}(\Sigma q\phi(a)), \end{aligned}$$

the result of the theorem.

It should be observed that here $\phi(A) = 0$, $\phi(B) = 1$. When a ϕ has been found, it may (after Theorem **83**) be replaced by any $\alpha\phi + \beta$.

[a] E_A and E_B are extreme cases of functions of $\mathfrak{D}(A, B)$: if F belongs to $\mathfrak{D}(A, B)$, then $E_A \geqq F \geqq E_B$ for all x.

It has still to be proved that $\psi(t)$ is strictly increasing and continuous.

(1) If
$$0 \leqq t_1 < t_2 \leqq 1,$$
then
$$(1-t_1) E_A + t_1 E_B \geqq (1-t_2) E_A + t_2 E_B$$
for all x, with inequality for some x. Hence, by [2],
$$\psi(t_1) = \mathfrak{M}[(1-t_1) E_A + t_1 E_B] < \mathfrak{M}[(1-t_2) E_A + t_2 E_B] = \psi(t_2).$$

(2) Suppose, if possible, that $\psi(t)$ has a discontinuity on the right at t_0, where $0 \leqq t_0 < 1$. Then we can find a ξ such that
$$\psi(t_0) < \xi < \psi(t_0 + \epsilon)$$
for arbitrarily small ϵ, and
$$E_{\psi(t_0)} \geqq E_\xi \geqq E_{\psi(t_0+\epsilon)}$$
for all x, with inequality for some x. Hence, by [2],

(6.22.1)
$$\mathfrak{M}[\tfrac{1}{2}E_{\psi(t_0)} + \tfrac{1}{2}E_{\psi(t)}] < \mathfrak{M}[\tfrac{1}{2}E_\xi + \tfrac{1}{2}E_{\psi(t)}] < \mathfrak{M}[\tfrac{1}{2}E_{\psi(t_0+\epsilon)} + \tfrac{1}{2}E_{\psi(t)}]$$
for any t of $(0, 1)$. But if s and t lie in $(0, 1)$, then, by [1],
$$\psi(s) = \mathfrak{M}[E_{\psi(s)}] = \mathfrak{M}[(1-s) E_A + sE_B],$$
and similarly for t; and, by [3],
$$\mathfrak{M}[\tfrac{1}{2}E_{\psi(s)} + \tfrac{1}{2}E_{\psi(t)}] = \mathfrak{M}\left[\frac{(1-s) E_A + sE_B}{2} + \frac{(1-t) E_A + tE_B}{2}\right]$$
$$= \mathfrak{M}\left[\left(1 - \frac{s+t}{2}\right) E_A + \frac{s+t}{2} E_B\right] = \psi\left(\frac{s+t}{2}\right).$$
Combining this with (6.22.1), we see that
$$\psi\left(\frac{t_0+t}{2}\right), \quad \psi\left(\frac{t_0+t+\epsilon}{2}\right)$$
are separated by a number, viz. $\mathfrak{M}[\tfrac{1}{2}E_\xi + \tfrac{1}{2}E_{\psi(t)}]$, which is independent of ϵ; and so, making $\epsilon \to 0$,
$$\psi\left(\frac{t_0+t}{2}\right) < \psi\left(\frac{t_0+t}{2} + 0\right).$$
Hence ψ has a discontinuity at $\tfrac{1}{2}(t_0+t)$, for all t of an interval; and this is impossible, because the discontinuities of a monotonic function are at most enumerable.

It follows that $\psi(t)$ has no right-hand discontinuity. Similarly, it has no left-hand discontinuity. It is therefore continuous, and this completes the proof.

We have confined our attention to finite distribution functions, so that all the functions F which have been considered are step-functions, and the means are the means of Ch. III. There is a similar theorem in which both hypothesis and conclusion are stronger in that they apply to a class of functions more extensive than $\mathfrak{D}(A, B)$. Let us denote by $\mathfrak{D}^*(A, B)$ the class of functions which have the properties (ii) and (iii) of § 6.19 and also satisfy (6.19.4). We can then prove a theorem which differs from Theorem **215** only in the substitution of $\mathfrak{D}^*$ for $\mathfrak{D}$. The proof is very much the same, but is slightly more elaborate in its final stages. See de Finetti (**1**).

MISCELLANEOUS THEOREMS AND EXAMPLES

216. 'Velocity averaged by time is less than velocity averaged by distance.'

[This is $\left(\int \frac{ds}{dt}dt\right)^2 < \int dt \int \frac{ds}{dt} ds = \int dt \int \left(\frac{ds}{dt}\right)^2 dt$, a case of Theorem **181**.]

217. If the kinetic energy of a mass M of moving homogeneous incompressible fluid is E, and the average velocity of its particles is V, then $E > \frac{1}{2}MV^2$, unless all particles have the same velocity.

[If ρ is the density, v the velocity of an element $dS = dx\,dy\,dz$, then

$$M = \rho\int dS, \quad V\int dS = \int v\,dS, \quad E = \tfrac{1}{2}\rho\int v^2 dS,$$

and the result follows from Theorem **181** (for triple integrals).]

218. A unit electric current passes through a closed plane circuit enclosing an area A, and exerts a force F on a unit magnetic pole P in the plane of and interior to the circuit. Then

$$2AF^2 > (2\pi)^3$$

unless the circuit is a circle whose centre is P.

[Suppose, for simplicity, that the circuit is 'star shaped' with respect to P (i.e. that every point of the line from P to any point of the circuit lies inside the circuit). Then, using polar coordinates r, θ about P, and integrals from 0 to 2π,

$$2\pi = \int d\theta = \int \left(\frac{1}{r}\right)^{\frac{2}{3}} (r^2)^{\frac{1}{3}}\, d\theta < \left(\int \frac{d\theta}{r}\right)^{\frac{2}{3}} \left(\int r^2 d\theta\right)^{\frac{1}{3}} = F^{\frac{2}{3}}(2A)^{\frac{1}{3}},$$

unless r is constant.]

219. If $f_\nu(x, y)$ and $g_\nu(x, y)$ are two (finite or infinite) sets of functions of x and y, then

$$(\Sigma \iint fg\,dx\,dy)^2 < \Sigma \iint f^2 dx\,dy\, \Sigma \iint g^2 dx\,dy,$$

unless there are two constants a and b, not both zero, such that

$$af_\nu(x, y) \equiv bg_\nu(x, y),$$

for every ν.

[From Theorems **7** (for infinite series) and **181** (for double integrals), or directly, by the second method of § 2.4. The theorem illustrates the following principle. The inequality

$$\text{(i)} \qquad (\Sigma\Sigma\Sigma\, uv)^2 \leqq \Sigma\Sigma\Sigma\, u^2\, \Sigma\Sigma\Sigma\, v^2,$$

where u and v are functions of three integral variables m, n and p, does not differ materially from the ordinary form of Cauchy's inequality; but we can derive materially different inequalities from (i) by replacing different selections of the signs of summation by signs of integration.]

220. Suppose that the a are positive, and that q_r is defined by

$$\frac{1}{(1-a_1x)(1-a_2x)\dots(1-a_nx)} = 1 + nq_1x + \dots + \binom{n+r-1}{r} q_r x^r + \dots.$$

Then
$$q_r^2 < q_{r-1}q_{r+1} \quad (r=1, 2, \dots)$$
unless all the a are equal.

221.
$$q_1 < q_2^{\frac{1}{2}} < q_3^{\frac{1}{3}} < \dots$$
unless all the a are equal.

[Theorems **220** and **221** were communicated to us by Prof. I. Schur. The q are means of homogeneous products of the a, like the p of § 2.22, but now the a in a product are not necessarily different. In particular $q_1 = p_1$.

Theorem **221** follows from Theorem **220** as Theorem **52** followed from Theorem **51**. To prove Theorem **220** we observe that

(i) $$q_r = (n-1)!\iint\dots\int (a_1x_1 + a_2x_2 + \dots + a_nx_n)^r\, dx_1 \dots dx_{n-1},$$

where $x_n = 1 - x_1 - x_2 - \dots - x_{n-1}$ and the domain of integration is defined by $x_1 > 0$, ..., $x_{n-1} > 0$, $x_n > 0$. We obtain Theorem **220** by applying Theorem **181** (for multiple integrals) to (i).

The formula (i) leads to a more complete theorem. If the a are real (but not necessarily positive) then the quadratic form $\Sigma q_{r+s}y_ry_s$ is strictly positive; and if the a are positive, then the form $\Sigma q_{r+s+1}y_ry_s$ is strictly positive; except (in both cases) when all the a are equal.]

222. If $p > 1$, f is L^p in $(0, a)$, and

$$F(x) = \int_0^x f(t)\,dt,$$

then
$$F(x) = o(x^{1/p'})$$
for small x.

[By Theorem **189**,

$$F^p \leqq \int_0^x f^p dt \left(\int_0^x dt\right)^{p-1} = x^{p-1}\int_0^x f^p dt,$$

and the second factor tends to 0.]

223. If $p > 1$ and f is L^p in $(0, \infty)$, then $F(x) = o(x^{1/p'})$ both for small and for large x.

[For small x, by Theorem **222**. To prove the result for large x, choose X so that

$$\int_X^\infty f^p dx < \epsilon^p$$

and suppose $x > X$. Then

$$(F(x)-F(X))^p = \left(\int_X^x f\,dt\right)^p \leqq (x-X)^{p-1}\int_X^x f^p\,dt < \epsilon^p x^{p-1},$$

$$F(x) < F(X) + \epsilon x^{1/p'} < 2\epsilon x^{1/p'}$$

for sufficiently large x.]

224. If y is an integral except perhaps at $x=0$ and xy'^2 is integrable in $(0, a)$, then

$$y = o\left\{\left(\log\frac{1}{x}\right)^{\frac{1}{2}}\right\}$$

for small x.

$$\left[\int_\epsilon^a y'\,dx \leqq \left(\int_\epsilon^a \frac{dx}{x}\int_\epsilon^a xy'^2\,dx\right)^{\frac{1}{2}}.\right]$$

225. If y is an integral except perhaps at 0 and 1, and $x(1-x)y'^2$ is integrable in $(0, 1)$, then y is L^2 and

$$0 \leqq \int_0^1 y^2\,dx - \left(\int_0^1 y\,dx\right)^2 \leqq \tfrac{1}{2}\int_0^1 x(1-x)\,y'^2\,dx.$$

[That y is L^2 follows from Theorem **224**. The first inequality is included in Theorem **181**. For the second, we have

$$\int_0^1 y^2\,dx - \left(\int_0^1 y\,dx\right)^2 = \tfrac{1}{2}\int_0^1\int_0^1 \{y(u)-y(v)\}^2\,du\,dv$$

$$= \int_0^1 du\int_u^1 dv\left(\int_u^v y'(t)\,dt\right)^2 \leqq \int_0^1 du\int_u^1 (v-u)\,dv\int_u^v (y'(t))^2\,dt$$

$$= \int_0^1 (y'(t))^2\,dt\int_0^t du\int_t^1 (v-u)\,dv = \tfrac{1}{2}\int_0^1 t(1-t)\,y'^2\,dt.$$

Of the two inequalities, the first can reduce to an equality only if y is constant, the second only if y is linear.]

226. If $m > 1$, $n > -1$, and f is positive and an integral, then

$$\text{(i)} \qquad \int_0^\infty x^n f^m\,dx \leqq \frac{m}{n+1}\left\{\int_0^\infty x^{\frac{m(n+1)}{m-1}} f^m\,dx\right\}^{\frac{m-1}{m}}\left(\int_0^\infty |f'|^m\,dx\right)^{\frac{1}{m}},$$

with equality only when $f = B\exp\{-Cx^{(m+n)/(m-1)}\}$, where $B \geqq 0$, $C > 0$.

In particular

$$\text{(ii)} \qquad \int_0^\infty f^2\,dx < 2\left(\int_0^\infty x^2 f^2\,dx\right)^{\frac{1}{2}}\left(\int_0^\infty f'^2\,dx\right)^{\frac{1}{2}},$$

unless $f = Be^{-Cx^2}$; and this inequality holds whether f be positive or not, and also for the range $(-\infty, \infty)$.

[The most interesting case is (ii), which is due to Weyl (**1**, 345), and is useful in quantum-mechanics.

Assume that the integrals on the right-hand side of (i) are finite. Since f is continuous, and $n < m(n+1)/(m-1)$, that on the left-hand side is also finite. Hence

$$\varliminf_{x\to\infty} x^{n+1} f^m = 0,$$

and so, integrating by parts over $(0, x_k)$, where (x_k) is an appropriate sequence which tends to infinity with k,

$$\int_0^\infty x^n f^m\,dx = -\frac{m}{n+1}\lim_{k\to\infty}\int_0^{x_k} x^{n+1}f^{m-1}f'\,dx.$$

But, by Theorem **189**,

$$\int_0^\infty x^{n+1}f^{m-1}|f'|\,dx < \left\{\int_0^\infty x^{\frac{m(n+1)}{m-1}}f^m\,dx\right\}^{\frac{m-1}{m}}\left(\int_0^\infty |f'|^m\,dx\right)^{\frac{1}{m}},$$

unless $f' \leqq 0$ and f' and $x^{(n+1)/(m-1)}f$ are effectively proportional. This hypothesis leads to the form of f stated.]

227. If ϕ increases,

$$(\int fg\,d\phi)^2 < \int f^2\,d\phi \int g^2\,d\phi,$$

unless f and g are effectively proportional (in the sense of § 6.17).

[Included in Theorem **210**; wanted in Theorem **228**.]

228. If $a \geqq 0$, $b \geqq 0$, $a \neq b$, and ϕ is non-negative and decreasing, then

$$\left(\int_0^\infty x^{a+b}\phi\,dx\right)^2 < \left\{1-\left(\frac{a-b}{a+b+1}\right)^2\right\}\int_0^\infty x^{2a}\phi\,dx\int_0^\infty x^{2b}\phi\,dx,$$

unless $\phi = C$, where $C > 0$, in $(0, \xi)$, and $\phi = 0$ in (ξ, ∞).

[ξ may be 0. The inequality is stronger than that resulting from a direct application of Theorem **181**. It follows from Theorem **227** if we reduce the integrals to the form considered there by partial integration. The case $a = 0$, $b = 2$ was mentioned by Gauss in connection with the Theory of Errors: see Gauss (**1**, IV, 12) and Pólya and Szegö (**1**, II, 114, 318).]

229. If $a \geqq 0$, $b \geqq 0$, $a \neq 1$, and ϕ is non-negative and increasing, then

$$\left(\int_0^1 x^{a+b}\phi\,dx\right)^2 > \left\{1-\left(\frac{a-b}{a+b+1}\right)^2\right\}\int_0^1 x^{2a}\phi\,dx\int_0^1 x^{2b}\phi\,dx,$$

unless $\phi = C$.

[See Pólya and Szegö (**1**, I, 57, 214). In this case Theorem **181** gives a reversed inequality, with the factor 1 on the right-hand side.]

230. If $\qquad 0 < a \leqq f \leqq A < \infty, \quad 0 < b \leqq g \leqq B < \infty,$

then
$$\int f^2\,dx\int g^2\,dx \leqq \left[\frac{1}{2}\left\{\sqrt{\left(\frac{AB}{ab}\right)}+\sqrt{\left(\frac{ab}{AB}\right)}\right\}\int fg\,dx\right]^2.$$

[Analogue of Theorem **71**: see Pólya and Szegö (**1**, I, 57, 214).]

231. If we consider the closed or open intervals (in general four in number) with end-points $-a$, b, where $a \geqq 0$, $b \geqq 0$, and suppose each of a and b zero, positive and finite, or infinite, we obtain in all 34 types of intervals I. Assign to each interval I a function $f(x)$ defined for $0 < x < 1$ and such that $\log \mathfrak{M}_r(f)$, where $\mathfrak{M}_r(f)$ is formed for the interval $(0, 1)$ and with $q = 1$, is finite just for the values of r in I.

[Examples:

I is $-a<r\leqq b$; $\quad f(x)=x^{1/a}(1-x)^{-1/b}\left(\log\frac{2}{1-x}\right)^{-2/b}$:

I is $-\infty\leqq r\leqq\infty$; $\quad f(x)=1+x^2$:

I is the single point 0; $\quad f(x)=\exp(-x^{-\frac{1}{2}}+(1-x)^{-\frac{1}{2}})$:

I is empty; $\quad f(x)=\exp(-x^{-1}+(1-x)^{-1})$.

This contains part of the proof of what is stated near the end of § 6.11.]

232. *Geometrical interpretation of Minkowski's inequality.* Suppose that a point in functional space is defined as a function of L^2, two functions defining the same point if and only if their difference is null; and that the distance between two points f and g is defined by

$$\delta(f,g)=\sqrt{\{\int(f-g)^2\,dx\}}.$$

Then (i) the distance between two distinct points is positive; and (ii)

$$\delta(f,h)\leqq\delta(f,g)+\delta(g,h).$$

[If we define distance by

$$\delta(f,g)=(\int|f-g|^r\,dx)^{1/r}\quad(r\geqq 1),$$

we obtain similar results in 'functional space L^r'.]

233. The shortest distance between two given points in Euclidean space is the straight line.

[A curve in space is given by

$$x=x(t),\quad y=y(t),\quad z=z(t).$$

We may suppose that t increases from 0 to 1 on the arc in question. If we assume that x, y, z are integrals of functions of L^2, then the length l is given by

$$l^2=[\int(x'^2+y'^2+z'^2)^{\frac{1}{2}}dt]^2=\mathfrak{M}_{\frac{1}{2}}(x'^2+y'^2+z'^2)\geqq\mathfrak{M}_{\frac{1}{2}}(x'^2)+\mathfrak{M}_{\frac{1}{2}}(y'^2)+\mathfrak{M}_{\frac{1}{2}}(z'^2),$$

by Theorem **198**; and this is not less than

$$(\int x'\,dt)^2+(\int y'\,dt)^2+(\int z'\,dt)^2=(x_1-x_0)^2+(y_1-y_0)^2+(z_1-z_0)^2.$$

If there is equality, $Ax'\equiv By'\equiv Cz'$, and the curve is a straight line.]

234. If $0<p<1$ and

$$\int fg\,dx\geqq A\,(\int g^{p'}dx)^{1/p'}$$

for all g, then $\int f^p\,dx\geqq A^p$.

[Compare Theorem **70**. If $f>0$ for all x, define g by $fg=f^p$. If $f>0$ in E, $f=0$ in CE, and the measure of CE is finite, define g by $fg=f^p$ in E and by $g=G$ in CE, and proceed as in the proof of Theorem **70**. If the measure of CE is infinite, take (for example)

$$g=Ge^{x^2}$$

in CE. Then

$$\int f^p\,dx=\int fg\,dx\geqq A\left(\int_E f^p\,dx+G^{p'}\int_{CE}e^{p'x^2}dx\right)^{1/p'},$$

and the result again follows when $G\to\infty$.]

235. Suppose that f and p are positive and that f has the period 2π; that

$$F(x)=\int_0^{2\pi} f(x+t)\,p(t)\,dt \Big/ \int_0^{2\pi} p(t)\,dt;$$

and that the means $\mathfrak{M}_r$ refer to the interval $(0, 2\pi)$ and a constant weight-function. Then

$$\mathfrak{M}_r(F) \geqq \mathfrak{M}_r(f) \quad (0 \leqq r \leqq 1), \quad \mathfrak{M}_r(F) \leqq \mathfrak{M}_r(f) \quad (r \geqq 1).$$

[This may be deduced from Theorem **204**, or proved directly (supposing for example $r \geqq 1$) as follows:

$$\begin{aligned}\mathfrak{M}_r^r(F) &= \frac{1}{2\pi}\int dx \left[\frac{\int f(x+t)\,p(t)\,dt}{\int p(t)\,dt}\right]^r \\ &\leqq \frac{1}{2\pi}\int dx \frac{\int f^r(x+t)\,p(t)\,dt\,(\int p(t)\,dt)^{r-1}}{(\int p(t)\,dt)^r} \\ &= \frac{1}{2\pi}\frac{\int p(t)\,dt \int f^r(x+t)\,dx}{\int p(t)\,dt} = \frac{1}{2\pi}\int f^r(u)\,du = \mathfrak{M}_r^r(f).\end{aligned}$$

For the case $r=0$, see Pólya and Szegö (**1**, I, 56, 212).]

236. We say that $f(x, y, \dots)$ and $g(x, y, \dots)$ are *similarly ordered* if

$$\{f(x_1, y_1, \dots) - f(x_2, y_2, \dots)\}\{g(x_1, y_1, \dots) - g(x_2, y_2, \dots)\} \geqq 0,$$

oppositely ordered if f and $-g$ are similarly ordered. Prove that

$$\iint \dots f\,dx\,dy \dots \iint \dots g\,dx\,dy \dots \leqq \iint \dots dx\,dy \dots \iint \dots fg\,dx\,dy \dots,$$

if f and g are similarly ordered, while the sign is reversed if f and g are oppositely ordered. The integration is extended over any common part of the regions of definition of f and g.

[Analogue of Theorem **43** (with $r=1$), due in substance to Tchebychef (who considers only monotonic functions of one variable).]

237. If ϕ and ψ satisfy the conditions of Theorem **156**, and

$$\Phi(x)=\int_0^x \phi(t)\,dt, \quad \Psi(x)=\int_0^x \psi(t)\,dt,$$

then
$$\int fg\,dx \leqq \int \Phi(f)\,dx + \int \Psi(g)\,dx.$$

238. If f and g are positive, and k a positive constant, and $f\log^+ f$ and e^{kg} are integrable, then fg is integrable.

[By Theorem **63**, $kfg \leqq f\log^+ f + e^{kg-1}$.]

239. If f is positive, then

$$\int_0^a f\log\frac{1}{x}\,dx \leqq 2\int_0^a f\log^+ f\,dx + \frac{4\sqrt{a}}{e}.$$

[Take $g=\frac{1}{2}\log\frac{1}{x}$, $k=1$, in the inequality used in proving Theorem **238**.]

240. If f is positive and L in $(0, a)$, and

$$F(x)=\int_0^x f\,dt,$$

then
$$\int_0^a f(x)\log\frac{1}{x}\,dx=\int_0^a \frac{F(x)}{x}\,dx+F(a)\log\frac{1}{a},$$
whenever either integral is finite.

241. Suppose that a is positive and finite; that $B=B(a)$ denotes generally a number depending on a only; that $f(x)\geqq 0$; and that

$$F(x)=\int_0^x f(t)\,dt,$$

$$J=\int_0^a f\log^+ f\,dx,\quad K=\int_0^a \frac{F}{x}\,dx.$$

Then (i) if J is finite K is also finite, and

$$K<BJ+B:$$

(ii) when f is a decreasing function the converse is also true; if K is finite then J is finite, and
$$J<BK\log^+ K+B.$$

[For the last two theorems see Hardy and Littlewood (8).]

242. If f is positive and L in $(0, a)$ and

$$g=\int_x^a \frac{f(t)}{t}\,dt,$$

then g is L and
$$\int_0^a g(x)\,dx=\int_0^a f(x)\,dx.$$

[Integrate by parts; or substitute for g and change the order of integration.]

243. We define $\mathfrak{M}_\phi(f)$, where ϕ is a continuous and strictly increasing function, by

$$\mathfrak{M}_\phi(f)=\phi^{-1}\{\int\phi(f)\,q\,dx\}.$$

Then, in order that
$$\mathfrak{M}_\phi(f)\leqq\mathfrak{M}_\psi(f)$$
for all f, it is necessary and sufficient that ψ should be convex with respect to ϕ.

244. In order that

$$\mathfrak{M}_{\phi_n}^{x_n}\ldots\mathfrak{M}_{\phi_1}^{x_1}(f)\leqq\mathfrak{M}_{\psi_n}^{x_n}\ldots\mathfrak{M}_{\psi_1}^{x_1}(f)$$

for all $f=f(x_1, x_2, \ldots, x_n)$, it is necessary and sufficient that every ψ_ν be convex with respect to the corresponding ϕ_ν.

245. In order that

$$\mathfrak{M}_{p_n}^{x_n}\ldots\mathfrak{M}_{p_1}^{x_1}(f)\leqq\mathfrak{M}_{q_{\nu_n}}^{x_{\nu_n}}\ldots\mathfrak{M}_{q_{\nu_1}}^{x_{\nu_1}}(f)$$

for all f, it is necessary and sufficient that (i) $q_\nu\geqq p_\nu$ and (ii) $q_\mu\geqq p_\nu$ when $\mu>\nu$ and the permutation by which $\nu_1, \nu_2, \ldots, \nu_n$ is derived from $1, 2, \ldots, n$ involves an inversion of μ and ν.

[For the last three theorems, which correspond to Theorems **92, 93** and **137**, see Jessen (**2, 3**).]

246. Hölder's inequality may be deduced from

$$\mathfrak{M}_1{}^y\,\mathfrak{M}_0{}^x(f) \leqq \mathfrak{M}_0{}^x\,\mathfrak{M}_1{}^y(f)$$

(Theorem **203**), by taking

$$f(x,y)=f_1{}^p(y)\ \left(0\leqq x<\frac{1}{p}\right),\quad f(x,y)=f_2{}^{p'}(y)\ \left(\frac{1}{p}\leqq x\leqq 1\right).$$

[See Jessen (**3**).]

247. If (i) $\phi(x,t)$ is positive, continuous, and convex in x, for $x_1 \leqq x \leqq x_2$, $t>0$; (ii) $p(t)\geqq 0$; (iii) the integral

$$I(x)=\int_0^\infty \phi(x,\,t)\,p(t)\,dt$$

is finite for $x=x_1$ and $x=x_2$; then $I(x)$ is continuous and convex for

$$x_1<x<x_2.$$

[That $I(x)$ is bounded and convex follows immediately from the convexity of ϕ; that it is continuous, from Theorem **111**.]

248. If $f(x)$ and $\phi(x)$ are positive and $\phi(x)$ convex for positive x, and

$$I(x)=x\int_0^\infty \phi\left\{\frac{f(t)}{x}\right\}dt$$

is finite for $x=x_1$ and $x=x_2$, then $I(x)$ is continuous and convex for

$$x_1<x<x_2.$$

[By Theorem **119**, $x\phi(1/x)$ and

$$f(t)\,.\,\frac{x}{f(t)}\,\phi\left\{\frac{f(t)}{x}\right\}$$

are convex, and we can apply Theorem **247**. More general results can be derived from Theorem **120**.]

249. In order that

$$\int_a^b \phi(g(x))\,dx \leqq \int_a^b \phi(f(x))\,dx$$

should be true for every convex and continuous ϕ, it is necessary and sufficient that

$$\int_a^b g(x)\,dx=\int_a^b f(x)\,dx$$

and

$$\int_a^b (g(x)-y)^+\,dx \leqq \int_a^b (f(x)-y)^+\,dx$$

for all y.

[Here a^+ means $\operatorname{Max}(a,0)$, as in § 6.1.]

250. If f and g are increasing functions, then an equivalent condition is

$$\int_\xi^b g(x)\,dx \leqq \int_\xi^b f(x)\,dx$$

for $a\leqq\xi\leqq b$.

[For the two last theorems, which embody analogues for integrals of parts of Theorem **108**, see Hardy, Littlewood, and Pólya (**2**).]

251. If $f_1(t), f_2(t), \ldots, f_l(t)$ are real and integrable in $(0, 1)$, then *either* (i) there is a function $x(t)$ such that

$$\int_0^1 f_1(t)\,x(t)\,dt > 0, \ldots, \int_0^1 f_l(t)\,x(t)\,dt > 0,$$

or (ii) there are non-negative numbers $y_1, y_2, \ldots, y_l$, not all zero, such that

$$y_1 f_1(t) + y_2 f_2(t) + \ldots + y_l f_l(t) \equiv 0.$$

252. If $f_1(t), f_2(t), \ldots, f_m(t)$ are real and continuous in $(0, 1)$, then *either* (i) there are real numbers $x_1, x_2, \ldots, x_m$ such that

$$x_1 f_1(t) + x_2 f_2(t) + \ldots + x_m f_m(t)$$

is non-negative for all, and positive for some, t of $(0, 1)$, *or* (ii) there is a positive and continuous function $y(t)$ such that

$$\int_0^1 f_1(t)\,y(t)\,dt = 0, \ldots, \int_0^1 f_m(t)\,y(t)\,dt = 0.$$

[Theorems **251** and **252** are both integral analogues of an important theorem of Stiemke (**1**) concerning systems of linear inequalities. Suppose that

$$a_{\lambda\mu} \quad (\lambda = 1, 2, \ldots, l;\ \mu = 1, 2, \ldots, m)$$

is a rectangular array of l rows and m columns, and that

$$L_\lambda(x) = a_{\lambda 1} x_1 + a_{\lambda 2} x_2 + \ldots + a_{\lambda m} x_m,$$
$$M_\mu(x) = a_{1\mu} y_1 + a_{2\mu} y_2 + \ldots + a_{l\mu} y_l;$$

and consider the two problems:

(i) to find a real set (x) for which

$$L_1(x) > 0,\ L_2(x) > 0, \ldots,\ L_l(x) > 0;$$

(ii) to find a non-negative and non-null set (y) for which

$$M_1(y) = 0,\ M_2(y) = 0, \ldots, M_m(y) = 0.$$

Since
$$\Sigma y L(x) = \Sigma x M(y),$$

the two problems cannot both be soluble for the same set (a), and Stiemke's theorem asserts that one is soluble whatever the set (a).

Theorems **251** and **252** state analogues of Stiemke's theorem in which the m columns or l rows are replaced by a continuous infinity of columns or rows. These theorems, and further references to the theory of systems of linear inequalities, which we have excluded from our programme only on account of its algebraical and geometrical preliminaries, will be found in Haar (**1**) and Dines (**1**).]

CHAPTER VII

SOME APPLICATIONS OF THE CALCULUS OF VARIATIONS

7.1. Some general remarks. The 'simplest problem of the Calculus of Variations' is that of determining a maximum or minimum value of

$$J(y)=\int_{x_0}^{x_1} F(x,y,y')\,dx$$

for all functions $y=y(x)$ for which

(1) $y_0=y(x_0)$, $y_1=y(x_1)$ are given,

(2) y' is continuous.

Let us denote this class of functions by $\mathfrak{K}$. Then our object is to find a function

$$y=Y(x)$$

of $\mathfrak{K}$, such that either

$$J(y)<\int_{x_0}^{x_1} F(x,Y,Y')\,dx=J(Y),$$

or $J(y)>J(Y)$, for all y of $\mathfrak{K}$ other than Y. The general theory tells us that, if such a function Y exists, it must satisfy 'Euler's equation'

$$\text{(E)} \qquad \frac{\partial F}{\partial y}=\frac{d}{dx}\left(\frac{\partial F}{\partial y'}\right).$$

Let us consider some simple examples.

(i) Suppose that $\quad J(y)=\int_0^1 y'^2\,dx$

and $y_0=0$, $y_1=1$. Then (E) is $y''=0$, and the only solution satisfying the conditions is $y=x$. It is easy to verify that $Y=x$ does in fact give a minimum for $J(y)$. For $J(Y)=1$ and

$$1=\left(\int_0^1 y'\,dx\right)^2<\int_0^1 dx\int_0^1 y'^2\,dx=J(y),$$

by Theorem **181**, unless $y'\equiv 1$, $y=x$; so that $J>1$ for all y other than Y.

We have in fact proved more than the problem as stated demands, since the proof is valid *whenever y is an integral*. This last hypothesis is however essential, since there are functions y for which

$$y(0)=0, \quad y(1)=1, \quad y' \equiv 0, \quad J(y)=0. \tag{7.1.1}$$

In order that y should be an integral, i.e. in order that there should be an integrable function $f(x)$ such that

$$y(x)=\int_a^x f(u)\,du,$$

it is necessary and sufficient that $y(x)$ should be 'absolutely continuous'. It is necessary, but not sufficient, that $y(x)$ should have bounded variation. In particular it is not sufficient that $y(x)$ should be monotonic; there are *increasing* functions y which satisfy (7.1.1).

If y is the integral of f, then $y' \equiv f$; an integral is the integral of its derivative. All this is expounded in detail in books on the theory of functions of a real variable[a]. The main theorem needed in this chapter is the theorem of integration by parts, stated in § 6.3 (a).

These remarks lead us to lay down the following convention. Throughout this chapter it will be assumed that, *whenever y and y' occur in an enunciation or a proof, y is an integral* (and so the integral of y'). A similar assumption will be made about y' and y'' (if y'' occurs in the problem); and the assumption naturally applies also to letters other than y. Without this assumption, all the problems of this chapter would lose their significance.

(ii)[b] Suppose that

$$J(y)=\int_0^1 (y'^2+y'^3)\,dx,$$

and $y_0=y_1=0$. The only solution of (E) satisfying the conditions is $y=0$. If $Y=0$, $J(Y)=0$, but Y does not give a maximum or minimum of $J(y)$. It is in fact easy to construct a y of $\mathfrak{K}$ for which

[a] See for example de la Vallée Poussin (**2**), Hobson (**1**), Titchmarsh (**1**).

[b] This and the next example are due to Weierstrass and are of great historical importance.

$J(y)$ is as large, positively or negatively, as we please. Thus, if $f(x)$ is any function for which $f(0)=f(1)=0$ and

$$\int_0^1 f'^3 dx > 0,$$

and $y=Cf(x)$, then $J(y)$ is large when C is large and has the sign of C.

(iii) Suppose that

$$J(y)=\int_0^1 xy'^2 dx,$$

$y_0=0, y_1=1$. Here $J(y)>0$ for all y of $\mathfrak{K}$; but $y=x^m$ gives $J=\frac{1}{2}m$, so that there are y of $\mathfrak{K}$ for which $J(y)$ is as small as we please. For y of $\mathfrak{K}$, $J(y)$ has an unattained lower bound 0. The same is true for classes of y more general than $\mathfrak{K}$ (for example, the class of integrals). On the other hand, $J(y)$ attains its bound 0 for the function mentioned under (i) above, and also for the discontinuous function which is 0 for $x=0$ and 1 for $x>0$.

7.2. Object of the present chapter. The Calculus of Variations might be expected to provide a very powerful weapon for the proof of integral inequalities. There are however hardly any instances of its application to inequalities of the types important in general analysis. This may be explained on two grounds. In the first place, the Calculus of Variations is concerned avowedly with *attained* maxima or minima, while many of the most important integral inequalities assert unattained upper or lower bounds. Secondly, the 'continuity' hypotheses of the classical theory are very restrictive. It is often more troublesome to extend an inequality, proved by variational methods for a special class of functions, to the most general classes for which the inequality is required, than to construct a direct proof of the full result. For these reasons the Calculus of Variations has been almost ignored in this chapter of analysis.

The ideas of the Calculus are however often very useful, and we apply them here to a number of special inequalities. When, as in example (i) above, or Theorems **254** and **256** below, the bound asserted by the inequality is attained, and attained by an extremal, that is to say by a solution of Euler's equation, these ideas are obviously relevant, and the result may well be one which it would be difficult to obtain in any other way. We shall however find that they are sometimes effective even when the bound is unattained and the final result lies outside the scope of the theory.

Our arguments will not demand any detailed knowledge of the theory; except in § 7.8, we shall require only its simplest formal ideas[a].

7.3. Example of an inequality corresponding to an unattained extremum. As a first example of the use of variational methods, we select a special case of a theorem which was first proved in an entirely different manner, and to which we shall return in § 9.8.

253. *If y' belongs to $L^2(0,\infty)$, $y_0=0$, and y is not always zero, then*

$$J(y)=\int_0^\infty \left(4y'^2-\frac{y^2}{x^2}\right)dx>0.$$

It is necessary for our present purpose to consider the more general integral

$$J(y)=\int_0^\infty \left(\mu y'^2-\frac{y^2}{x^2}\right)dx \quad (\mu \geqq 4). \tag{7.3.1}$$

Euler's equation is

$$x^2y''+\lambda y=0 \quad (\lambda=1/\mu \leqq \tfrac{1}{4}).$$

Its solution is $$y=Ax^m+Bx^n,$$

where $$m=\tfrac{1}{2}+\sqrt{(\tfrac{1}{4}-\lambda)}, \quad n=\tfrac{1}{2}-\sqrt{(\tfrac{1}{4}-\lambda)},$$

if $\mu>4$, and is $$y=x^{\frac{1}{2}}(A+B\log x)$$

if $\mu=4$. In neither case is there a solution (other than $y=0$) for which y' is L^2.

For this reason it is necessary to modify the problem before we attempt to apply variational ideas. We consider

$$J(y)=\int_0^1 \left(\mu y'^2-\frac{y^2}{x^2}\right)dx$$

[a] Euler's equation and Hilbert's invariant integral. Anything which we assume will be found without difficulty in the books of Bliss (1) or Bolza (1).

with $y_0=0$, $y_1=1$, $\mu>4$. There is then one[a] extremal satisfying the conditions, viz.

$$y=Y=x^m=x^{\frac{1}{2}+a}, \tag{7.3.2}$$

where

$$a=\sqrt{\left(\frac{1}{4}-\frac{1}{\mu}\right)}, \quad \mu=\frac{1}{\frac{1}{4}-a^2}. \tag{7.3.3}$$

A simple calculation gives

$$J(Y)=\frac{2}{1-2a}, \tag{7.3.4}$$

and this suggests the following theorem.

254.[b] *If $\mu>4$, $y(0)=0$, $y(1)=1$, and y' is L^2, then*

$$J(y)=\int_0^1\left(\mu y'^2-\frac{y^2}{x^2}\right)dx \geqq \frac{2}{1-2a}, \tag{7.3.5}$$

where a is defined by (7.3.3). *The only case of equality is that defined by* (7.3.2).

7.4. First proof of Theorem 254. We give two proofs of Theorem **254**. The first demands no knowledge of the Calculus of Variations, though the transformations which we use are suggested by our knowledge of the form of the extremal Y.

If

$$y=x^{\frac{1}{2}+a}+\eta=Y+\eta, \tag{7.4.1}$$

then

$$J(y)=J(Y)+J(\eta)+K(Y,\eta), \tag{7.4.2}$$

where
$$K(Y,\eta)=2\int_0^1\left(\mu Y'\eta'-\frac{Y\eta}{x^2}\right)dx.$$

Since Y' and so η' are L^2, $\eta=o(x^{\frac{1}{2}})$ for small x;[c] and so

$$K=2\lim_{\delta\to 0}\left(-\int_\delta^1\frac{Y\eta}{x^2}dx+\Big[\mu Y'\eta\Big]_\delta^1-\int_\delta^1\mu Y''\eta\,dx\right)$$

$$=-2\lim_{\delta\to 0}\int_\delta^1\eta\left(\mu Y''+\frac{Y}{x^2}\right)dx=0.$$

[a] The extremal
$$y=\lambda x^{\frac{1}{2}+a}+(1-\lambda)x^{\frac{1}{2}-a}$$
gives $y_0=0$, $y_1=1$ for any λ; but y' is not L^2, and $J(y)$ diverges, unless $\lambda=1$.

[b] For this and some later theorems in this chapter see Hardy and Littlewood (**10**).

[c] Theorem **222**.

Hence (7.4.2) gives

$$J(y) = \frac{2}{1-2a} + J(\eta); \tag{7.4.3}$$

and it is sufficient to prove that

$$J(\eta) > 0.$$

Here η' is L^2, and η is not null but vanishes at the ends of the interval.

We now write $\eta = Y\zeta$.

Then

$$\begin{aligned} J_\delta(\eta) &= \int_\delta^1 \left(\mu\eta'^2 - \frac{\eta^2}{x^2}\right) dx \\ &= \mu\int_\delta^1 (Y\zeta' + Y'\zeta)^2 dx - \int_\delta^1 \frac{Y^2\zeta^2}{x^2} dx \\ &= \mu\int_\delta^1 Y^2\zeta'^2 dx + \int_\delta^1 \left(\mu Y'^2 - \frac{Y^2}{x^2}\right)\zeta^2 dx + 2\mu\int_\delta^1 YY'\zeta\zeta' dx. \end{aligned} \tag{7.4.4}$$

But

$$2\mu\int_\delta^1 YY'\zeta\zeta' dx = -\mu(YY'\zeta^2)_\delta - \mu\int_\delta^1 (Y'^2 + YY'')\zeta^2 dx. \tag{7.4.5}$$

Combining (7.4.4) and (7.4.5), and observing that Y is a solution of

$$\mu Y'' + \frac{Y}{x^2} = 0,$$

we obtain

$$J_\delta(\eta) = -\mu(YY'\zeta^2)_\delta + \mu\int_\delta^1 (Y\zeta')^2 dx. \tag{7.4.6}$$

But

$$YY'\zeta^2 = (\tfrac{1}{2}+a)\frac{(Y\zeta)^2}{x} = (\tfrac{1}{2}+a)\frac{\eta^2}{x} \to 0$$

when $x \to 0$. Hence, when we make $\delta \to 0$ in (7.4.6), we obtain

$$J(\eta) = \mu\int_0^1 (Y\zeta')^2 dx, \tag{7.4.7}$$

which is positive unless the integrand is null, i.e. unless $\zeta = 0$.

This proves the theorem. The condition $\mu > 4$ was required to make Y' belong to L^2. We have however reduced the theorem to dependence on the identity (7.4.7). Since

$$Y\zeta' = \eta' - \frac{\frac{1}{2}+a}{x}\eta = y' - \frac{\frac{1}{2}+a}{x}y,$$

(7.4.7) and (7.4.3) give

$$(7.4.8)\quad J(y)=\int_0^1\left(\mu y'^2-\frac{y^2}{x^2}\right)dx=\frac{2}{1-2a}+\mu\int_0^1\left(y'-\frac{\frac{1}{2}+a}{x}y\right)^2dx;$$

and here Y has disappeared. Since both sides of (7.4.8) are continuous in μ, we may now include the case $\mu=4$, $a=0$. The identity may be verified directly by partial integration when its form has been discovered (though some care is required about convergence at the lower limit).

We are now in a position to prove Theorem **253**. If we write

$$x=X/\xi,\quad cy(x)=Y(X),$$

and then replace X, Y again by x, y, we obtain

$$(7.4.9)\quad \int_0^\xi\left(\mu y'^2-\frac{y^2}{x^2}\right)dx=\frac{2}{1-2a}\frac{c^2}{\xi}+\mu\int_0^\xi\left(y'-\frac{\frac{1}{2}+a}{x}y\right)^2dx,$$

where now $y(0)=0$, $y(\xi)=c$. If y' is L^2 in $(0,\infty)$, $c=o(\xi^{\frac{1}{2}})$ for large ξ,[a] and the first term on the right tends to 0 when $\xi\to\infty$. Making $\xi\to\infty$, and supposing $\mu=4$, we obtain

$$\int_0^\infty\left(4y'^2-\frac{y^2}{x^2}\right)dx=4\int_0^\infty\left(y'-\frac{y}{2x}\right)^2dx.$$

This formula, which makes Theorem **253** intuitive, is valid whenever y' is L^2, and may of course be verified directly[b].

7.5. Second proof of Theorem 254. In our second proof we make explicit the variational theory which underlies the first.

Suppose that $y=Y(x)$, or E, is the extremal through the end-points P_0 and P_1, and that

$$(7.5.1)\quad y=y(x,\alpha),$$

or $E(\alpha)$, is a family of extremals containing E and depending on a parameter α. Suppose further either that

(i) $E(\alpha)$ covers up a region surrounding E in a $(1,1)$ manner, so that just one extremal passes through every point of the region, and α is a one-valued function of x and y; or that

(ii) every curve of $E(\alpha)$ passes through P_0, so that $y(x_0,\alpha)$ is independent of α, but condition (i) is satisfied in all other respects.

[a] Theorem **223**.

[b] Grandjot (**1**) gives a number of somewhat similar identities for series.

In these circumstances[a] (7.5.1) is said to define a *field* of extremals including E.

The slope
$$y'(x,\alpha)$$
of the extremal through a point P of the field may be expressed as a one-valued function
$$p=p(x,y)$$
of x and y. Hilbert's 'invariant integral' is

$$J^*(C)=\int_C\{(F-pF_p)\,dx+F_p\,dy\}.$$

Here F and F_p are the values of $F(x,y,y')$ and $F_{y'}(x,y,y')$ when y' is replaced by p, and the integral is taken along any curve C which lies in the region covered by the field.

The fundamental properties of Hilbert's integral are as follows.

(i) $J^*(C)$ depends only on the ends Q, R of C; in other words

$$(F-pF_p)\,dx+F_p\,dy=dW$$

is a perfect differential, and

$$J^*(C)=W_R-W_Q.$$

(ii) If C is the extremal E, then

$$J^*(E)=\int_E F\,dx=J(E),$$

say. It follows that, if C runs from P_0 to P_1, then

$$\begin{aligned}J(C)-J(E)=J(C)-J^*(E)&=J(C)-J^*(C)\\&=\int_C F(x,y,y')\,dx\\&\quad-\int_C\{(F(x,y,p)-pF_p(x,y,p))\,dx+F_p(x,y,p)\,dy\}\\&=\int_C \mathcal{E}(x,y,p,y')\,dx,\end{aligned}$$

where

$$\mathcal{E}(x,y,p,y')=F(x,y,y')-F(x,y,p)-(y'-p)\,F_p(x,y,p).$$

[a] With the addition of certain conditions concerning the differentiability of $\alpha(x,y)$ which it is unnecessary to repeat here: see Bolza (**1**, 95–105).

Here y' is the slope of C at any point and p the slope of the extremal through the point; and $\mathcal{E}$ is Weierstrass's 'excess-function'. If $\mathcal{E} > 0$ whenever $y' \neq p$, then

$$J(E) < J(C)$$

and E gives a true minimum of J.

In the present case we take

$$y = \alpha x^{\frac{1}{2}+a}$$

as $E(\alpha)$. We find

$$p = \alpha(\tfrac{1}{2}+a)x^{-\frac{1}{2}+a} = (\tfrac{1}{2}+a)\frac{y}{x},$$

$$F = \mu p^2 - \frac{y^2}{x^2}, \quad F_p = \frac{2y}{(\frac{1}{2}-a)x}, \quad F - pF_p = -\frac{y^2}{(\frac{1}{2}-a)x^2},$$

$$J^* = \frac{1}{\frac{1}{2}-a}\int\left(-\frac{y^2}{x^2}dx + \frac{2y}{x}dy\right) = \int dW,$$

where
$$W = \frac{y^2}{(\frac{1}{2}-a)x}.$$

Here
$$\mathcal{E} = \mu y'^2 - \mu p^2 - (y'-p)\,2\mu p = \mu(y'-p)^2 > 0$$

unless $y' = p$. The identity

$$J(C) - J(E) = \int_C \mathcal{E}\,dx$$

reduces to

$$\int_0^1\left(\mu y'^2 - \frac{y^2}{x^2}\right)dx - \frac{2}{1-2a} = \mu\int_0^1\left\{y' - \left(\frac{1}{2}+a\right)\frac{y}{x}\right\}^2 dx,$$

which is (7.4.8).

This argument shows the genesis of (7.4.8), but does not prove it, for two reasons. In the first place, F has a singularity, and the theory of the field breaks down, for $x = 0$. Secondly, the theory presupposes the continuity of y'.

In order to dispose of the first difficulty, we may take P_0 and P_1 to be $(\delta, \delta^{\frac{1}{2}+a})$ and $(1, 1)$. The theory then gives the identity

$$\int_\delta^1\left(\mu y'^2 - \frac{y^2}{x^2}\right)dx = \frac{2(1-\delta^{2a})}{1-2a} + \mu\int_\delta^1\left\{y' - \left(\frac{1}{2}+a\right)\frac{y}{x}\right\}^2 dx;$$

and we obtain (7.4.8), for continuous y', by making δ tend to zero.

When (7.4.8) is proved for continuous y', it may be extended to general y' of L^2 by standard processes of approximation. We deal with this point in the next section, in a different problem where we have no alternative elementary proof.

Similar considerations lead to the identity

$$\int_0^1 \left(\mu y'^k - \frac{y^k}{x^k}\right) dx = \frac{1}{(k-1)(1-\lambda)} + \mu \int_0^1 \left\{ y'^k - \left(\frac{\lambda y}{x}\right)^k - k\left(y' - \frac{\lambda y}{x}\right)\left(\frac{\lambda y}{x}\right)^{k-1} \right\} dx. \quad (7.5.2)$$
[a]

Here

$$y(0)=0, \quad y(1)=1, \quad y' \geqq 0, \quad k>1, \quad \mu > \left(\frac{k}{k-1}\right)^k = K,$$

and λ is the (unique) root of

$$\mu(k-1)\lambda^{k-1}(\lambda-1)+1=0 \quad (7.5.3)$$

which lies between $1/k'$ and 1. When the form of (7.5.2) has been determined we may put

$$\mu = K,$$

where $\lambda = 1/k'$ and $\mu\lambda^k = 1$. We thus obtain

$$\int_0^1 \left(K y'^k - \frac{y^k}{x^k}\right) dx = \frac{k}{k-1} + K \int_0^1 \left\{ y'^k - \left(\frac{\lambda y}{x}\right)^k - k\left(y' - \frac{\lambda y}{x}\right)\left(\frac{\lambda y}{x}\right)^{k-1} \right\} dx.$$

It may be verified directly, by partial integration, that this is true whenever y' is L^k; and we can prove as in § 7.4 that the identity remains true when the upper limit 1 is replaced by ∞ and the term $k/(k-1)$ is omitted. Since, by Theorem **41**,

$$a^k - b^k > k(a-b)b^{k-1}$$

for all positive a, b, we thus obtain a proof of a theorem (Theorem **327**) which will be stated explicitly, and proved in an entirely different manner, in § 9.8.

Incidentally we obtain

255. *If* $\quad k>1, \quad \mu > \left(\frac{k}{k-1}\right)^k = K,$

[a] The theory of the field gives the form of the identity, which may then be verified independently. The limitation to curves for which $y' \geqq 0$ would introduce another slight complication into a properly variational proof.

$y(0)=0$, $y(1)=1$, and y' is L^k, then

$$J=\int_0^1\left(\mu y'^k-\frac{y^k}{x^k}\right)dx \geqq \frac{1}{(k-1)(1-\lambda)},$$

where λ is the root of (7.5.3) *between $1/k'$ and* 1.

7.6. Further examples illustrative of variational methods. It is difficult to distinguish at all precisely between 'elementary' and 'variational' proofs, since there are many proofs of intermediate types. We give a selection of such proofs, worked out with varying degrees of detail, in this and the succeeding sections.

(I). **256.** *If $y(0)=0$ and $2k$ is an even positive integer, then*

$$\int_0^1 y^{2k}\,dx \leqq C\int_0^1 y'^{2k}\,dx, \tag{7.6.1}$$

where

$$C=\frac{1}{2k-1}\left(\frac{2k}{\pi}\sin\frac{\pi}{2k}\right)^{2k}. \tag{7.6.2}$$

There is equality only for a certain hyperelliptic curve.

(i) We suppose first that $y(1)\neq 0$, in which case we may take $y(1)=1$, and consider

$$J(y)=\int_0^1 (Cy'^{2k}-y^{2k})\,dx.$$

Euler's equation is

$$(2k-1)\,Cy'^{2k-2}y''+y^{2k-1}=0,$$

which gives

$$(2k-1)\,Cy'^{2k}=C'-y^{2k},$$

where C' is a constant of integration.

There is one extremal which passes through $(0, 0)$ and $(1, 1)$ and cuts $x=1$ at right angles. In fact, if we take $C'=1$, then y' vanishes when $y=1$. Also

$$x=\{(2k-1)\,C\}^{1/2k}\int\frac{dy}{(1-y^{2k})^{1/2k}}$$

and, since

$$\int_0^1\frac{dy}{(1-y^{2k})^{1/2k}}=\frac{1}{2k}\int_0^1(1-u)^{-1/2k}u^{1/2k-1}\,du=\frac{\pi}{2k}\operatorname{cosec}\frac{\pi}{2k},$$

there is an extremal of this type which passes through $(0,0)$ and $(1, 1)$. If we denote this extremal by Y, then

$$J(Y)=\Big[CYY'^{2k-1}\Big]_0^1-\int_0^1 Y\{(2k-1)CY'^{2k-2}Y''+Y^{2k-1}\}\,dx=0,$$

since $Y'(1)=0$.

To prove the theorem, we must show that $y=Y$ gives a strong minimum; and this follows easily from the general theory. The extremal is a rising curve of the same general form as the curve $y=\sin\frac{1}{2}\pi x$, to which it reduces when $k=1$. The curve $y=\alpha Y$ is also an extremal; the family $y=\alpha Y$ defines a field in the sense of §7.5; and the excess-function

$$\mathcal{E}=y'^{2k}-p^{2k}-2k(y'-p)p^{2k-1}$$

is positive. Hence the standard conditions for a minimum are satisfied. This proof is genuinely 'variational', and (in view of the trouble of calculating the slope-function p explicitly) it might be difficult to find a more elementary proof.

There is however one point in the proof which demands an additional remark. The 'general theory' assumes that y' is continuous, and it may not be obvious how its conclusions, in particular in regard to the uniqueness of the solution, are extended to the more general y considered here.

Let us denote by I^{2p} the class of integrals y of functions of L^{2p}, by I^* the class of integrals y^* of continuous functions. The general theory shows that

$$J(y^*)>J(Y) \tag{7.6.3}$$

for a y^* different from Y, while we require the same result for any y of I^{2p}. We can approximate to a y of I^{2p}, different from Y, by a sequence of functions y^*, in such a manner that

$$J(y)=\lim J(y^*);$$

but all that then follows from (7.6.3) is

$$J(y)\geqq J(Y),$$

the strict sign of inequality being lost in the passage to the limit.

The difficulty disappears if we look at the question differently. The general theory proves not only the inequality (7.6.3) but also the *identity*

$$J(y^*)-J(Y)=\int\mathcal{E}(y^*)\,dx,$$

where $\mathcal{E}(y^*)$ is the excess-function corresponding to y^* and the field $y = \alpha Y$. Approximating to y by an appropriate sequence of y^*, we replace this identity by

$$J(y) - J(Y) = \int \mathcal{E}(y)\,dx,$$

and the integral is positive unless $y = Y$.

(ii) The case in which $y(1) = 0$ may be discussed similarly, since $y = 0$ is then an extremal satisfying the conditions, and is included in the field $y = \alpha Y$ used in (i).

The proof might have been arranged differently if we had made no hypothesis about the value of $y(1)$. The problem is then one with a 'variable end-point', that of minimising $J(y)$ for curves drawn from the origin to meet the line $x = 1$. The extremals cut this line 'transversally' (in this case orthogonally), and all the curves $y = \alpha Y$ satisfy this condition. The general theory shows that all the extremals give the same value of $J(y)$, and this value must be 0, since it is 0 when $\alpha = 0$.[a]

7.7. Further examples: Wirtinger's inequality. (II) Let us consider more particularly the case $k = 1$ of (I). Changing the limits, the result is that

$$\int_0^{\frac{1}{2}\pi} y^2\,dx < \int_0^{\frac{1}{2}\pi} y'^2\,dx \tag{7.7.1}$$

if $y(0) = 0$ and y is not a multiple of $\sin x$.

The general theory suggests that there is an identity of the type

$$\int_0^{\frac{1}{2}\pi} (y'^2 - y^2)\,dx = \int_0^{\frac{1}{2}\pi} \{y' - y\psi(x)\}^2\,dx$$

or

$$\int_0^{\frac{1}{2}\pi} \{y^2(1+\psi^2) - 2yy'\psi\}\,dx = 0.$$

This will plainly be true if

$$y^2(1+\psi^2)\,dx - 2y\psi\,dy$$

is an exact differential dz, and z vanishes at the limits; and this requires

$$-\psi' = 1 + \psi^2, \quad \psi = -\tan(x+k),$$

[a] We owe these remarks to Prof. Bliss.

in which case $z=-\psi y^2$. If we take $k=\frac{1}{2}\pi$, $\psi=\cot x$, then z vanishes when $x=\frac{1}{2}\pi$; and since y' is L^2 and so[a] $y=o(x^{\frac{1}{2}})$, z also vanishes when $x=0$. We thus obtain

$$(7.7.2) \qquad \int_0^{\frac{1}{2}\pi} (y'^2-y^2)\,dx=\int_0^{\frac{1}{2}\pi} (y'-y\cot x)^2\,dx,$$

which makes (7.7.1) intuitive.

A slight modification of (7.7.2) leads to

257. *If $y(0)=y(\pi)=0$ and y' is L^2, then*

$$\int_0^{\pi} y^2\,dx<\int_0^{\pi} y'^2\,dx$$

unless $$y=C\sin x.$$

For $y=o(x^{\frac{1}{2}})$ for small x, and $y=o\{(\pi-x)^{\frac{1}{2}}\}$ for x near π,[a] so that $y^2\cot x$ vanishes at both limits. Hence

$$(7.7.3) \qquad \int_0^{\pi} (y'^2-y^2)\,dx=\int_0^{\pi} (y'-y\cot x)^2\,dx.$$

Another modification of (7.7.2) leads to a more interesting theorem due to Wirtinger[b].

258. *If y has the period 2π, y' is L^2, and*

$$(7.7.4) \qquad \int_0^{2\pi} y\,dx=0,$$

then $$\int_0^{2\pi} y^2\,dx<\int_0^{2\pi} y'^2\,dx$$

unless $$y=A\cos x+B\sin x.$$

We cannot write down at once an identity similar to (7.7.2) or (7.7.3), but with 0, 2π as limits, because $y\cot x$ will usually have infinities in the range of integration. We may however argue as follows[c].

[a] Theorem **222**.

[b] See Blaschke (**1**, 105). The most immediate proof is by an application of Parseval's Theorem to the Fourier developments

$$y\sim\tfrac{1}{2}a_0+\Sigma\,(a_n\cos nx+b_n\sin nx), \quad y'\sim\Sigma\,(nb_n\cos nx-na_n\sin nx)$$

(with $a_0=0$).

[c] The proof which follows was communicated to us by Dr Hans Lewy.

The function $$z(x)=y(x+\pi)-y(x)$$

has opposite signs for the arguments x and $x+\pi$, and therefore vanishes at least once in $(0, \pi)$. We suppose that $z(\alpha)=0$, where $0 \leqq \alpha < \pi$, and write $y(\alpha)=a$. Since y' is L^2, $(y-a)^2 \cot(x-\alpha)$ vanishes for $x=\alpha$ and $x=\alpha+\pi$;[a] and

$$\int_0^{2\pi} [y'^2-(y-a)^2-\{y'-(y-a)\cot(x-\alpha)\}^2]\,dx$$
$$=\Big[(y-a)^2 \cot(x-\alpha)\Big]_0^{2\pi}=0.$$

Hence, using (7.7.4), we obtain

$$\int_0^{2\pi} (y'^2-y^2)\,dx=2\pi a^2+\int_0^{2\pi} \{y'-(y-a)\cot(x-\alpha)\}^2 dx,$$

which is positive unless $a=0$ and

$$y' \equiv y \cot(x-\alpha), \quad y=C \sin(x-\alpha).$$

There is a special interest in Theorem **258** because the proof of the classical isoperimetric property of the circle may be based upon it. We consider a simple closed curve C whose area is A and whose perimeter is L, and take

$$\phi=\frac{2\pi s}{L},$$

where s is the arc of the curve, as parameter, so that

$$x=x(\phi), \quad y=y(\phi) \qquad (0 \leqq \phi \leqq 2\pi).$$

We suppose for simplicity that x' and y' are continuous; the proof is valid for more general x, y. We may also suppose without loss of generality that the centre of gravity of the perimeter lies on the axis of x, so that

$$\int_0^{2\pi} y\,d\phi=0.$$

We have then

$$\left(\frac{dx}{d\phi}\right)^2+\left(\frac{dy}{d\phi}\right)^2=\left\{\left(\frac{dx}{ds}\right)^2+\left(\frac{dy}{ds}\right)^2\right\}\frac{L^2}{4\pi^2}=\frac{L^2}{4\pi^2},$$

[a] Using Theorem **222** as in § 7.4 and above.

$$\frac{L^2}{2\pi}-2A=\int_0^{2\pi}\left\{\left(\frac{dx}{d\phi}\right)^2+\left(\frac{dy}{d\phi}\right)^2\right\}d\phi+2\int_0^{2\pi}y\frac{dx}{d\phi}d\phi\ [a]$$

$$=\int_0^{2\pi}\left(\frac{dx}{d\phi}+y\right)^2d\phi+\int_0^{2\pi}\left\{\left(\frac{dy}{d\phi}\right)^2-y^2\right\}d\phi$$

$$\geqq\int_0^{2\pi}\left\{\left(\frac{dy}{d\phi}\right)^2-y^2\right\}d\phi\geqq 0,$$

by Theorem **258**. There will be inequality unless

$$y=A\cos\phi+B\sin\phi$$

and also $\quad x=-\int y\,d\phi=-A\sin\phi+B\cos\phi+C,$

when the curve is a circle[b].

7.8. An example involving second derivatives. (III) It is known[c] that, if f has a second derivative for $x\geqq 0$, and μ_0, μ_1, μ_2 are the upper bounds of $|f|$, $|f'|$, $|f''|$, then

$$\mu_1^2\leqq 4\mu_0\mu_2.$$

This suggests that there may be a corresponding relation between the integrals

$$J_0=\int_0^\infty |f|^p dx,\quad J_1=\int_0^\infty |f'|^p dx,\quad J_2=\int_0^\infty |f''|^p dx,$$

where $p\geqq 1$. The next theorem settles this question in the case $p=2$.

259. *If y and y'' are L^2 in $(0,\infty)$[d], then*

$$\left(\int_0^\infty y'^2dx\right)^2<4\int_0^\infty y^2dx\int_0^\infty y''^2dx,$$

unless $y=AY(Bx)$, where

$$Y=e^{-\frac{1}{2}x}\sin(x\sin\gamma-\gamma)\quad(\gamma=\tfrac{1}{3}\pi),$$

when there is equality.

[a] Or $-2\int_0^{2\pi}y\frac{dx}{d\phi}d\phi$, according to the sense of the variation of s as we pass in a given sense round the curve.

[b] The proof is in principle that of Hurwitz (**2**), but differs (*a*) in that we do not use the theory of Fourier series and (*b*) in our unsymmetrical treatment of x and y.

[c] Landau (**2**, **3**).

[d] In accordance with the convention of § 7.1, y' is the integral of y'' and y of y'.

If we consider the 'isoperimetrical' problem of the Calculus of Variations defined by

$$\text{`}J_1=\int_0^\infty y'^2dx \textit{ maximum, } J_0=\int_0^\infty y^2dx \textit{ and } J_2=\int_0^\infty y''^2dx \textit{ given'},$$

we have to form Euler's equation for

$$\int_0^\infty (y'^2-\lambda y^2-\mu y''^2)\,dx.$$

It is a linear equation of the form

$$ay''''+by''+cy=0,$$

whose solutions are linear combinations of real or complex exponentials. When we try to choose the parameters in the most advantageous way, we are led to consider the function Y.

It seems difficult to complete the proof on these lines by use of the general theory. We shall deduce Theorem **259** from the simpler theorem which follows.

260. *Under the conditions of Theorem* **259**,

$$J(y)=\int_0^\infty (y^2-y'^2+y''^2)\,dx>0$$

unless $y=AY$, *when there is equality.*

We give several proofs of this theorem to illustrate differences of method. The first two are, as they stand, elementary; the third, of which we give only an outline, makes explicit the variational theory which lies behind the other two. We begin by an observation which is necessary in any case, viz. that J_1 is finite.

To prove this, we have

$$(7.8.1)\qquad \int_0^X y'^2dx=\Big[yy'\Big]_0^X-\int_0^X yy''\,dx.$$

Since J_0 and J_2 are finite, the last integral tends to a finite limit[a] when $X\to\infty$. If J_1 were infinite, yy' and *a fortiori*

$$y^2=2\int yy'\,dx$$

would tend to infinity, which is impossible on account of the convergence of J_0. Hence J_1 is finite, and all three terms in (7.8.1)

[a] By Theorem **181**.

tend to limits. In particular yy' tends to a limit, which can only be 0 (again on account of the convergence of J_0).

(1) Our first proof proceeds rather on the lines of § 7.4. It is easily verified that

$$Y+Y'+Y''=0, \quad Y(0)+Y'(0)=0, \quad Y''(0)=0$$

and $$J(Y)=0.^{a}$$

We now write $$y=z+cY,$$

choosing c so that $z(0)=0$. Then z, z', and z'' are L^2, z and z' are $o(x^{\frac{1}{2}})$ for large x,[b] and $zz'\to 0$.

Now $$J(y)=J(z)+2cK(Y,z)+c^2J(Y),$$

where

$$K=\int_0^\infty (Yz-Y'z'+Y''z'')\,dx$$

$$=-\int_0^\infty (Y'+Y'')z\,dx-\int_0^\infty (Y+Y')z''\,dx-\int_0^\infty Y'z'\,dx$$

$$=\int_0^\infty (Y+Y'+Y'')z'\,dx=0.$$

Hence $J(y)=J(z)$, and it is enough to prove that $J(z)>0$ unless $z=0$. But, since $z(0)=0$ and $zz'\to 0$ when $x\to\infty$,

$$\int_0^\infty z'^2\,dx=-\int_0^\infty zz''\,dx,$$

and so $$J(z)=\int_0^\infty (z^2+zz''+z''^2)\,dx>0$$

unless $z=0$. This proves Theorem **260**.

(2) We may try (following the lines of § 7.5 and § 7.7) to reduce Theorem **260** to dependence upon an identity. For this, we make

$$\{y^2-y'^2+y''^2-(y''+\phi y+\psi y')^2\}\,dx \tag{7.8.2}$$

an exact differential; and the simplest choice of ϕ and ψ is $\phi=\psi=1$, when (7.8.2) reduces to

$$-d\,(y+y')^2.$$

Thus

$$\int_0^X \{y^2-y'^2+y''^2-(y+y'+y'')^2\}\,dx=-\Big[(y+y')^2\Big]_0^X.$$

[a] This requires a little calculation.

[b] By Theorem **223**.

Since J_0, J_1, and J_2 are finite, the left-hand side has a limit when $X \to \infty$.[a] Also $yy' \to 0$, and so $y^2 + y'^2$ tends to a limit, which can only be 0. It follows that

(7.8.3)
$$\int_0^\infty (y^2 - y'^2 + y''^2)\,dx = \{y(0) + y'(0)\}^2 + \int_0^\infty (y + y' + y'')^2\,dx,$$

which is positive unless

$$y'' + y' + y = 0$$

and $$y'(0) + y(0) = 0.$$

These conditions give $y = AY$.

(3) The underlying variational theory is a little more complex than that of § 7.5. If we put

(7.8.4) $$y' = z,$$

(7.8.5) $$J(y) = J(y, z) = \int_0^\infty (y^2 - y'^2 + z'^2)\,dx,$$

and suppose that

(7.8.6) $$y(0) = 1, \quad z(0) = \zeta, \quad y(\infty) = 0, \quad z(\infty) = 0,$$

then the problem is a 'Lagrange problem', viz. that of minimising $J(y, z)$ when subject to (7.8.4) and (7.8.6).

The extremals of the field, in space (x, y, z) are given[b] by

(7.8.7) $$\frac{\partial \Phi}{\partial y} - \frac{d}{dx}\left(\frac{\partial \Phi}{\partial y'}\right) = 0, \quad \frac{\partial \Phi}{\partial z} - \frac{d}{dx}\left(\frac{\partial \Phi}{\partial z'}\right) = 0,$$

where $$\Phi = F - \lambda(y' - z) = y^2 - y'^2 + z'^2 - \lambda(y' - z),$$

and λ is a function of x defined by the equations themselves. In this case the equations (7.8.7) reduce to

(7.8.8) $$2y + \frac{d}{dx}(2y' + \lambda) = 0, \quad \lambda = \frac{d}{dx}(2z'),$$

and from these and (7.8.4) we find

(7.8.9) $$y'''' + y'' + y = 0.$$

[a] The integrals

$$\int_0^\infty yy'\,dx, \quad \int_0^\infty y'y''\,dx, \quad \int_0 yy''\,dx$$

being convergent, by Theorem **181**.

[b] For an account of the general theory see Bliss (**2**).

The most general solution of (7.8.9), for which y and z vanish at infinity, is

(7.8.10) $$y = ae^{-\rho x} + \bar{a}e^{-\bar{\rho}x},$$

where $\rho = e^{\frac{1}{3}\pi i}$ and the bar denotes the conjugate.

The equations (7.8.10) and

(7.8.11) $$z = -a\rho e^{-\rho x} - \bar{a}\bar{\rho}e^{-\bar{\rho}x},$$

(7.8.12) $$\lambda = 2ae^{-\rho x} + 2\bar{a}e^{-\bar{\rho}x},$$

define a two-parameter 'field' of extremals through $(\infty, 0, 0)$. The 'slope functions' p, q, and the 'multiplier' λ of the field are the functions

$$p(x, y, z), \quad q(x, y, z), \quad \lambda(x, y, z),$$

obtained by expressing the slopes of the extremal, and the function λ defined by (7.8.8), in terms of x, y, z. A straightforward calculation gives

(7.8.13) $$p = z, \quad q = -y - z,$$

(7.8.14) $$\lambda = 2y.$$[a]

The analogue of the Hilbert integral is

$$J^* = \int [(\Phi - p\Phi_p - q\Phi_q)\,dx + \Phi_p\,dy + \Phi_q\,dz].$$

This integral has properties corresponding to those of §7.5; it is independent of the path between its end-points, and its value along an extremal is the same as that of

$$J = \int F\,dx.$$

Also, if E is the extremal, and C any other curve, joining the end-points, we have

(7.8.15) $$J_C - J_E = J_C - J_E^* = J_C - J_C^* = \int_C \mathcal{E}\,dx,$$

where $\mathcal{E}$ is the excess-function, defined here by

$$\mathcal{E} = \Phi(x, y, z, y', z', \lambda) - \Phi(x, y, z, p, q, \lambda) - (y' - p)\Phi_p - (z' - q)\Phi_q.$$

[a] p, q, λ are in the first instance functions of x and the parameters a, $\bar{a}$ of the extremal. Here, in particular,

$$y' = -a\rho e^{-\rho x} - \bar{a}\bar{\rho}e^{-\bar{\rho}x} = z, \quad z' = a\rho^2 e^{-\rho x} + \bar{a}\bar{\rho}^2 e^{-\bar{\rho}x} = -y - z,$$

$$\lambda = 2z'' = -2a\rho^3 e^{-\rho x} - 2\bar{a}\bar{\rho}^3 e^{-\bar{\rho}x} = 2y.$$

In this case it will be found that

$$(7.8.16) \qquad \mathcal{E}=(y+z+z')^2-(y'-z)^2$$

and

$$J_E=(1+\zeta)^2.$$

Since $\mathcal{E}$ reduces to $(y+y'+z)^2$ when $y'=z$, we obtain

$$\int_0^\infty (y^2-y'^2+z'^2)\,dx=(1+\zeta)^2+\int_0^\infty (y+y'+z')^2\,dx,$$

which is (7.8.3). We have thus proved Theorem **260** in three different ways.

We have supposed that $y(0)\neq 0$, so that we can take $y(0)=1$. The case in which $y(0)=0$ may be discussed similarly[a].

Our object has been only to illustrate a method and not to discuss a difficult general theory, and we have therefore presented the argument very shortly. The following remarks[b] may help to make the method intelligible.

(*a*) The integral J^* constructed from a two-parameter set of extremals is not necessarily invariant. Here we can verify the invariance of J^* directly; in fact

$$J^*=-\int d(y+z)^2.$$

This invariance ensures that our extremals form a 'field'. The 'reason' is to be found in the fact that they pass through a fixed point, y, z, and λ all vanishing for $x=\infty$.

(*b*) In this case Φ is quadratic, and

$$\mathcal{E}=\tfrac{1}{2}\{(y'-p)^2\Phi_{pp}+2(y'-p)(z'-q)\Phi_{pq}+(z'-q)^2\Phi_{qq}\},$$

which leads immediately to the formula (7.8.16).

(*c*) Suppose that E and C are defined as above, that E_0 is the positive axis of x, and L an arbitrary curve joining $(0, 1, \zeta)$ to the origin. Then

$$\begin{aligned} J_C=J_C-J_{E_0}&=J_C-J^\star_{E_0}=J^\star_L+J_C-J^\star_{L+E_0}\\ &=J^\star_L+J_C-J^\star_C=J^\star_L+\int_0^\infty \mathcal{E}\,dx\\ &=(1+\zeta)^2+\int_0^\infty (y+y'+y'')^2\,dx. \end{aligned}$$

This is (7.8.3); the argument avoids a direct calculation of J_E. Alternatively we may argue

$$J_C-J_E=\int_0^\infty \mathcal{E}\,dx,$$

$$J_E=J^\star_E=J^\star_L+J^\star_{E_0}=(1+\zeta)^2.$$

[a] Compare § 7.6.

[b] For which we are indebted to Prof. Bliss and Mr L. C. Young.

In order to deduce Theorem **259** from Theorem **260**, we apply the last theorem to $y(x/\rho)$ instead of to $y(x)$. We thus obtain

$$\rho^4 J_0 - \rho^2 J_1 + J_2 = \int_0^\infty (\rho^4 y^2 - \rho^2 y'^2 + y''^2)\, dx > 0.$$

Since this is true for all positive ρ, and in particular when $\rho^2 = J_1/2J_0$, it follows that $J_1{}^2 < 4J_0 J_2$ unless $y(x/B) = AY(x)$.

7.9. A simpler problem. It is interesting to observe that the corresponding theorem for the interval $(-\infty, \infty)$ is much more elementary and quite different.

261. *If y and y'' are L^2 in $(-\infty, \infty)$, then*

$$\left(\int_{-\infty}^{\infty} y'^2 dx\right)^2 < \int_{-\infty}^{\infty} y^2 dx \int_{-\infty}^{\infty} y''^2 dx$$

unless $y = 0$. The (unit) constant is the best possible.

In fact (as in the proof of Theorem **260**) $yy' \to 0$ and

$$J_1{}^2 = \left(\int_{-\infty}^{\infty} y'^2 dx\right)^2 = \left(-\int_{-\infty}^{\infty} yy'' dx\right)^2 < \int_{-\infty}^{\infty} y^2 dx \int_{-\infty}^{\infty} y''^2 dx = J_0 J_2.$$

To prove the constant best possible, we take $y = \sin x$ for $|x| \leqq n\pi$, $y = 0$ for $|x| > n\pi$, and round off the angles at $x = \pm n\pi$ so as to make y'' continuous. We can plainly do this with changes in the three integrals which are bounded when $n \to \infty$, and then each of them differs boundedly from $n\pi$, so that

$$J_1{}^2 > (1-\epsilon) J_0 J_2$$

if n is sufficiently large.

MISCELLANEOUS THEOREMS AND EXAMPLES

262. If $y(0) = y(1) = 0$ and y' is L^2, then

$$\int_0^1 \frac{y^2}{x(1-x)} dx < \tfrac{1}{2} \int_0^1 y'^2 dx,$$

unless $y = cx(1-x)$.

[If

$$J(y) = \int_0^1 \left(\tfrac{1}{2} y'^2 - \frac{y^2}{x(1-x)}\right) dx,$$

then (E) is

$$x(1-x)y'' + 2y = 0.$$

and $y = \alpha x(1-x)$ is an extremal satisfying the conditions whatever be α. By varying α we can define a field[a] round any particular extremal. It will be found that in this case $J(Y) = 0$ and

$$p = \frac{1-2x}{x(1-x)} y, \quad W = \frac{1-2x}{2x(1-x)} y^2, \quad \mathcal{E} = \tfrac{1}{2}\left\{y' - \frac{1-2x}{x(1-x)} y\right\}^2.$$

[a] The field differs slightly in character from those described in § 7.5, since each extremal passes through (0, 0) and (1, 0).

The underlying identity is

$$\int_0^1 \left(\tfrac{1}{2}y'^2 - \frac{y^2}{x(1-x)}\right) dx = \tfrac{1}{2}\int_0^1 \left(y' - \frac{1-2x}{x(1-x)}y\right)^2 dx.$$

The theorem should be compared with Theorem **225**. For fuller details, and proofs of both theorems by means of Legendre functions, see Hardy and Littlewood (**10**).]

263. If $J=\int_0^\infty y^2\,dx$, $K=\int_0^\infty y'^2\,dx$, then

$$4JK > \{y(0)\}^4$$

unless $y = ae^{-bx}$.

264. If

(*a*) $$y(-1)=-1,\quad y(1)=1,\quad y'(-1)=y'(1)=0$$

and k is a positive integer, then

$$\int_{-1}^1 (y'')^{2k}\,dx \geqq 2\left(\frac{4k-1}{2k-1}\right)^{2k-1},$$

with inequality unless

$$y = \frac{4k-1}{2k}x - \frac{2k-1}{2k}x^{(4k-1)/(2k-1)}.$$

[This is an example of the theory of § 7.8, simpler than that in the text. In this case

$$\mathcal{E} = z'^{2k} - q^{2k} - 2k(z'-q)q^{2k-1}.]$$

265. If y satisfies (*a*), and has a second derivative y'' for every x of $(-1, 1)$, then $|y''(x)|>2$ for some x.

[This theorem, which is easily proved directly, corresponds to the limiting case $k=\infty$ of Theorem **264**. The extremal curve of Theorem **264** reduces to

$$y = 2x - x^2 \operatorname{sgn} x.$$

For this curve $y' = 2(1-|x|)$, and $y'' = -2\operatorname{sgn} x$ except for $x=0$. There is no second derivative at the origin.]

266. If y is L^2, $z'=y$, and

$$y(0)=y(2\pi)=z(0)=z(2\pi)=0,$$

then

$$\int_0^{2\pi} (y'^2 - y^2)\,dx = \int_0^{2\pi}\left\{y' + \frac{(x\cos x - \sin x)\,y + (1-\cos x)\,z}{2 - 2\cos x - x\sin x}\right\}^2 dx.$$

[This identity, which gives another (though less simple) proof of Theorem **258**, is the result of treating Wirtinger's inequality as a case of Lagrange's problem, on the lines of § 7.8.]

267.

$$\int_0^\infty (y^2 + 2y'^2 + y''^2)\,dx > \tfrac{3}{2}\{y(0)\}^2$$

unless

$$y = Ce^{-x}(x+2).$$

268. If $k \geqq 1$ and y and y'' are L^k in $(-\infty, \infty)$ or $(0, \infty)$, then

$$\left(\int |y'|^k\,dx\right)^2 \leqq K(k)\int |y|^k\,dx \int |y''|^k\,dx,$$

the integrals being taken over the interval in question.[a]

[a] Hardy, Littlewood and Landau (1).

269. If $k > 1$ and y and y'' are L^k and $L^{k'}$ respectively, then

$$\int_{-\infty}^{\infty} y'^2\,dx < \left(\int_{-\infty}^{\infty} |y|^k\,dx\right)^{1/k} \left(\int_{-\infty}^{\infty} |y''|^{k'}\,dx\right)^{1/k'},$$

unless y is null.

270. If y' is L^2, then

(i) $$\int_0^\infty \frac{y^4}{x^3}\,dx < \tfrac{3}{2}\left(\int_0^\infty y'^2\,dx\right)^2$$

unless

(ii) $$y = \frac{x}{ax+b},$$

where a and b are positive, in which case there is equality. More generally, if

(iii) $$l > k > 1, \quad r = \frac{l}{k} - 1,$$

and y' is positive and L^k, then

(iv) $$\int_0^\infty \frac{y^l}{x^{l-r}}\,dx < K\left(\int_0^\infty y'^k\,dx\right)^{l/k},$$

where

(v) $$K = \frac{1}{l-r-1}\left[\frac{r\Gamma(l/r)}{\Gamma(1/r)\,\Gamma\{(l-1)/r\}}\right],$$

unless

(vi) $$y = \frac{x}{(ax^r+b)^{1/r}}.$$

[It is easy to prove an inequality of the type of (i) but with a less favourable constant. Thus, if we denote the integrals in (i) by K and J respectively, we have

$$y^2 = \left(\int_0^x y'\,dt\right)^2 \leqq Jx$$

and so

$$K = \int_0^\infty \frac{y^4}{x^3}\,dx \leqq J\int_0^\infty \frac{y^2}{x^2}\,dx < 4J^2,$$

by Theorem **253**. We know no elementary proof of the full result. For details of the variational proof, which is much more difficult than that of Theorem **260**, we must refer to Bliss (**3**).]

271. If $\alpha > 1$, and $\mathfrak{G}(f)$ is the geometric mean of f over $(0, x)$, then

$$\int_0^\infty x^{\alpha-1}\,\mathfrak{G}^\alpha(f)\,dx < \frac{1}{\alpha-1}\left\{\Gamma\left(\frac{\alpha}{\alpha-1}\right)\right\}^{1-\alpha}\left(\int_0^\infty f\,dx\right)^\alpha,$$

unless $$f \equiv C\exp(-Bx^{\alpha-1}).$$

[See Hardy and Littlewood (**7**). The limiting case $\alpha = 1$ corresponds to Theorem **335**.]

272. In the problem

$$\text{'}\int_{-\infty}^{\infty} y^2\,dx \textit{ maximum}, \int_{-\infty}^{\infty} x^2y^2\,dx \textit{ and } \int_{-\infty}^{\infty} y'^2\,dx \textit{ given'},$$

the Euler equation is of the form

$$y'' + (a+bx^2)\,y = 0,$$

and is soluble by parabolic cylinder functions. It has a solution $y = e^{-cx^2}$ if $b = -a^2 = -(2c)^2$.

[This gives the variational basis for Weyl's inequality (see Theorem **226**).]

CHAPTER VIII

SOME THEOREMS CONCERNING BILINEAR AND MULTILINEAR FORMS

8.1. Introduction. In this chapter we prove a number of general theorems concerning the maxima of bilinear and multilinear forms. In the early part of the chapter we consider forms in n sets of variables, but suppose the variables and coefficients positive. Later, we abandon this restriction, but suppose that $n=2$; and most of the latter part of the chapter is occupied by the proof of an important theorem of M. Riesz concerning bilinear forms with complex variables and coefficients.

POSITIVE MULTILINEAR FORMS

8.2. An inequality for multilinear forms with positive variables and coefficients. We suppose that

$$x_i,\ y_j,\ \dots,\ z_k$$

are n sets of variables, $i, j, \dots, k$ running from $-\infty$ to ∞; and that

$$\Sigma,\ \sum_i,\ \sum_i{}'$$

indicate respectively summation with respect to all suffixes, summation with respect to i only, and summation with respect to $j, \dots, k$ (all suffixes except i). The sum

$$S=\Sigma a_{ij\dots k}x_i y_j \dots z_k$$

is called a multilinear form in the variables $x, y, \dots, z$. When n is 1, 2, or 3, the form is said to be linear, bilinear, or trilinear.

If the series are absolutely convergent, then

$$S=\sum_i x_i \sum_i{}' a_{ij\dots k} y_j \dots z_k=\sum_j y_j \sum_j{}' a_{ij\dots k} x_i \dots z_k=\dots.$$

273. *Suppose that*

$$0<\alpha\leqq 1,\quad 0<\beta\leqq 1,\quad \dots,\quad 0<\gamma\leqq 1 \tag{8.2.1}$$

and

$$0\leqq\frac{\alpha+\beta+\dots+\gamma-1}{n-1}\leqq \operatorname{Min}(\alpha,\beta,\dots,\gamma); \tag{8.2.2}$$

that $\bar{\alpha}, \bar{\beta}, \ldots, \bar{\gamma}$ *are defined by*

$$(8.2.3)\qquad \alpha-\bar{\alpha}=\beta-\bar{\beta}=\ldots=\gamma-\bar{\gamma}=\frac{\alpha+\beta+\ldots+\gamma-1}{n-1}$$

(*so that* $0 \leqq \bar{\alpha} \leqq \alpha, \ldots, 0 \leqq \bar{\gamma} \leqq \gamma$), *and that* $\rho, \sigma, \ldots, \tau$ *are positive and*

$$(8.2.4)\qquad \bar{\alpha}\rho+\bar{\beta}\sigma+\ldots+\bar{\gamma}\tau=1.$$

Suppose further that

$$x_i \geqq 0, \quad y_j \geqq 0, \quad \ldots, \quad z_k \geqq 0, \quad a_{ij\ldots k} \geqq 0,$$

$$(8.2.5)\qquad \sum_i x_i^{1/\alpha} \leqq X, \quad \sum_j y_j^{1/\beta} \leqq Y, \quad \ldots, \quad \sum_k z_k^{1/\gamma} \leqq Z,$$

and

$$(8.2.6)\qquad \sum_i{}' a_{ij\ldots k}^{\rho} = A_i \leqq A, \quad \sum_j{}' a_{ij\ldots k}^{\sigma} = B_j \leqq B, \quad \ldots, \quad \sum_k{}' a_{ij\ldots k}^{\tau} = C_k \leqq C.$$

Then $\quad S=\Sigma a_{ij\ldots k}x_i y_j \ldots z_k \leqq A^{\bar{\alpha}} B^{\bar{\beta}} \ldots C^{\bar{\gamma}} X^{\alpha} Y^{\beta} \ldots Z^{\gamma}.$

We have in fact

$$1-\bar{\alpha}-\bar{\beta}-\ldots-\bar{\gamma}=\frac{\alpha+\beta+\ldots+\gamma-1}{n-1}=\alpha-\bar{\alpha}=\beta-\bar{\beta}=\ldots,$$

and so, by Theorem **11**,[a]

$$\begin{aligned} S &= \Sigma\,(a^{\rho}x^{1/\alpha})^{\bar{\alpha}}(a^{\sigma}y^{1/\beta})^{\bar{\beta}}\ldots(a^{\tau}z^{1/\gamma})^{\bar{\gamma}}(x^{1/\alpha}y^{1/\beta}\ldots z^{1/\gamma})^{1-\bar{\alpha}-\bar{\beta}-\ldots-\bar{\gamma}} \\ &\leqq (\Sigma a^{\rho}x^{1/\alpha})^{\bar{\alpha}}(\Sigma a^{\sigma}y^{1/\beta})^{\bar{\beta}}\ldots(\Sigma a^{\tau}z^{1/\gamma})^{\bar{\gamma}}(\Sigma x^{1/\alpha}y^{1/\beta}\ldots z^{1/\gamma})^{1-\bar{\alpha}-\bar{\beta}-\ldots-\bar{\gamma}} \\ &= (\sum_i x^{1/\alpha}\sum_i{}' a^{\rho})^{\bar{\alpha}}\ldots(\sum_k z^{1/\gamma}\sum_k{}' a^{\tau})^{\bar{\gamma}}(\sum_i x^{1/\alpha}\ldots\sum_k z^{1/\gamma})^{1-\bar{\alpha}-\ldots-\bar{\gamma}} \\ &\leqq (AX)^{\bar{\alpha}}\ldots(CZ)^{\bar{\gamma}}(XY\ldots Z)^{1-\bar{\alpha}-\ldots-\bar{\gamma}} \\ &= A^{\bar{\alpha}}B^{\bar{\beta}}\ldots C^{\bar{\gamma}}X^{\alpha}Y^{\beta}\ldots Z^{\gamma}. \end{aligned}$$

We note some special cases.

(1) If
$$\alpha+\beta+\ldots+\gamma=1,$$
then
$$\bar{\alpha}=\alpha, \quad \bar{\beta}=\beta, \quad \ldots, \quad \bar{\gamma}=\gamma,$$
and the statement of the theorem becomes simpler.

(2) When $n=2$, the second of the conditions (8.2.2) is satisfied automatically. If we write[b]
$$\alpha=\frac{1}{p}, \quad \beta=\frac{1}{q},$$
then
$$\bar{\alpha}=\frac{1}{q'}, \quad \bar{\beta}=\frac{1}{p'}.$$

[a] Extended to infinite series: we shall not usually repeat this remark.

[b] In the preceding chapters the letters p and q have been reserved for the weights of mean values. In this chapter they are not required for this purpose, and we use them as indices.

Exchanging ρ and σ, and A and B, we obtain

274. *If*

$$p \geqq 1, \quad q \geqq 1, \quad \frac{1}{p}+\frac{1}{q} \geqq 1, \quad \rho > 0, \quad \sigma > 0, \quad \frac{\rho}{p'}+\frac{\sigma}{q'} = 1,$$

$$\sum_i a_{ij}^{\rho} \leqq A, \quad \sum_j a_{ij}^{\sigma} \leqq B, \quad \sum_i x_i^p \leqq X, \quad \sum_j y_j^q \leqq Y,$$

then
$$S = \Sigma a_{ij} x_i y_j \leqq A^{1/p'} B^{1/q'} X^{1/p} Y^{1/q}.$$

(3) An interesting and still more special case is that in which $\rho = \sigma = 1$, $q = p'$.

275. *If*
$$p > 1, \quad q > 1, \quad \frac{1}{p}+\frac{1}{q} = 1,$$

$$\sum_i a_{ij} \leqq A, \quad \sum_j a_{ij} \leqq B, \quad \sum_i x_i^p \leqq X, \quad \sum_j y_j^q \leqq Y,$$

then
$$S = \Sigma\Sigma a_{ij} x_i y_j \leqq (BX)^{1/p} (AY)^{1/q}.$$

For the case $p = q = 2$, see Frobenius (**1**) and Schur (**1**).

8.3. A theorem of W. H. Young. Another specialisation of Theorem **274** leads to an inequality of W. H. Young which is very important in the theory of Fourier series.

Suppose that $\sigma = \rho > 1$, so that

$$\frac{1}{p'}+\frac{1}{q'} = \frac{1}{\rho} < 1, \quad \frac{1}{p}+\frac{1}{q} > 1;$$

and take $a_{ij} = a_{i+j}$. Then

$$\sum_i a_{ij}^{\rho} = \sum_j a_{ij}^{\rho} = \Sigma a_n^{\rho} = A,$$

say, for every j and i respectively. Hence, if we write

$$z_n = \sum_{i+j=n} x_i y_j, \tag{8.3.1}$$

we have, by Theorem **274**,

$$\Sigma a_n z_n = \Sigma a_{i+j} x_i y_j \leqq A^{1/\rho} X^{1/p} Y^{1/q}. \tag{8.3.2}$$

Since (8.3.2) is true for all a_n for which $\Sigma a_n^{\rho} = A$, it follows, by Theorem **15**, that
$$\Sigma z_n^{\rho'} \leqq X^{\rho'/p} Y^{\rho'/q}.$$

This must be replaced, when $\rho = 1$, by $z_n \leqq X^{1/p} Y^{1/q}$.

We have thus proved (apart from the specification of the cases of equality) the following theorem of Young[a].

[a] Young (**3**. **4**, **6**). Young does not consider the question of equality.

276. *If* $p>1, \quad q>1, \quad \frac{1}{p}+\frac{1}{q}>1,$

and z_n *is defined by* (8.3.1), *then*

$$\Sigma z_n^{\frac{pq}{p+q-pq}} \leqq (\Sigma x_i^p)^{\frac{q}{p+q-pq}} (\Sigma y_j^q)^{\frac{p}{p+q-pq}}. \tag{8.3.3}$$

Equality can occur only if all the x, or all the y, or all the x but one and all the y but one, are zero.

We add a more direct proof which enables us to settle the question of equality. If we write $1/p=1-\lambda$, $1/q=1-\mu$, then $\lambda>0$, $\mu>0$, $\lambda+\mu<1$, and we can enunciate the theorem in the following form.

277. *If* $\lambda>0, \quad \mu>0, \quad \lambda+\mu<1,$

z is defined by (8.3.1), *and* $\mathfrak{S}_r(x)$ *as in* §2.10, *then*

$$\mathfrak{S}_{1/(1-\lambda-\mu)}(z) \leqq \mathfrak{S}_{1/(1-\lambda)}(x)\,\mathfrak{S}_{1/(1-\mu)}(y), \tag{8.3.4}$$

with equality only in the cases specified in Theorem **276.**

Let $\nu=1-\lambda-\mu$. It follows from Theorem **11** that

$$(\Sigma uv)^{\frac{1}{\nu}} = (\Sigma u^{\frac{\lambda}{\lambda+\nu}} v^{\frac{\mu}{\mu+\nu}} u^{\frac{\nu}{\lambda+\nu}} v^{\frac{\nu}{\mu+\nu}})^{\frac{1}{\nu}} \leqq (\Sigma u^{\frac{1}{\lambda+\nu}})^{\frac{\lambda}{\nu}} (\Sigma v^{\frac{1}{\mu+\nu}})^{\frac{\mu}{\nu}} \Sigma u^{\frac{1}{\lambda+\nu}} v^{\frac{1}{\mu+\nu}}.$$

Applying this inequality, with

$$u=y_{n-i}=y_j, \quad v=x_i,$$

to (8.3.1), we obtain

$$z_n^{\frac{1}{\nu}} \leqq (\sum_i x_i^{\frac{1}{\mu+\nu}})^{\frac{\mu}{\nu}} (\sum_j y_j^{\frac{1}{\lambda+\nu}})^{\frac{\lambda}{\nu}} \sum_{i+j=n} x_i^{\frac{1}{\mu+\nu}} y_j^{\frac{1}{\lambda+\nu}}, \tag{8.3.5}$$

$$z_n^{\frac{1}{\nu}} \leqq \mathfrak{S}_{\frac{1}{\mu+\nu}}^{\frac{\mu}{\nu(\mu+\nu)}}(x)\, \mathfrak{S}_{\frac{1}{\lambda+\nu}}^{\frac{\lambda}{\nu(\lambda+\nu)}}(y) \sum_{i+j=n} x_i^{\frac{1}{\mu+\nu}} y_j^{\frac{1}{\lambda+\nu}}. \tag{8.3.6}$$

Hence

$$\sum_n z_n^{\frac{1}{\nu}} \leqq \mathfrak{S}_{\frac{1}{\mu+\nu}}^{\frac{\mu}{\nu(\mu+\nu)}}(x)\, \mathfrak{S}_{\frac{1}{\lambda+\nu}}^{\frac{\lambda}{\nu(\lambda+\nu)}}(y) \sum_n \sum_{i+j=n} x_i^{\frac{1}{\mu+\nu}} y_j^{\frac{1}{\lambda+\nu}}.$$

Since the double sum here is equal to

$$\sum_i \sum_j x_i^{\frac{1}{\mu+\nu}} y_j^{\frac{1}{\lambda+\nu}} = \mathfrak{S}_{\frac{1}{\mu+\nu}}^{\frac{1}{\mu+\nu}}(x)\, \mathfrak{S}_{\frac{1}{\lambda+\nu}}^{\frac{1}{\lambda+\nu}}(y),$$

we obtain
$$\sum_n z_n^{\frac{1}{\nu}} \leqq \mathfrak{S}_{\frac{1}{\mu+\nu}}^{\frac{1}{\nu}}(x)\, \mathfrak{S}_{\frac{1}{\lambda+\nu}}^{\frac{1}{\nu}}(y),$$

and this is (8.3.4).

We can dispose of the question of equality as follows.

If not all x, and not all y, are zero, there are n such that $x_i y_j > 0$ for some i, j for which $i+j=n$. We call the lattice point corresponding to such a pair i, j a point P_n. Then (8.3.5) is true, for such an n, if, in the first two sums on the right, we limit ourselves to values of i and j corresponding to points P_n. If there is equality in (8.3.5), then, for such i, j the ratios

$$x_i^{\frac{1}{\mu+\nu}} : y_j^{\frac{1}{\lambda+\nu}} : x_i^{\frac{1}{\mu+\nu}} y_j^{\frac{1}{\lambda+\nu}}$$

do not depend on i and j, and the corresponding x_i, and the corresponding y_j, are all equal. It follows that there can, for any n, be only a finite number of points P_n.

Suppose that all these conditions are satisfied for a certain n. Then equality will still be excluded in the next inequality (8.3.6), unless the i and j corresponding to the P_n exhaust *all* i and j for which $x_i > 0$ and $y_j > 0$. It follows that the total number of positive x_i and y_j is finite. There is therefore a single point for which $x_i y_j > 0$ and $n = i+j$ is a minimum. For this n there is a *unique* P_n and, if there is equality in (8.3.6) for this n, then $x_i = 0$ and $y_j = 0$ except for the corresponding pair i, j.

8.4. Generalisations and analogues. Theorems **276** and **277** have many interesting specialisations, generalisations, and analogues. We state a number of these without proof.

278. *If* $\lambda > 0, \mu > 0, \ldots, \nu > 0, \lambda + \mu + \ldots + \nu < 1$, *and*

$$w_n = \sum_{i+j+\ldots+k=n} x_i y_j \ldots z_k,$$

then

$$\mathfrak{S}_{\frac{1}{1-\lambda-\mu-\ldots-\nu}}(w) \leqq \mathfrak{S}_{\frac{1}{1-\lambda}}(x)\,\mathfrak{S}_{\frac{1}{1-\mu}}(y) \ldots \mathfrak{S}_{\frac{1}{1-\nu}}(z),$$

unless all numbers of one set, or all but one of every set, are zero.

279. *If*
$$c_n = \sum_{i_1+i_2+\ldots+i_k=n} a_{i_1} a_{i_2} \ldots a_{i_k},$$

then
$$\Sigma c_n{}^2 \leqq \left(\Sigma a_n^{\frac{2k}{2k-1}}\right)^{2k-1},$$

unless all a but one are zero.

Theorems **277–279**, being 'homogeneous in Σ' in the sense of § 1.4, have integral analogues.

280. *If* $\lambda>0,\ \mu>0,\ \lambda+\mu<1$,

$$\mathfrak{J}_r(f)=\left(\int_{-\infty}^{\infty} f^r\,dx\right)^{1/r}$$

and
$$h(x)=\int_{-\infty}^{\infty} f(t)\,g(x-t)\,dt,$$

then
$$\mathfrak{J}_{\frac{1}{1-\lambda-\mu}}(h)<\mathfrak{J}_{\frac{1}{1-\lambda}}(f)\,\mathfrak{J}_{\frac{1}{1-\mu}}(g),$$

unless f or g is null.

281. *If* $\lambda>0,\ \mu>0,\ \lambda+\mu<1$,

$$\mathfrak{J}_r(f)=\int_0^{\infty} f^r\,dx,$$

and
$$h(x)=\int_0^{x} f(t)\,g(x-t)\,dt,$$

then
$$\mathfrak{J}_{\frac{1}{1-\lambda-\mu}}(h)<\mathfrak{J}_{\frac{1}{1-\lambda}}(f)\,\mathfrak{J}_{\frac{1}{1-\mu}}(g),$$

unless f or g is null.

282. *If k is an integer and*

$$\phi(x)=\int_{-\infty}^{\infty}\cdots\int_{-\infty}^{\infty} f(x_1)f(x_2)\ldots f(x_{k-1})$$
$$\times f(x-x_1-\ldots-x_{k-1})\,dx_1\,dx_2\ldots dx_{k-1},$$

then
$$\int_{-\infty}^{\infty}\phi^2(x)\,dx\leqq\left(\int_{-\infty}^{\infty} f^{\frac{2k}{2k-1}}(x)\,dx\right)^{2k-1}.$$

283. *If*

$$\phi(x)=\int f(x_1)\ldots f(x_{k-1})f(x-x_1-\ldots-x_{k-1})\,dx_1\ldots dx_{k-1},$$

the integration being defined by

$$x_i\geqq 0,\quad \Sigma x_i\leqq x,$$

then
$$\int_0^{\infty}\phi^2(x)\,dx\leqq\left(\int_0^{\infty} f^{\frac{2k}{2k-1}}(x)\,dx\right)^{2k-1}.$$

284. *If $f(x)$ has the period 2π, and*

$$\phi(x)=\frac{1}{(2\pi)^{k-1}}\int_{-\pi}^{\pi}\cdots\int_{-\pi}^{\pi}f(x_1)f(x_2)\ldots f(x_{k-1}) \times f(x-x_1-\ldots-x_{k-1})\,dx_1dx_2\ldots dx_{k-1},$$

then
$$\frac{1}{2\pi}\int_{-\pi}^{\pi}\{\phi(x)\}^2dx \leqq \left(\frac{1}{2\pi}\int_{-\pi}^{\pi}f^{\frac{2k}{2k-1}}(x)\,dx\right)^{2k-1}.$$

8.5. Applications to Fourier series. Theorems **279** and **284** have important applications in the theory of Fourier series. Here we are concerned with functions and coefficients which are not positive, but the theorems which we have proved are sufficient for the applications.

Suppose first that $f(x)$ and $g(x)$ are complex functions of L^2, and that

$$\sum_{-\infty}^{\infty}a_n e^{nix}, \quad \sum_{-\infty}^{\infty}b_n e^{nix}$$

are their complex Fourier series. Then it is well known that

$$\Sigma a_n\bar{b}_n=\frac{1}{2\pi}\int_{-\pi}^{\pi}f(x)\overline{g(x)}\,dx \tag{8.5.1}$$

(the bar denoting the conjugate). In particular, if $f(x)$ is L^2,

$$\Sigma\,|\,a_n\,|^2=\frac{1}{2\pi}\int_{-\pi}^{\pi}|\,f\,|^2\,dx. \tag{8.5.2}$$

Conversely, if $\Sigma\,|\,a_n\,|^2$ is convergent, there is an $f(x)$ of L^2 which has the a_n as its Fourier constants and satisfies (8.5.2).

These theorems ('Parseval's Theorem' and 'the Riesz-Fischer Theorem') were generalised by Young and Hausdorff. We write

$$\mathfrak{S}_p(a)=(\Sigma\,|\,a_n\,|^p)^{1/p}, \quad \mathfrak{J}_p(f)=\left(\frac{1}{2\pi}\int_{-\pi}^{\pi}|\,f(x)\,|^p\,dx\right)^{1/p}; \tag{8.5.3}$$

so that $\mathfrak{S}_p(a)$ is $\mathfrak{S}_p(|\,a\,|)$ as defined in § 2.10, and (8.5.2) may be written

$$\mathfrak{S}_2(a)=\mathfrak{J}_2(f). \tag{8.5.4}$$

Young and Hausdorff proved that, if

$$1<p\leqq 2, \tag{8.5.5}$$

then

$$\mathfrak{J}_{p'}(f)\leqq\mathfrak{S}_p(a) \tag{8.5.6}$$

and

$$\mathfrak{S}_{p'}(a)\leqq\mathfrak{J}_p(f). \tag{8.5.7}$$

The limitation on p is essential. The theorems were proved first by Young (**3**, **4**, **6**) for a special sequence of values of p and p', viz.

$$p=\frac{2k}{2k-1}, \quad p'=2k \quad (k=1,2,3,\ldots), \tag{8.5.8}$$

and then generally by Hausdorff (**2**).

We confine ourselves here to the case (8.5.8), considered by Young. In this case (8.5.6) and (8.5.7) are corollaries of Theorems **279** and **284** respectively. For example, the c_n of Theorem **279**[a] is the Fourier constant of $\psi = f^k$, and so

$$\mathfrak{J}_{2k}^{2k}(f) = \frac{1}{2\pi}\int_{-\pi}^{\pi} |f|^{2k}\,dx = \frac{1}{2\pi}\int_{-\pi}^{\pi} |\psi|^2\,dx$$
$$= \Sigma\,|c_n|^2 \leqq \mathfrak{S}_{2k/(2k-1)}^{2k}(a),$$

which is (8.5.6). Similarly $a_n{}^k$ is the Fourier constant of the $\phi(x)$ of Theorem **284**, and so (8.5.7) may be deduced from Theorem **284**.

The proof of (8.5.6) and (8.5.7) for general p is decidedly more difficult: see § 8.17.

It is interesting to observe (as another application of Hölder's inequality) how (8.5.7) may be *deduced* from (8.5.6). Write

$$b_n = |a_n|^{p'-1}\operatorname{sgn} a_n = |a_n|^{p'}/\bar{a}_n$$

if $a_n \neq 0$ and $|n| \leqq N$, and $b_n = 0$ otherwise; and let

$$g(x) = \Sigma b_n e^{nix}.$$

Then
$$\sum_{-N}^{N} |a_n|^{p'} = \sum_{-N}^{N} a_n \bar{b}_n = \frac{1}{2\pi}\int_{-\pi}^{\pi} f(x)\overline{g(x)}\,dx,$$

since $\bar{g}$ is a trigonometrical polynomial. Hence, using Hölder's inequality and (8.5.6), we obtain

$$\sum_{-N}^{N} |a_n|^{p'} \leqq \mathfrak{J}_p(f)\,\mathfrak{J}_{p'}(g) \leqq \mathfrak{J}_p(f)\,\mathfrak{S}_p(b)$$
$$= \mathfrak{J}_p(f)\left(\sum_{-N}^{N} |a_n|^{(p'-1)p}\right)^{1/p} = \mathfrak{J}_p(f)\left(\sum_{-N}^{N} |a_n|^{p'}\right)^{1/p}.$$

Transposing the last factor, and then making N tend to infinity, we obtain (8.5.7).

8.6. The convexity theorem for positive multilinear forms.

In this section we prove a simple but important property of multilinear forms with positive variables and coefficients. The theorem which we prove is a mere corollary of Hölder's inequality, but it is useful, and will serve as an introduction to the deeper theorem of § 8.13.

285.[b] *Suppose that $a \geqq 0$, $x \geqq 0$, ..., $z \geqq 0$, and that*

$$M_{\alpha,\beta,\ldots,\gamma}$$

is the upper bound of

$$S = \Sigma a_{ij\ldots k} x_i y_j \ldots z_k,$$

for all $x, y, \ldots, z$ for which

$$\Sigma x^{1/\alpha} \leqq 1, \quad \Sigma y^{1/\beta} \leqq 1, \quad \ldots, \quad \Sigma z^{1/\gamma} \leqq 1;$$

[a] Now, of course, formed from complex a.

[b] M. Riesz (**1**): Riesz has $n = 2$.

then $\log M_{\alpha,\beta,\ldots,\gamma}$ *is a convex function of* $\alpha, \beta, \ldots, \gamma$ *in the region* $\alpha > 0, \beta > 0, \ldots, \gamma > 0$.

By a convex function of n variables $\alpha, \beta, \ldots, \gamma$ we mean (§3.12) a function convex along any straight line in the space of $\alpha, \beta, \ldots, \gamma$.

We have to verify that if

$$t_1 \geqq 0, \quad t_2 \geqq 0, \quad t_1 + t_2 = 1,$$

$$\alpha = t_1\alpha_1 + t_2\alpha_2, \quad \beta = t_1\beta_1 + t_2\beta_2, \quad \ldots, \quad \gamma = t_1\gamma_1 + t_2\gamma_2,$$

then

$$M_{\alpha,\beta,\ldots,\gamma} \leqq M^{t_1}_{\alpha_1,\beta_1,\ldots,\gamma_1} M^{t_2}_{\alpha_2,\beta_2,\ldots,\gamma_2}. \tag{8.6.1}$$

Now

$$S = \Sigma axy\ldots z = \Sigma\,(ax^{\alpha_1/\alpha}y^{\beta_1/\beta}\ldots z^{\gamma_1/\gamma})^{t_1}(ax^{\alpha_2/\alpha}y^{\beta_2/\beta}\ldots z^{\gamma_2/\gamma})^{t_2}$$
$$\leqq (\Sigma ax^{\alpha_1/\alpha}y^{\beta_1/\beta}\ldots z^{\gamma_1/\gamma})^{t_1}(\Sigma ax^{\alpha_2/\alpha}y^{\beta_2/\beta}\ldots z^{\gamma_2/\gamma})^{t_2}.$$

Since $\qquad \Sigma\,(x^{\alpha_1/\alpha})^{1/\alpha_1} = \Sigma x^{1/\alpha} \leqq 1, \ldots,$

the first sum on the right does not exceed $M_{\alpha_1,\beta_1,\ldots,\gamma_1}$; and similarly the second does not exceed $M_{\alpha_2,\beta_2,\ldots,\gamma_2}$. This proves (8.6.1).

The theorem may be extended to the closed region $\alpha \geqq 0$, $\beta \geqq 0, \ldots$, if we replace the conditions $\Sigma x^{1/\alpha} \leqq 1$, $\Sigma y^{1/\beta} \leqq 1, \ldots$ by $x \leqq 1$, $y \leqq 1, \ldots$ when $\alpha, \beta, \ldots$ are zero.

Suppose for example that $n = 2$ and

$$\sum_i a_{ij} \leqq A, \quad \sum_j a_{ij} \leqq B.$$

Then $M_{0,1} \leqq A$ and $M_{1,0} \leqq B$, so that $M_{\alpha,1-\alpha} \leqq B^\alpha A^{1-\alpha}$ for $0 < \alpha < 1$. If $p > 1$, $q = p'$, we may take $\alpha = 1/p$, $1 - \alpha = 1/q$; and then we obtain

$$M_{1/p,1/q} \leqq M^{1/p}_{1,0} M^{1/q}_{0,1} \leqq B^{1/p} A^{1/q},$$

which is equivalent to the result of Theorem **275**.

GENERALITIES CONCERNING BILINEAR FORMS

8.7. General bilinear forms. So far we have been occupied with 'positive' multilinear forms, i.e. forms whose variables and coefficients are non-negative. The most important multilinear forms are bilinear, and the remainder of this chapter, and most of the next, is concerned, from one point of view or another, with bilinear forms, which will not generally be positive.

We shall denote the form whose coefficients are a_{ij} by A, and similarly with other letters. We change the convention of § 8.1 concerning the range of the suffixes; until the end of § 8.12, $i, j, \ldots$ run from 1 to ∞. We write

$$\mathfrak{S}_r(x) = \mathfrak{S}_r(|x|) = \left(\sum_1^\infty |x_i|^r\right)^{1/r}.$$

We also write

$$\sum_i a_{ij} x_i = X_j, \quad \sum_j a_{ij} y_j = Y_i. \tag{8.7.1}$$

When the form is positive

$$A = \sum_i \sum_j a_{ij} x_i y_j = \sum_i x_i Y_i = \sum_j y_j X_j, \tag{8.7.2}$$

the convergence of any one of these series involving the convergence of the others and the equality of the three. The equations (8.7.2) are true also for complex a, x, y when the form is finite.

We shall make repeated use of the following general theorems[a].

286. *Suppose that*

$$p > 1, \quad q > 1, \quad \frac{1}{p} + \frac{1}{p'} = 1, \quad \frac{1}{q} + \frac{1}{q'} = 1$$

(so that $p' > 1$, $q' > 1$*) and that* a, x, y *are real and non-negative. Then the three assertions*

$$|A(x,y)| \leqq K\mathfrak{S}_p(x)\mathfrak{S}_q(y) \tag{8.7.3}$$

for all x, y;[b]

$$\mathfrak{S}_{q'}(X) \leqq K\mathfrak{S}_p(x) \tag{8.7.4}$$

for all x;

$$\mathfrak{S}_{p'}(Y) \leqq K\mathfrak{S}_q(y) \tag{8.7.5}$$

for all y; *are equivalent.*

287. *The three assertions*

(i) *there is inequality in* (8.7.3), *unless either* (x_i) *or* (y_j) *is null;*

(ii) *there is inequality in* (8.7.4), *unless* (x_i) *is null;*

(iii) *there is inequality in* (8.7.5), *unless* (y_j) *is null;*

are also equivalent.

[a] For the case $p = q = 2$, see Hellinger and Toeplitz (**1**); for $q = p'$, F. Riesz (**1**). The substance of the general theorems is to be found in M. Riesz (**1**). The important cases are naturally those in which K has its least possible value, i.e. is the bound of A (§8.8).

[b] Here $A \geqq 0$; but we write $|A|$ for A in view of Theorem **288**.

288. *When the forms are finite, Theorems* **286** *and* **287** *are true also for forms with complex variables and coefficients.*

Theorem **286** is a simple corollary of Theorems **13** and **15**. It follows from (8.7.2) and Theorem **13** that

$$A = \sum_j y_j X_j \leqq \mathfrak{S}_q(y)\mathfrak{S}_{q'}(X), \tag{8.7.6}$$

so that (8.7.4) is a sufficient condition for the truth of (8.7.3); and Theorem **15** shows that it is also a necessary condition. Hence (8.7.3) and (8.7.4) are equivalent, and similarly (8.7.3) and (8.7.5) are equivalent.

There will be inequality in (8.7.3) unless (y_j) is null or there in equality in (8.7.4). Hence the second assertion of Theorem **287** implies the first. If the x_i, and so the X_j, are given, we can, by Theorem **13**, choose a non-null (y_j) so that there shall be equality in (8.7.6). Hence, if there is equality in (8.7.4) for a non-null (x_i), then there is equality in (8.7.3) for a non-null (x_i) and (y_j). Hence the two assertions are equivalent, and similarly the first and third are equivalent. This proves Theorem **287**.

Finally, the whole argument applies equally to complex a, x, y when the forms are finite[a]. We have only to use Theorem **14** instead of Theorem **13**.

The most important case is that in which $q = p'$, $q' = p$, when (8.7.3), (8.7.4), and (8.7.5) become

$$|A| \leqq K\mathfrak{S}_p(x)\mathfrak{S}_{p'}(y), \quad \mathfrak{S}_p(X) \leqq K\mathfrak{S}_p(x), \quad \mathfrak{S}_{p'}(Y) \leqq K\mathfrak{S}_{p'}(y).$$

BOUNDED BILINEAR FORMS

8.8. Definition of a bounded bilinear form. Throughout the rest of this chapter we suppose, except when there is an explicit statement to the contrary, that the variables and coefficients in the forms considered are arbitrary real or complex numbers.

We describe the aggregate of all sets (x) or $x_1, x_2, \ldots$, real or complex, for which

$$\mathfrak{S}_p(x) = \mathfrak{S}_p(|x|) = (\Sigma |x_i|^p)^{1/p} \leqq 1 \tag{8.8.1}$$

as *space* $[p]$. Here p is any positive number; but usually $p > 1$.

[a] Otherwise there are difficulties concerning the mode of summation of A.

Similarly we describe the aggregate of sets (x, y) for which

(8.8.2) $$\mathfrak{S}_p(x) \leqq 1, \quad \mathfrak{S}_q(y) \leqq 1$$

as *space* $[p, q]$. The most important case is that in which $p = q = 2$. The importance of this space was first recognised by Hilbert, and we may describe it shortly as *Hilbert space*.

In our general definition p or q may be ∞, if we interpret $\mathfrak{S}_\infty(x)$ as $\text{Max}\,|x|$. Thus space $[\infty, \infty]$ is the aggregate of sets (x, y) for which $|x| \leqq 1$, $|y| \leqq 1$.

A bilinear form

(8.8.3) $$A = A(x, y) = \Sigma\Sigma a_{ij} x_i y_j$$

is said to be *bounded in* $[p, q]$ if

(8.8.4) $$|A_n(x, y)| = \left| \sum_{i=1}^{n} \sum_{j=1}^{n} a_{ij} x_i y_j \right| \leqq M,$$

where M is independent of the x and y, and of n, for all points of $[p, q]$. We call A_n a *section* (*Abschnitt*) of A: a form is bounded if its sections are bounded.

It is plain that (8.8.4) will hold for all points of $[p, q]$ if it holds whenever

(8.8.5) $$\mathfrak{S}_p(x) = 1, \quad \mathfrak{S}_q(y) = 1.$$

In this case (8.8.4) may be written

(8.8.6) $$|A_n(x, y)| \leqq M \mathfrak{S}_p(x) \mathfrak{S}_q(y),$$

and here both sides are homogeneous of degree 1 in x and in y, so that the conditions (8.8.5) are immaterial. We might therefore have taken (8.8.6), with unrestricted x, y, in our definition of a bounded form.

So far M has been any number for which (8.8.4) or (8.8.6) is true; if so, we say that A is bounded by M, or M is *a bound* of A. It is natural to take M to be the smallest such bound, and then we say that M is *the bound* of A.[a]

[a] If M_n is the maximum of A_n, under the conditions

$$\sum_1^n |x_i|^p \leqq 1, \quad \sum_1^n |y_j|^q \leqq 1,$$

then plainly $M_n \leqq M_{n+1}$ and M_n is bounded in n. Hence

$$M = \lim_{n \to \infty} M_n$$

exists, and is the smallest bound of A.

An important special case is that in which $p=q$ and

$$a_{ij}=a_{ji},$$

when A is said to be *symmetrical*. In this case *a necessary and sufficient condition that A should be bounded is that the quadratic form*

$$A(x,x)=\Sigma\Sigma a_{ij}x_ix_j$$

should be bounded. When we say that $A(x, x)$ is bounded, we mean, naturally, that $A(x, y)$ is bounded when the x and y are the same, i.e. that

$$|A_n(x,x)|\leqq M$$

for all x for which $\mathfrak{S}_p(x)=1$.

In the first place, it is obvious that the condition is necessary, and that the bound of $A(x, x)$ does not exceed that of $A(x, y)$. That the condition is sufficient follows from the identity

$$(8.8.7)\quad A_n(x,y)=\tfrac{1}{4}A_n(x+y,x+y)-\tfrac{1}{4}A_n(x-y,x-y).$$

When $p=2$ we can go a little further; *the bounds of $A(x, x)$ and $A(x, y)$ are the same*. In fact, if M is the bound of $A(x, x)$, then (8.8.7) gives

$$|A_n(x,y)|\leqq\tfrac{1}{4}M\Sigma(|x+y|^2+|x-y|^2)=\tfrac{1}{2}M\Sigma(|x|^2+|y|^2)\leqq M.$$

It is evident that, when the coefficients a are positive, A is bounded if it is bounded for non-negative x and y, and that its bound may be defined with reference only to such x and y. If

$$A^*=\Sigma\Sigma\,|a_{ij}|\,x_iy_j$$

is bounded, A is said to be *absolutely bounded*.

8.9. Some properties of bounded forms in $[p, q]$. The theory of bounded forms is very important, but we cannot develop it systematically here. We prove enough to enable us to give an account of some special forms in which we shall be interested in the sequel.

We take $p>1$, $q>1$, and, as usual,

$$p'=\frac{p}{p-1},\quad q'=\frac{q}{q-1}.$$

289. *If A has the bound M in $[p, q]$, then*

$$\sum_i |a_{ij}|^{p'} \leqq M^{p'}, \quad \sum_j |a_{ij}|^{q'} \leqq M^{q'} \tag{8.9.1}$$

for every j and i respectively.

Take all the x zero except x_I, $x_I = 1$, and all the y zero except $y_1, y_2, \ldots, y_J$. By (8.8.6),

$$\left|\sum_1^J a_{Ij} y_j\right| \leqq M \left(\sum_1^J |y_j|^q\right)^{1/q}.$$

Since this is true for all y_j and all J, it follows, by Theorem **15**, that

$$\sum_j |a_{Ij}|^{q'} \leqq M^{q'}.$$

This is the second of (8.9.1), and the first is proved similarly.

Thus a *necessary* condition for boundedness in $[2, 2]$ is that

$$\sum_i |a_{ij}|^2 < \infty, \quad \sum_j |a_{ij}|^2 < \infty$$

for all j and i. The condition

$$\Sigma\Sigma |a_{ij}|^2 < \infty$$

is *sufficient*, since then

$$|A|^2 \leqq \sum_{i,j} |a_{ij}|^2 \sum_{i,j} |x_i^2 y_j^2| = \sum_{i,j} |a_{ij}|^2 \sum_i |x_i|^2 \sum_j |y_j|^2;$$

but this condition is by no means necessary, even when the coefficients are positive. Thus, as we shall see in § 8.12,

$$\Sigma\Sigma \frac{x_i y_j}{i+j}$$

is bounded.

290. *Any row or column of a bounded form is absolutely convergent.*

For

$$\sum_i |a_{ij} x_i y_j| \leqq |y_j| \left(\sum_i |x_i|^p\right)^{1/p} \left(\sum_i |a_{ij}|^{p'}\right)^{1/p'} \leqq M |y_j| \mathfrak{S}_p(x),$$

by Theorem **289**.

It is plain that, when $a_{ij} \geqq 0$, a necessary condition that A should be bounded is that

$$\Sigma\Sigma a_{ij} x_i y_j \tag{8.9.2}$$

should be convergent for all positive x and y in $[p, q]$. It is natural to ask whether this is also true for bounded forms with arbitrary real or complex coefficients, i.e. whether, when A is bounded, the series (8.9.2) is necessarily convergent (for the x and y of $[p, q]$) in any of the recognised senses. The answer is affirmative: if A is bounded, the series (8.9.2) is convergent (indeed uniformly) in the three standard senses, as a double series in Pringsheim's sense or as a repeated sum by rows or columns. But the idea of the convergence of the double series is irrelevant to our present purpose (and is not very important in the general theory), and we shall not prove these theorems. See Hellinger and Toeplitz (**1**) for the case [2, 2].

8.10. The Faltung of two forms in $[p, p']$. We suppose now that $q=p'$. If A and B are bounded in $[p, p']$, with bounds M and N, then, by Theorem **289**,

$$\sum_k |a_{ik}|^p \leqq M^p, \quad \sum_k |b_{kj}|^{p'} \leqq N^{p'},$$

and therefore, by Theorem **13**, the series

$$f_{ij} = \sum_k a_{ik} b_{kj} \tag{8.10.1}$$

is absolutely convergent. We call

$$F = F(A, B) = \sum\sum f_{ij} x_i y_j$$

the *Faltung* of A and B. The order of A and B is relevant, $F(A, B)$ and $F(B, A)$ being usually different forms.

291. *If A and B have the bounds M and N in $[p, p']$, then F is bounded in $[p, p']$, and its bound does not exceed MN.*

Suppose that $m \geqq n$ and that $x_i = 0$ for $i > n$. Then, since A is bounded by M in $[p, p']$, we have

$$|A_m| = \left| \sum_{k=1}^m y_k \sum_{i=1}^n a_{ik} x_i \right| \leqq M$$

for all x and y for which $\mathfrak{S}_p(x) \leqq 1$ and $\mathfrak{S}_{p'}(y) \leqq 1$. Hence, by Theorem **15**,

$$\sum_{k=1}^m \left| \sum_{i=1}^n a_{ik} x_i \right|^p \leqq M^p$$

for $m \geqq n$; and therefore

$$\sum_k \left| \sum_{i=1}^n a_{ik} x_i \right|^p \leqq M^p. \tag{8.10.2}$$

Similarly

$$\sum_k \left| \sum_{j=1}^n b_{kj} y_j \right|^{p'} \leqq N^{p'}. \tag{8.10.3}$$

But

(8.10.4)

$$\sum_{i=1}^n \sum_{j=1}^n f_{ij} x_i y_j = \sum_{i=1}^n \sum_{j=1}^n x_i y_j \sum_k a_{ik} b_{kj} = \sum_k \left(\sum_{i=1}^n a_{ik} x_i \right) \left(\sum_{j=1}^n b_{kj} y_j \right).$$

From (8.10.2), (8.10.3), (8.10.4), and Theorem **13**, it follows that

$$\left|\sum_{i=1}^{n}\sum_{j=1}^{n} f_{ij}x_i y_j\right| \leqq MN,$$

which proves the theorem.

It is plain that we can define the Faltung of A and B, whether A and B are bounded or not, whenever the series

$$\sum_k |a_{ik}|^p, \quad \sum_k |b_{kj}|^{p'}$$

are convergent.

8.11. Some special theorems on forms in [2, 2].[a] In this section we confine ourselves to the classical case $p=q=2$, and suppose the variables and coefficients real (though not generally positive). We suppose then that A is a real form, and denote by

$$A' = \Sigma\Sigma a_{ji} x_i y_j \tag{8.11.1}$$

the form obtained from A by exchanging suffixes in a_{ij}.

If

$$\sum_k a^2_{ik} < \infty \tag{8.11.2}$$

for all i, the series

$$c_{ij} = \sum_k a_{ik} a_{jk} \tag{8.11.3}$$

are absolutely convergent, and, by (8.10.1),

$$C(x,y) = \Sigma\Sigma c_{ij} x_i y_j \tag{8.11.4}$$

is the Faltung $F(A,A')$ of A and A'. In particular $C(x,x)$ is a quadratic form whose section C_n is given, after (8.10.4), by

$$C_n(x,x) = \sum_k \left(\sum_{i=1}^{n} a_{ik} x_i\right)^2. \tag{8.11.5}$$

We write

$$C(x,x) = N(A)$$

and call $C(x,x)$ the *norm* of A. When A satisfies (8.11.2), we say that *the norm of A exists*. The existence of $N(A)$ is, after Theorem **289**, a *necessary* condition for the boundedness of A.

If A is bounded, with bound M, then, by Theorem **291**, $N(A)$

[a] Hellinger and Toeplitz (**1**), Schur (**1**).

is bounded, and its bound P does not exceed M^2. On the other hand, whenever $N(A)$ exists,

$$|A_n(x,y)| = \left| \sum_{j=1}^{n} y_j \sum_{i=1}^{n} a_{ij} x_i \right| \leqq \left(\sum_{j=1}^{n} y_j^2 \right)^{\frac{1}{2}} \left\{ \sum_{j=1}^{n} \left(\sum_{i=1}^{n} a_{ij} x_i \right)^2 \right\}^{\frac{1}{2}}$$

$$\leqq \left(\sum_{j=1}^{n} y_j^2 \right)^{\frac{1}{2}} \{C_n(x,x)\}^{\frac{1}{2}}.$$

Hence, if $N(A)$ is bounded by P, A is bounded by $P^{\frac{1}{2}}$.

Collecting our results, we obtain

292. *A necessary and sufficient condition that a real form A should be bounded in* [2, 2] *is that the norm $N(A)$ of A should exist and be bounded. If M is the bound of A, and P that of $N(A)$, then*

$$P = M^2.$$

A useful corollary is

293. *If A, B, ... is a finite set of forms whose norms exist, and*

$$H(x,x) = N(A) + N(B) + \dots$$

is bounded, with bound P, then A, B, ... are bounded, with bounds which do not exceed $P^{\frac{1}{2}}$.

In fact, if $N_n(A)$, ... are sections of $N(A)$, ..., then $N_n(A)$, ... are non-negative[a], by (8.11.5), and

$$H_n(x,x) = N_n(A) + N_n(B) + \dots.$$

Hence $N_n(A)$, ... are bounded by $P^{\frac{1}{2}}$.

8.12. Application to Hilbert's forms. We now apply Theorem **293** to two very important special forms first studied by Hilbert.

294. *The forms*

$$A = \Sigma\Sigma \frac{x_i y_j}{i+j-1}, \quad B = \Sigma\Sigma' \frac{x_i y_j}{i-j},$$

where $i, j = 1, 2, \dots$ and the dash implies the omission of the terms in which $i=j$, are bounded in real space [2, 2], *with bounds not exceeding π.*

[a] That is to say, assume non-negative values only for real x. The phrase 'positive form' has been used in this chapter in a different sense, that of a form with non-negative coefficients and variables.

It is plain that each form satisfies condition (8.11.2). We write

$$N(A)=\Sigma\Sigma c_{ij}x_i x_j, \quad N(B)=\Sigma\Sigma d_{ij}x_i y_j$$

and calculate $c_{ij}+d_{ij}$.

If $i=j$, we have

(8.12.1)
$$c_{ii}+d_{ii}=\sum_{k=1}^{\infty}\frac{1}{(i+k-1)^2}+\sum_{k=1}^{\infty}{}'\frac{1}{(i-k)^2}=\sum_{-\infty}^{\infty}{}'\frac{1}{(i-k)^2}=\tfrac{1}{3}\pi^2.$$

If $i\neq j$, then

$$c_{ij}+d_{ij}=\sum_{k=1}^{\infty}\frac{1}{(i+k-1)(j+k-1)}+\sum_{k=1}^{\infty}{}'\frac{1}{(i-k)(j-k)}$$
$$=\sum_{k=-\infty}^{\infty}{}'\frac{1}{(i-k)(j-k)}=\frac{1}{i-j}\sum_{k=-\infty}^{\infty}{}'\left(\frac{1}{j-k}-\frac{1}{i-k}\right),$$

the dash here excluding the values $k=i$ and $k=j$. If K is greater than both $|i|$ and $|j|$, then

$$\sum_{k=-K}^{K}{}'\left(\frac{1}{j-k}-\frac{1}{i-k}\right)$$
$$=\frac{2}{i-j}+\left(\frac{1}{j-K}+\dots+\frac{1}{j+K}-\frac{1}{i-K}-\dots-\frac{1}{i+K}\right),$$

with two series unbroken except for the omission of terms with denominator 0, and the bracket tends to zero when $K\to\infty$.[a] Hence

(8.12.2)
$$c_{ij}+d_{ij}=\frac{2}{(i-j)^2}\quad (i\neq j).$$

From (8.12.1) and (8.12.2) it follows that

(8.12.3)
$$N(A)+N(B)=\frac{\pi^2}{3}\Sigma x_i^2+2\Sigma\Sigma'\frac{x_i x_j}{(i-j)^2}.$$

The first form here has the bound $\frac{1}{3}\pi^2$; and, since

$$\sum_i{}'\frac{1}{(i-j)^2}=\sum_j{}'\frac{1}{(i-j)^2}<\tfrac{1}{3}\pi^2,$$

the second satisfies the conditions of Theorem **275**, and has the bound $\frac{2}{3}\pi^2$. Hence $N(A)+N(B)$ has the bound π^2, and Theorem **294** follows from Theorem **293**.[b]

[a] All terms cancel except a number independent of K.

[b] The proof is that of Schur (**1**).

That A is bounded can be proved more simply: we give a number of proofs in Ch. IX.

A is absolutely bounded (§ 8.8), since its coefficients are positive. It is important to observe that this is not true of B. To prove this, it is enough to prove that

$$\Sigma\Sigma' \frac{x_i y_j}{|i-j|} = \infty$$

for a positive set (x, y) for which Σx_i^2 and Σy_j^2 are convergent. We take

$$x_i = i^{-\frac{1}{2}}(\log i)^{-1} \ (i>1), \quad y_j = j^{-\frac{1}{2}}(\log j)^{-1} \ (j>1)$$

and $x_1 = x_2$, $y_1 = y_2$. Then

$$\Sigma\Sigma' \frac{x_i y_j}{|i-j|} \geq \sum_{j=1}^{\infty}\sum_{k=1}^{\infty} k^{-1} x_{j+k} y_j \geq \sum_{k=1}^{\infty} k^{-1} \sum_{l=k+1}^{\infty} x_l y_l$$

$$= \sum_{k=1}^{\infty} \frac{1}{k} \sum_{l=k+1}^{\infty} \frac{1}{l(\log l)^2} \geq \sum_{k=1}^{\infty} \frac{1}{k} \int_{k+1}^{\infty} \frac{du}{u(\log u)^2} = \sum_{k=1}^{\infty} \frac{1}{k \log(k+1)},$$

and this series is divergent.

We shall see in Ch. IX that A is bounded in $[p, p']$. B is also bounded in $[p, p']$, but the proof is much more difficult: see M. Riesz (**1, 2**), Titchmarsh (**2, 3**).

THE THEOREM OF M. RIESZ

8.13. The convexity theorem for bilinear forms with complex variables and coefficients. We prove next a very important theorem due to M. Riesz[a]. This, like Theorem 285, asserts the convexity of $\log M_{\alpha,\beta}$, where $M_{\alpha,\beta}$ is the upper bound of a form of the type A; but in Riesz's theorem the form is bilinear, the a, x, y are general complex numbers, and convexity is proved only in a restricted domain of α and β.

It is essential to Riesz's argument that $M_{\alpha,\beta}$ should be an attained maximum and not merely an upper bound; and we therefore consider a finite bilinear form

$$A = \sum_{i=1}^{m}\sum_{j=1}^{n} a_{ij} x_i y_j. \tag{8.13.1}$$

295. *Suppose that $M_{\alpha,\beta}$ is the maximum of A for*

$$\sum_1^m |x_i|^{1/\alpha} \leq 1, \quad \sum_1^n |y_j|^{1/\beta} \leq 1, \tag{8.13.2}$$

it being understood that, if $\alpha = 0$ or $\beta = 0$, these inequalities are

[a] M. Riesz (**1**). Our proof is substantially that of Riesz. An alternative proof has been given (not quite completely) by Paley (**2, 4**).

replaced by $|x_i| \leqq 1$ *or* $|y_j| \leqq 1$. *Then* $\log M_{\alpha,\beta}$ *in convex in the triangle*

(8.13.3) $$0 \leqq \alpha \leqq 1, \quad 0 \leqq \beta \leqq 1, \quad \alpha + \beta \geqq 1.$$

We have to prove that, if (α_1, β_1) and (α_2, β_2) are two points of the triangle (8.13.3), $0 < t < 1$, and

(8.13.4) $$\alpha = \alpha_1 t + \alpha_2 (1-t), \quad \beta = \beta_1 t + \beta_2 (1-t),$$

then

(8.13.5) $$M_{\alpha,\beta} \leqq M^t_{\alpha_1,\beta_1} M^{1-t}_{\alpha_2,\beta_2}.$$

We need the fact that $M_{\alpha,\beta} \to M_{0,\beta}$ as $\alpha \to 0$. Let $(\bar{x}, \bar{y})$ be a maximal set for (α, β), $(\bar{x}_0, \bar{y}_0)$ one for $(0, \beta)$. Then $(\bar{x}, \bar{y})$ is a 'permissible' set for $(0, \beta)$ and $(m^{-\alpha}\bar{x}_0, \bar{y}_0)$ for (α, β). Hence $m^{-\alpha} M_{0,\beta} \leqq M_{\alpha,\beta} \leqq M_{0,\beta}$ and $M_{\alpha,\beta} \to M_{0,\beta}$. Also $M_{\alpha,\beta} \to M_{\alpha,0}$ as $\beta \to 0$. There is thus some continuity of $M_{\alpha,\beta}$ at the corners $(0, 1)$ and $(1, 0)$ of the triangle (8.13.3), at other points of which $M_{\alpha,\beta}$ is obviously continuous. Hence, after Theorem **88**, it is enough to prove that (8.13.5) is true, when $\alpha_1, \beta_1, \alpha_2, \beta_2$ are given, for *some* t for which $0 < t < 1$. It is, moreover, enough to consider $M_{\alpha,\beta}$ for $\alpha > 0$, $\beta > 0$, in which case we can define p, q, p', q' by

(8.13.6) $$\alpha = \frac{1}{p}, \quad \beta = \frac{1}{q}, \quad \frac{1}{p} + \frac{1}{p'} = 1, \quad \frac{1}{q} + \frac{1}{q'} = 1.$$

We may then write (8.13.2) in the form

(8.13.7) $$\mathfrak{S}_p(x) \leqq 1, \quad \mathfrak{S}_q(y) \leqq 1;$$

and the inequalities (8.13.3) are equivalent to either of

(8.13.8) $$q' \geqq p \geqq 1$$

and

(8.13.9) $$p' \geqq q \geqq 1.$$

We shall also write, as in § 8.7,

(8.13.10) $$X_j = X_j(x) = \sum_i a_{ij} x_i, \quad Y_i = Y_i(y) = \sum_j a_{ij} y_j,$$

so that

(8.13.11) $$A = \sum_j X_j y_j = \sum_i x_i Y_i$$

or simply

(8.13.12) $$A = \Sigma X y = \Sigma x Y.$$

Theorem **286** enables us to give another definition of $M_{\alpha,\beta}$ which is more convenient for our present purpose. It is plain that A attains its maximum for a set (x, y) in which

$$\mathfrak{S}_p(x) = 1, \quad \mathfrak{S}_q(y) = 1; \tag{8.13.13}$$

and $M_{\alpha,\beta}$ is the least number K satisfying

$$|A| \leqq K\mathfrak{S}_p(x)\mathfrak{S}_q(y) \tag{8.13.14}$$

for all such (x, y). Since both sides are homogeneous of degree 1 in x and y, the restrictions (8.13.13) are now irrelevant, and $M_{\alpha,\beta}$ may be defined as the least K satisfying (8.13.14) for all (x, y).[a]

By Theorem **286**, this is also the least K satisfying

$$\mathfrak{S}_{q'}(X) \leqq K\mathfrak{S}_p(x) \tag{8.13.15}$$

for all x, or

$$\mathfrak{S}_{p'}(Y) \leqq K\mathfrak{S}_q(y) \tag{8.13.16}$$

for all y. We may therefore define $M_{\alpha,\beta}$ by

$$M_{\alpha,\beta} = \operatorname{Max} \frac{\mathfrak{S}_{q'}(X)}{\mathfrak{S}_p(x)} = \operatorname{Max} \frac{\mathfrak{S}_{p'}(Y)}{\mathfrak{S}_q(y)}, \tag{8.13.17}$$

the maxima being taken for all non-null sets x or y.

8.14. Further properties of a maximal set (x, y). Suppose that (x^*, y^*) is a set of (x, y), subject to (8.13.7), for which $|A|$ attains its maximum, and that X^*, Y^* are the corresponding values of X, Y. It is obvious (as we have observed already) that

$$\mathfrak{S}_p(x^*) = 1, \quad \mathfrak{S}_q(y^*) = 1. \tag{8.14.1}$$

Also, as in (8.7.6),

$$|A| \leqq \mathfrak{S}_{q'}(X)\mathfrak{S}_q(y), \quad |A| \leqq \mathfrak{S}_{p'}(Y)\mathfrak{S}_p(x). \tag{8.14.2}$$

There must be equality in each of (8.14.2) when x, y have the values x^*, y^*; for otherwise we could increase $|A|$ by leaving the x, X unaltered and changing the y, or by leaving the y, Y unaltered and changing the x. Hence

$$M_{\alpha,\beta} = |A(x^*, y^*)| = \mathfrak{S}_{q'}(X^*)\mathfrak{S}_q(y^*) = \mathfrak{S}_{p'}(Y^*)\mathfrak{S}_p(x^*).$$

Further, by Theorem **14**,

$$|X_j^*|^{q'} = \omega^{q'} |y_j^*|^q,$$

or

$$|X_j^*| = \omega |y_j^*|^{q-1}, \tag{8.14.3}$$

[a] This is merely a repetition of an argument used already in § 8.8.

where ω is positive and independent of j; and

$$\arg X_j^* y_j^*$$

is independent of j. Hence

$$M_{\alpha,\beta} = |A(x^*, y^*)| = |\Sigma X^* y^*| = \Sigma |X^* y^*| = \omega \Sigma |y^*|^q = \omega.$$

Substituting in (8.14.3), and adding the corresponding result for Y_i^*, we obtain

$$(8.14.4) \qquad |X_j^*| = M_{\alpha,\beta} |y_j^*|^{q-1}, \quad |Y_i^*| = M_{\alpha,\beta} |x_i^*|^{p-1}.$$

8.15. Proof of Theorem 295. In what follows we suppose (x, y) a maximal set (for the indices α, β), omitting the asterisks. We write $p_1 = 1/\alpha_1$, and so on, and M, M_1, M_2 for $M_{\alpha.\beta}, M_{\alpha_1,\beta_1}, M_{\alpha_2,\beta_2}$. Our use of $p, p_1, \dots$ excludes the points (0, 1), (1, 0) of the triangle, but, as we remarked in § 8.13, this will not impair the proof.

By (8.14.4)

$$M\mathfrak{S}_{(p-1)p_1'}^{p-1}(x) = (M^{p_1'} \Sigma |x_i|^{(p-1)p_1'})^{1/p_1'}$$
$$= (\Sigma |Y_i|^{p_1'})^{1/p_1'} = \mathfrak{S}_{p_1'}(Y).$$

Comparing this with (8.13.17), we obtain

$$(8.15.1) \qquad M\mathfrak{S}_{(p-1)p_1'}^{p-1}(x) \leqq M_1 \mathfrak{S}_{q_1}(y).$$

Similarly

$$(8.15.2) \qquad M\mathfrak{S}_{(q-1)q_2'}^{q-1}(y) \leqq M_2 \mathfrak{S}_{p_2}(x).$$

Hence, if $0 < t < 1$, we have

(8.15.3)

$$M\mathfrak{S}_{(p-1)p_1'}^{(p-1)t}(x)\,\mathfrak{S}_{(q-1)q_2'}^{(q-1)(1-t)}(y) \leqq M_1^t M_2^{1-t}\,\mathfrak{S}_{q_1}^t(y)\,\mathfrak{S}_{p_2}^{1-t}(x).$$

Let us assume provisionally that there is *some* t between 0 and 1 which satisfies

$$(8.15.4) \qquad \frac{1}{p} = \frac{t}{p_1} + \frac{1-t}{p_2}, \quad \frac{1}{q} = \frac{t}{q_1} + \frac{1-t}{q_2},$$

that is to say the equations (8.13.4), and that

$$(8.15.5) \quad \mathfrak{S}_{p_2}^{1-t}(x) \leqq \mathfrak{S}_{(p-1)p_1'}^{(p-1)t}(x), \quad \mathfrak{S}_{q_1}^t(y) \leqq \mathfrak{S}_{(q-1)q_2'}^{(q-1)(1-t)}(y).$$

Then (8.15.3) and (8.15.5) will give

$$M \leqq M_1^t M_2^{1-t}$$

for *some* t, and after Theorem **88** the theorem will follow.

It remains to justify the assumptions expressed by (8.15.4) and (8.15.5). Let us assume further that there are numbers μ and ν such that

(8.15.6) $$0<\mu\leqq 1, \quad 0<\nu\leqq 1,$$

(8.15.7)

$$p_2=(p-1)p_1'\mu+p(1-\mu), \quad q_1=(q-1)q_2'\nu+q(1-\nu).$$

By Theorem **18**,[a] $r\log\mathfrak{S}_r(x)=\log\mathfrak{S}_r^r(x)$ is a convex function of r; and the x, being a maximal set, satisfy (8.13.13). Hence

(8.15.8) $$\mathfrak{S}_{p_2}^{p_2}(x)\leqq\mathfrak{S}_{(p-1)p_1'}^{(p-1)p_1'\mu}(x)\,\mathfrak{S}_p^{p(1-\mu)}(x)=\mathfrak{S}_{(p-1)p_1'}^{(p-1)p_1'\mu}(x),$$

and similarly

(8.15.9) $$\mathfrak{S}_{q_1}^{q_1}(y)\leqq\mathfrak{S}_{(q-1)q_2'}^{(q-1)q_2'\nu}(y)\,\mathfrak{S}_q^{q(1-\nu)}(y)=\mathfrak{S}_{(q-1)q_2'}^{(q-1)q_2'\nu}(y).$$

If finally

(8.15.10) $$\frac{p_1'\mu}{p_2}=\frac{t}{1-t}, \quad \frac{q_2'\nu}{q_1}=\frac{1-t}{t},$$

then (8.15.8) and (8.15.9) will be equivalent to (8.15.5).

In order to complete the proof, it is necessary to show that (8.15.4), (8.15.6), (8.15.7), and (8.15.10) are consistent. These conditions contain six equations to be satisfied by the four numbers p, q, μ, ν, and two inequalities. The first equation (8.15.10) gives

$$(p_1'-1)\mu+1=\frac{(p_1'-1)p_2}{p_1'}\frac{t}{1-t}+1=\frac{p_2}{p_1}\frac{t}{1-t}+1$$

and

$$p_2+p_1'\mu=p_2\left(1+\frac{t}{1-t}\right)=\frac{p_2}{1-t};$$

and the first equation (8.15.7) gives

$$\frac{1}{p}=\frac{(p_1'-1)\mu+1}{p_2+p_1'\mu}=\frac{p_2t+p_1(1-t)}{p_1p_2}=\frac{t}{p_1}+\frac{1-t}{p_2},$$

which agrees with (8.15.4).

A similar argument applies to the equations involving q, so that (8.15.4) is a consequence of (8.15.7) and (8.15.10). Given p_1, q_1, p_2, q_2, and t, we can find μ, ν from (8.15.10) and p, q from (8.15.7), and these numbers will satisfy the six equations.

[a] Strictly, by Theorem **18** restated as Theorem **17** was restated in Theorem **87**.

It remains only to examine the inequalities (8.15.6). If μ and ν satisfy (8.15.10), and $0<t<1$, the inequalities are equivalent to

$$\frac{q_1}{q_2'} \leq \frac{t}{1-t} \leq \frac{p_1'}{p_2}. \tag{8.15.11}$$

Since (α_1, β_1) and (α_2, β_2) lie in the triangle (8.13.3), we have, by (8.13.8) and (8.13.9),

$$q_1 \leq p_1', \quad q_2' \geq p_2,$$

and *a fortiori*

$$\frac{q_1}{q_2'} \leq \frac{p_1'}{p_2}.$$

We can therefore choose t so as to satisfy (8.15.11), and then all our conditions are satisfied.

It will be observed that it is only in the last paragraph that we use the essential inequality $\alpha+\beta \geq 1$. When the form is positive, this inequality is irrelevant; $\log M_{\alpha,\beta}$ is then, by Theorem **285**, convex in the whole of the positive quadrant of (α, β).

8.16. Applications of the theorem of M. Riesz. (i) Theorem **295** is easily transformed into another theorem of very different appearance.

296. *Suppose that*

$$X_j(x) = \sum_{i=1}^{m} a_{ij} x_i \quad (j=1, 2, \ldots, n) \tag{8.16.1}$$

and that $M^*_{\alpha,\gamma}$ is the maximum of

$$\left(\sum_1^n |X_j|^{1/\gamma}\right)^\gamma$$

for

$$\sum_1^m |x_i|^{1/\alpha} \leq 1.$$

Then $\log M^*_{\alpha,\gamma}$ is convex in the triangle

$$0 \leq \gamma \leq \alpha \leq 1. \tag{8.16.2}$$

In fact, by (8.13.17),

$$M_{\alpha,\beta} = \operatorname{Max} \frac{\mathfrak{S}_{q'}(X)}{\mathfrak{S}_p(x)},$$

while

$$M^*_{\alpha,\gamma} = \operatorname{Max} \frac{\mathfrak{S}_{1/\gamma}(X)}{\mathfrak{S}_{1/\alpha}(x)} = \operatorname{Max} \frac{\mathfrak{S}_{1/\gamma}(X)}{\mathfrak{S}_p(x)}.$$

Hence

$$M_{\alpha,\beta} = M^*_{\alpha,\gamma}$$

if $\gamma = 1/q' = 1-\beta$; and then the conditions (8.16.2) are equivalent to (8.13.8) or to (8.13.3). Thus $M^*_{\alpha,\gamma}$ is a convex function of $(\alpha, 1-\gamma)$, or, what is the same thing, of (α, γ).

(ii) **297.** *Suppose that X_j is defined by* (8.16.1); *that*

(8.16.3) $$\sum_1^n |X_j|^2 \leqq \sum_1^m |x_i|^2$$

for all x; and that

(8.16.4) $$1 \leqq p \leqq 2.$$

Then

(8.16.5) $$\mathfrak{S}_{p'}(X) \leqq \mathfrak{m}^{(2-p)/p}\, \mathfrak{S}_p(x),$$

where

(8.16.6) $$\mathfrak{m} = \mathrm{Max}\,|a_{ij}|.$$

To deduce Theorem **297** from Theorem **296**, we write $\alpha = 1/p$, as before, and consider the line from $(\frac{1}{2}, \frac{1}{2})$ to $(1, 0)$ in the plane (α, γ). This line lies entirely in the triangle (8.16.2); and hence, by Theorem **296**,

$$M^*_{\alpha,1-\alpha} \leqq (M^*_{\frac{1}{2},\frac{1}{2}})^{2(1-\alpha)} (M^*_{1,0})^{2\alpha-1}$$

for $\frac{1}{2} < \alpha < 1$. It is plain from (8.16.3) that $M^*_{\frac{1}{2},\frac{1}{2}} \leqq 1$; and

$$M^*_{1,0} = \frac{\mathrm{Max}\,|X|}{\Sigma\,|x|} \leqq \mathfrak{m}.$$

Hence $$M^*_{\alpha,1-\alpha} \leqq \mathfrak{m}^{2\alpha-1} = \mathfrak{m}^{(2-p)/p},$$

which is equivalent to (8.16.5).

The condition (8.16.3) is certainly satisfied (with equality) if (8.16.1) is a 'unitary' substitution, i.e. a substitution which leaves $\Sigma\,|x|^2$ unaltered[a]. This case of the theorem was found by F. Riesz (4), and the general theorem by M. Riesz (**1**).

8.17. Applications to Fourier series. Out of many other important applications of Riesz's theorems, we select an application of Theorem **297** to the proof of Hausdorff's theorems[b].

[a] In this case $n = m$. A real unitary substitution is orthogonal.

[b] See § 8.5. Riesz deduces these theorems in a different manner, and gives a number of other applications.

(i) Suppose that m is an odd integer and

$$f_m(\theta)=\sum_{-\frac{1}{2}m}^{\frac{1}{2}m} e^{\mu\theta i}x_\mu,\text{[a]}$$

$$X_\nu=\sum_{-\frac{1}{2}m}^{\frac{1}{2}m} a_{\mu\nu}x_\mu=\frac{1}{\sqrt{m}}f_m\left(\frac{2\pi\nu}{m}\right),$$

$$a_{\mu\nu}=m^{-\frac{1}{2}}e^{2\mu\nu\pi i/m}.$$

The substitution is unitary, so that

$$\Sigma\,|X_\nu|^2=\Sigma\,|x_\mu|^2.$$

Also $\mathfrak{m}=m^{-\frac{1}{2}}$. Hence, by Theorem **297**,

$$\left(\frac{1}{m}\Sigma\left|f_m\left(\frac{2\pi\nu}{m}\right)\right|^{p'}\right)^{1/p'}\leqq(\Sigma\,|x_\mu|^p)^{1/p}. \tag{8.17.1}$$

The left-hand side being an approximation to $\mathfrak{J}_{p'}(f)$, we may deduce Hausdorff's theorem (8.5.6) by passages to the limit[b].

(ii) If m is again an odd integer,

$$x_\mu=f_m\left(\frac{2\pi\mu}{m}\right)=\sum_{r=-\frac{1}{2}m}^{\frac{1}{2}m} a_r e^{2r\mu\pi i/m}$$

and

$$X_\nu=\frac{1}{\sqrt{m}}\sum_{-\frac{1}{2}m}^{\frac{1}{2}m} e^{-2\nu\mu\pi i/m}x_\mu,$$

then simple calculations show that

$$X_\nu=m^{\frac{1}{2}}a_\nu,$$

and

$$\Sigma\,|X_\nu|^2=\Sigma\,|x_\mu|^2$$

as before. In this case Theorem **297** gives

$$(\Sigma\,|a_\nu|^{p'})^{1/p'}\leqq\left(\frac{1}{m}\Sigma\left|f_m\left(\frac{2\pi\mu}{m}\right)\right|^p\right)^{1/p};$$

and Hausdorff's theorem (8.5.7) follows by appropriate passages to the limit.

We can also, as we showed in § 8.5, deduce the second theorem from the first.

[a] We now write μ, ν for i, j, and extend the summations over the range

$$-\tfrac{1}{2}m<\mu<\tfrac{1}{2}m.$$

[b] If $f(\theta)$ is a polynomial

$$\sum_{-\frac{1}{2}M}^{\frac{1}{2}M} x_\mu e^{\mu\theta i},$$

then $f_m(\theta)=f(\theta)$ for $m\geqq M$, and the theorem for $f(\theta)$ follows immediately from (8.17.1). The extension to an arbitrary $f(\theta)$ depends on the theory of 'strong convergence'.

MISCELLANEOUS THEOREMS AND EXAMPLES

298. If $p>1$, and $a(x,y)$ is measurable and positive, then the three assertions

(i) $$\int_0^\infty\int_0^\infty a(x,y)f(x)g(y)\,dx\,dy \leqq K\left(\int_0^\infty f^p dx\right)^{1/p}\left(\int_0^\infty g^{p'}dy\right)^{1/p'},$$

for all non-negative f, g;

(ii) $$\int_0^\infty dy\left(\int_0^\infty a(x,y)f(x)\,dx\right)^p \leqq K^p\int_0^\infty f^p dx,$$

for all non-negative f;

(iii) $$\int_0^\infty dx\left(\int_0^\infty a(x,y)g(y)\,dy\right)^{p'} \leqq K^{p'}\int_0^\infty g^{p'}dy,$$

for all non-negative g, are equivalent. The assertions 'there is inequality in (i) unless f or g is null', 'there is inequality in (ii) unless f is null', 'there is inequality in (iii) unless g is null', are also equivalent.

[Analogue of Theorems **286** and **287**, with $q=p'$. There is a more general form with both p and q arbitrary.]

299. The forms

$$A=\Sigma\Sigma\frac{x_i y_j}{i+j-1+\lambda},\quad B=\Sigma\Sigma'\frac{x_i y_j}{i-j+\lambda}$$

where $\lambda>0$, and the dash is required only if λ is integral, are bounded in [2, 2], and have bounds π if λ is integral, $\pi\,|\operatorname{cosec}\lambda\pi\,|$ if λ is non-integral.

[Schur (**1**), Pólya and Szegö (**1**, I, 117, 290).]

300. If $p>1$ and $A=\Sigma\Sigma a_{ij}x_i y_j$ has a bound M in $[p,p']$, and

$$h_{ij}=\int f_i(t)\,g_j(t)\,dt,$$

where $$\int|f_i|^p dt\leqq\mu^p,\quad \int|g_j|^{p'}dt\leqq\nu^{p'},$$

then $$A^*=\Sigma\Sigma a_{ij}h_{ij}x_i y_j$$

has a bound $M\mu\nu$.

[For
$$\begin{aligned}|A^*|&=|\textstyle\int\{\Sigma\Sigma a_{ij}x_i f_i(t)y_j g_j(t)\}\,dt\,|\\
&\leqq M\textstyle\int\{\Sigma\,|x_i f_i(t)\,|^p\}^{1/p}\{\Sigma\,|\,y_j g_j(t)\,|^{p'}\}^{1/p'}dt\\
&\leqq M\,\{\textstyle\int\Sigma\,|x_i|^p\,|f_i(t)\,|^p dt\}^{1/p}\{\textstyle\int\Sigma|y_j|^{p'}\,|\,g_j(t)\,|^{p'}dt\}^{1/p'}\\
&\leqq M\mu\nu(\Sigma\,|x_i\,|^p)^{1/p}(\Sigma\,|\,y_j\,|^{p'})^{1/p'}.\end{aligned}$$
For the case $p=p'=2$, see Schur (**1**).]

301. $\Sigma\Sigma'\dfrac{(ij)^{\frac{1}{2}(\mu-1)}}{i^\mu-j^\mu}x_i y_j$ is bounded in [2, 2].

302. $\Sigma\Sigma'\dfrac{\sin(i-j)\theta}{i-j}x_i y_j$ is bounded in [2, 2], for any real θ. If $0\leqq\theta\leqq\pi$, the bound does not exceed $\operatorname{Max}(\theta,\pi-\theta)$.

[For the last two theorems see Schur (**1**).]

303. If a_n is the Fourier sine coefficient of an odd bounded function, or the Fourier cosine coefficient of an even bounded function, then the forms

$$\Sigma\Sigma a_{i+j} x_i y_j, \quad \Sigma\Sigma a_{i-j} x_i y_j$$

are bounded in [2, 2].

[Toeplitz (**1**). Suppose that i and j run from 1 to n; that x and y are real and

$$\Sigma x_i^2 = \Sigma y_j^2 = 1;$$

that $X = \Sigma x_i \cos i\theta, \quad X' = \Sigma x_i \sin i\theta, \quad Y = \Sigma y_j \cos j\theta, \quad Y' = \Sigma y_j \sin j\theta,$

and that (for example)

$$a_n = \frac{2}{\pi}\int_0^\pi f(\theta) \sin n\theta \, d\theta,$$

where $|f| \leqq M$. A simple calculation gives

$$\sum_1^n \sum_1^n a_{i-j} x_i y_j = \frac{2}{\pi}\int_0^\pi (X'Y - XY') f(\theta)\, d\theta.$$

Since

$$\left|\int_0^\pi XY' f(\theta)\, d\theta\right| \leqq \tfrac{1}{2}M \int_0^\pi (X^2 + Y'^2)\, d\theta = \tfrac{1}{4}M\pi(\Sigma x^2 + \Sigma y^2) = \tfrac{1}{2}M\pi,$$

we find the upper bound $2M$ for A. Similarly in the other cases.

If for example $f(\theta)$ is odd and equal to $\frac{1}{2}(\pi - \theta)$ for $0 < \theta < \pi$, then $M = \frac{1}{2}\pi$ and $a_n = n^{-1}$. We thus obtain the result of Theorem **294** concerning B.]

304. If (i) $\Sigma\Sigma a_{ij} x_i y_j$ is bounded in $[p, q]$, (ii) $k > 1$, $l > 1$, and (iii) (u_i), (v_j) are given sets for which $\mathfrak{S}_{pk'}(u) < \infty$, $\mathfrak{S}_{ql'}(v) < \infty$, then

$$A = \Sigma\Sigma a_{ij} u_i v_j x_i y_j$$

is bounded in $[pk, ql]$.

[For $\quad |A| \leqq M(\Sigma|ux|^p)^{1/p}(\Sigma|vy|^q)^{1/q}$

$$\leqq M(\Sigma|u^{pk'}|)^{1/pk'}(\Sigma|v^{ql'}|)^{1/ql'}(\Sigma|x|^{pk})^{1/pk}(\Sigma|y|^{ql})^{1/ql}.]$$

305. The form

$$\Sigma\Sigma' \frac{u_i v_j}{i-j} x_i y_j,$$

where u_i and v_j are given sets of numbers satisfying

$$\Sigma|u_i|^2 \leqq 1, \quad \Sigma|v_j|^2 \leqq 1,$$

is bounded, but not necessarily absolutely bounded, in $[\infty, \infty]$.

[Take $p = q = 2$, $k = l = \infty$ in Theorem **304**.

If the form were always absolutely bounded, then Hilbert's form B of § 8.12 would be absolutely bounded, which is untrue.]

306. If $\quad \mathfrak{M}_2(a) \geqq A_2 H, \quad \mathfrak{M}_4(a) \leqq A_4 H,$

then $$\frac{A_2^3}{A_4^2} H \leqq \mathfrak{M}_1(a) \leqq A_4 H.$$

[By Theorems **16** and **17**. The result is required in the proof of the next theorem.]

307. If perpendiculars are drawn from the corners $(\pm 1, \pm 1, \ldots, \pm 1)$ of the 'unit cube' in space of m dimensions, upon any linear $[m-1]$ through the centre of the cube, then the mean of the perpendiculars lies between two constants A and B, independent of m and the position of the $[m-1]$.

308. If
$$b_j = \left(\sum_i |a_{ij}|^2\right)^{\frac{1}{2}}, \quad c_i = \left(\sum_j |a_{ij}|^2\right)^{\frac{1}{2}},$$
then
$$P = (\Sigma\Sigma |a_{ij}|^{\frac{4}{3}})^{\frac{3}{4}} \leqq K(\Sigma b_j + \Sigma c_i) = K(B+C),$$
where K is an absolute constant.

309. A necessary condition that a form $A = \Sigma\Sigma a_{ij} x_i y_j$, with real or complex coefficients, should be bounded in $[\infty, \infty]$, with bound M, is that, in the notation of the last theorem, B, C, P should be less than KM.

[For the last five theorems see Littlewood (**2**).]

310. If
$$p \geqq 2, \ q \geqq 2, \ \frac{1}{p} + \frac{1}{q} \leqq \frac{1}{2},$$
$$\lambda = \frac{pq}{pq - p - q}, \quad \mu = \frac{4pq}{3pq - 2p - 2q};$$
and A is bounded in $[p, q]$, with bound M, then
$$(\Sigma b_j^{\lambda})^{1/\lambda} \leqq KM, \quad (\Sigma c_i^{\lambda})^{1/\lambda} \leqq KM, \quad (\Sigma\Sigma |a_{ij}|^{\mu})^{1/\mu} \leqq KM,$$
where b_j and c_i are defined as in Theorem **308**, and K depends on p and q only.

311. If
$$\frac{1}{2} \leqq \frac{1}{p} + \frac{1}{q} < 1,$$
but the conditions of Theorem **310** are satisfied in other respects, then
$$(\Sigma b_j^{\lambda})^{1/\lambda} \leqq KM, \quad (\Sigma c_i^{\lambda})^{1/\lambda} \leqq KM, \quad (\Sigma\Sigma |a_{ij}|^{\lambda})^{1/\lambda} \leqq KM.$$

312. If
$$p < 2 < q, \ \frac{1}{p} + \frac{1}{q} < 1,$$
and the conditions of Theorem **310** are satisfied otherwise, then
$$(\Sigma\Sigma |a_{ij}|^{\lambda})^{1/\lambda} \leqq KM.$$

313. If
$$p > 1, \ q > 1, \ \frac{1}{p} + \frac{1}{q} < 1,$$
$$a_{ij} \geqq 0,$$
A is bounded in $[p, q]$, with bound M, and
$$\beta_j = \left(\sum_i a_{ij}^{p'}\right)^{1/p'}, \quad \gamma_i = \left(\sum_j a_{ij}^{q'}\right)^{1/q'},$$
then
$$(\Sigma \beta_j^{\lambda})^{1/\lambda} \leqq M, \quad (\Sigma \gamma_i^{\lambda})^{1/\lambda} \leqq M, \quad (\Sigma\Sigma a_{ij}^{\lambda})^{1/\lambda} \leqq M.$$

[For the last four theorems see Hardy and Littlewood (**13**).]

314. *Hilbert's forms in* $[p, p']$. It will be proved in Ch. IX that the form A of Theorem **294** is bounded in $[p, p']$. The corresponding theorem for B lies a good deal deeper. We have to show that

$$\text{(i)} \qquad \left|\sum_1^n \sum_1^{n}{}' \frac{x_i y_j}{i-j}\right| \leqq K \mathfrak{S}_p(x)\, \mathfrak{S}_{p'}(y),$$

where $K = K(p)$ depends only on p; or, what is, after Theorem **286**, the same thing, that

$$\text{(ii)} \qquad \sum_i \left|\sum_j{}' \frac{y_j}{i-j}\right|^{p'} \leqq K^{p'} \sum |y_j|^{p'}.$$

It is enough, after Theorem **295**, to prove (i) or (ii) for *even integral* values of p' (or for some subsequence of these values). This demands some special device, the most natural, from our present point of view, being that used by Titchmarsh (**2**).

CHAPTER IX

HILBERT'S INEQUALITY AND ITS ANALOGUES AND EXTENSIONS

9.1. Hilbert's double series theorem. The researches of which we give an account in this chapter originate in a remarkable bilinear form which was first studied by Hilbert, and which we have already encountered in § 8.12, viz. the form

$$\Sigma\Sigma \frac{a_m b_n}{m+n},$$

where m and n run from 1 to ∞. Our first theorem is Theorem **315** below, which we state with its integral analogue and with a complement of a type which will occur frequently in this chapter.

315. *If* $\quad p>1, \quad p'=p/(p-1),$

and $\quad \Sigma a_m{}^p \leqq A, \quad \Sigma b_n{}^{p'} \leqq B,$

the summations running from 1 *to* ∞, *then*

$$\Sigma\Sigma \frac{a_m b_n}{m+n} < \frac{\pi}{\sin(\pi/p)} A^{1/p} B^{1/p'} \tag{9.1.1}$$

unless (a) *or* (b) *is null.*

316. *If* $\quad p>1, \quad p'=p/(p-1),$

and $\quad \int_0^\infty f^p(x)\,dx \leqq F, \quad \int_0^\infty g^{p'}(y)\,dy \leqq G,$

then

$$\int_0^\infty \int_0^\infty \frac{f(x)\,g(y)}{x+y}\,dx\,dy < \frac{\pi}{\sin(\pi/p)} F^{1/p} G^{1/p'}, \tag{9.1.2}$$

unless $f \equiv 0$ *or* $g \equiv 0$.

317. *The constant* $\pi \operatorname{cosec}(\pi/p)$ *is the best possible constant in each of Theorems* **315** *and* **316**.

The case $p=p'=2$ of Theorem **315** is 'Hilbert's double series theorem', and was proved first (apart from the exact determination of the constant) by Hilbert in his lectures on integral equations. Hilbert's proof was published by Weyl (2). The determination of the constant, and the integral analogue, are due to Schur (**1**), and the extensions to general p to Hardy and M. Riesz: see Hardy (**3**). Other proofs, of the whole or of parts of the theorems, and generalisations in different directions, have been given by Fejér and F. Riesz (**1**), Francis and Littlewood (**1**), Hardy (**2**), Hardy, Littlewood, and Pólya (**1**), Mulholland (**1, 3**), Owen (**1**), Pólya and Szegö (**1**, I, 117, 290), Schur (**1**), and F. Wiener (**1**). A number of these generalisations will be proved or quoted later.

The inequality (9.1.1) is of the same type as the general inequality discussed in § 8.2; but Theorem **315** is not included in Theorem **275**, since

$$\sum_m \frac{1}{m+n}$$

is divergent. It is to be observed that $\pi \operatorname{cosec} \alpha\pi$, where $\alpha = 1/p$, is (in accordance with Theorem **295**) convex for $0 < \alpha < 1$.

9.2. A general class of bilinear forms.

We shall deduce Theorem **315** from the following more general theorem[a].

318. *Suppose that $p > 1$, $p' = p/(p-1)$, and that $K(x,y)$ has the following properties:*

(i) *K is non-negative, and homogeneous of degree -1:*

(ii) $$\int_0^\infty K(x,1)\, x^{-1/p}\, dx = \int_0^\infty K(1,y)\, y^{-1/p'}\, dy = k:$$

and **either** (iii) *$K(x,1)\, x^{-1/p}$ is a strictly decreasing function of x, and $K(1,y)\, y^{-1/p'}$ of y*: **or**, *more generally,* (iii′) *$K(x,1)\, x^{-1/p}$ decreases from $x=1$ onwards, while the interval* (0, 1) *can be divided into two parts,* $(0, \xi)$ *and* $(\xi, 1)$, *of which one may be nul, in the first of which it decreases and in the second of which it increases; and $K(1,y)\, y^{-1/p'}$ has similar properties. Finally suppose that, when only the less stringent condition* (iii′) *is satisfied,*

(iv) $$K(x,x) = 0.$$

Then

(a) $$\sum\sum K(m,n)\, a_m b_n < k \left(\sum a_m^{\,p}\right)^{1/p} \left(\sum b_n^{\,p'}\right)^{1/p'}$$

unless (*a*) *or* (*b*) *is null;*

(b) $$\sum_n \left(\sum_m K(m,n)\, a_m\right)^p < k^p \sum a_m^{\,p}$$

unless (*a*) *is null;*

[a] Hardy, Littlewood, and Pólya (**1**). The case $p=2$ of the theorem is due in substance to Schur (**1**): Schur supposes $K(x, y)$ a decreasing function of both variables.

(c) $$\sum_m \left(\sum_n K(m,n)\, b_n\right)^{p'} < k^{p'} \sum b_n^{p'}$$

unless (b) *is null.*

In each case the summations are from 1 to ∞. Theorems **286** and **287** show that the three conclusions (a), (b), (c) are equivalent.

We may elucidate the hypotheses by the following remarks.

(1) The convergence and equality of the two integrals in condition (ii) is a consequence of the convergence of either, because of the homogeneity of K.

(2) The words 'decreasing', ... are to be interpreted in the strict sense throughout the theorem.

(3) ξ may be 0 or 1, one of the intervals $(0, \xi)$ and $(\xi, 1)$ then disappearing.

(4) In the most important application, in which

$$K(x,y) = \frac{1}{x+y},$$

condition (iii) is satisfied. An interesting case in which condition (iii′) is required is

$$K(x,y) = \frac{1}{(x+y)^{1-\alpha}\,|x-y|^{\alpha}} \quad (0<\alpha<1).^{[a]}$$

In this case $K(x, 1)$ has an infinity at $x=1$. In such cases condition (iv) is required in order to exclude equal pairs (m, m) from the summation.

It is easy to see that, if m and n are positive integers, and the summations are over $r = 1, 2, \ldots$, then

$$(9.2.1)\quad \sum K\left(\frac{r}{n}, 1\right)\left(\frac{r}{n}\right)^{-1/p}\frac{1}{n} < \int_0^\infty K(x,1)\, x^{-1/p}\, dx = k,$$

$$(9.2.2)\quad \sum K\left(1, \frac{r}{m}\right)\left(\frac{r}{m}\right)^{-1/p'}\frac{1}{m} < \int_0^\infty K(1,y)\, y^{-1/p'}\, dy = k.$$

For, if (iii) is satisfied, then

$$(9.2.3)\quad K\left(\frac{r}{n}, 1\right)\left(\frac{r}{n}\right)^{-1/p}\frac{1}{n} < \int_{(r-1)/n}^{r/n} K(x,1)\, x^{-1/p}\, dx,$$

and (9.2.1) follows by summation. If only (iii′) is satisfied, we use (9.2.3) for $r > n$ and for $r \leqq \xi n$, and

$$K\left(\frac{r}{n}, 1\right)\left(\frac{r}{n}\right)^{-1/p}\frac{1}{n} < \int_{r/n}^{(r+1)/n} K(x,1)\, x^{-1/p}\, dx$$

[a] See Hardy, Littlewood, and Pólya (1).

for $\xi n < r < n$; and the result again follows by summation when we observe that $K(1,1)=0$. The proof of (9.2.2) is similar.

Hence

$$\Sigma\Sigma K(m,n)\,a_m b_n = \Sigma\Sigma a_m K^{1/p}\left(\frac{m}{n}\right)^{1/pp'} b_n K^{1/p'}\left(\frac{n}{m}\right)^{1/pp'}$$
$$\leqq P^{1/p} Q^{1/p'},$$

where
$$P = \sum_m a_m{}^p \sum_n K(m,n)\left(\frac{m}{n}\right)^{1/p'}$$
$$= \sum_m a_m{}^p \sum_n K\left(1,\frac{n}{m}\right)\left(\frac{n}{m}\right)^{-1/p'}\frac{1}{m} < k\,\Sigma a_m{}^p,$$

by (9.2.2), unless (a) is null; and similarly

$$Q < k\,\Sigma b_n{}^{p'}$$

unless (b) is null. This proves the theorem.

If we take
$$K(x,y) = \frac{1}{x+y}$$

we obtain Theorem **315**. It may be shown that the k of Theorem **318** is a best possible constant, but in this direction we shall not go beyond proving Theorem **317**.[a]

9.3. The corresponding theorem for integrals. The theorem for integrals corresponding to Theorem **318** is

319. *Suppose that $p > 1$, that $K(x,y)$ is non-negative and homogeneous of degree -1, and that*

$$\int_0^\infty K(x,1)\,x^{-1/p}\,dx = \int_0^\infty K(1,y)\,y^{-1/p'}\,dy = k.$$

Then (a)

$$\int_0^\infty\int_0^\infty K(x,y)f(x)g(y)\,dx\,dy \leqq k\left(\int_0^\infty f^p\,dx\right)^{1/p}\left(\int_0^\infty g^{p'}\,dy\right)^{1/p'},$$

(b)
$$\int_0^\infty dy\left(\int_0^\infty K(x,y)f(x)\,dx\right)^p \leqq k^p\int_0^\infty f^p\,dx,$$

(c)
$$\int_0^\infty dx\left(\int_0^\infty K(x,y)g(y)\,dy\right)^{p'} \leqq k^{p'}\int_0^\infty g^{p'}\,dy.$$

If $K(x,y)$ is positive, then there is inequality in (b) *unless $f \equiv 0$, in* (c) *unless $g \equiv 0$, and in* (a) *unless either $f \equiv 0$ or $g \equiv 0$.*

[a] See § 9.5. The constant k is (again in accordance with Theorem **295**) convex in $\alpha = 1/p$.

The theorem may be proved by the method of § 9.2, which naturally goes rather more simply in this case. We have

$$\iint K(x,y) f(x) g(y)\, dx\, dy$$
$$= \iint f(x) K^{1/p} \left(\frac{x}{y}\right)^{1/pp'} g(y)\, K^{1/p'} \left(\frac{y}{x}\right)^{1/pp'} dx\, dy \leqq P^{1/p} Q^{1/p'},$$

where $$P = \int f^p(x)\, dx \int K(x,y) \left(\frac{x}{y}\right)^{1/p'} dy = k \int f^p\, dx$$

and $$Q = k \int g^{p'}\, dy.$$

If $K > 0$, and there is equality, then

$$A f^p(x) \left(\frac{x}{y}\right)^{1/p'} \equiv B g^{p'}(y) \left(\frac{y}{x}\right)^{1/p} \tag{9.3.1}$$

for almost all y.[a] If we give y a value for which $g(y)$ is positive and finite, and for which the equivalence holds, we see that $f^p(x)$ is equivalent to a function Cx^{-1}, and this is inconsistent with the convergence of $\int f^p\, dx$. Hence either f or g is null. It may also be shown that the constant is the best possible.

There is another interesting method of proof due to Schur[b]. We have

$$\int_0^\infty f(x)\, dx \int_0^\infty K(x,y) g(y)\, dy = \int_0^\infty f(x)\, dx \int_0^\infty x K(x, xw) g(xw)\, dw$$
$$= \int_0^\infty f(x)\, dx \int_0^\infty K(1,w) g(xw)\, dw = \int_0^\infty K(1,w)\, dw \int_0^\infty f(x) g(xw)\, dx$$

(if any of the integrals is convergent). Applying Theorem **189** to the inner integral, and observing that

$$\int g^{p'}(xw)\, dx = \frac{1}{w} \int g^{p'}(y)\, dy,$$

we obtain (a); and (b) and (c) are corollaries, by Theorem **191**.

The case $K(x, y) = 1/(x+y)$ gives Theorem **316**. We shall return later (§ 9.9) to other applications.

[a] That is to say it is true, for almost all y, that the two sides of (9.3.1) are equal for almost all x. See § 6.3 (d).

[b] Schur (**1**). Schur supposes $p = 2$.

9.4. Extensions of Theorems 318 and 319. (1) The following theorem is in some ways more and in others less general than Theorem **318**.

320. *Suppose that $K(x,y)$ is a strictly decreasing function of x and y, and also satisfies the conditions* (i) *and* (ii) *of Theorem* **318**; *that $\lambda_m > 0$, $\mu_n > 0$,*

$$\Lambda_m = \lambda_1 + \lambda_2 + \ldots + \lambda_m, \quad \mathrm{M}_n = \mu_1 + \mu_2 + \ldots + \mu_n;$$

and that $p > 1$. Then

$$\Sigma\Sigma K(\Lambda_m, \mathrm{M}_n) \lambda_m{}^{1/p'} \mu_n{}^{1/p} a_m b_n < k (\Sigma a_m{}^p)^{1/p} (\Sigma b_n{}^{p'})^{1/p'}$$

unless (a) *or* (b) *is null*[a].

The special case $\Lambda_m = m$, $\mathrm{M}_n = n$ is also a special case of Theorem **318**.

We deduce Theorem **320** from Theorem **319** by a process which has many applications[b]. We interpret Λ_0 and M_0 as 0, and take, in Theorem **319**,

$$f(x) = \lambda_m{}^{-1/p} a_m \quad (\Lambda_{m-1} \leqq x < \Lambda_m),$$

$$g(y) = \mu_n{}^{-1/p'} b_n \quad (\mathrm{M}_{n-1} \leqq y < \mathrm{M}_n).$$

If we observe that

$$\int_{\Lambda_{m-1}}^{\Lambda_m} \int_{\mathrm{M}_{n-1}}^{\mathrm{M}_n} K(x,y) f(x) g(y)\, dx\, dy > \lambda_m{}^{1/p'} \mu_n{}^{1/p} a_m b_n K(\Lambda_m, \mathrm{M}_n)$$

unless $a_m = 0$ or $b_n = 0$, we obtain Theorem **320**.

If $K(x,y) = 1/(x+y)$ we obtain[c]

321. $$\Sigma\Sigma \frac{\lambda_m{}^{1/p'} \mu_n{}^{1/p}}{\Lambda_m + \mathrm{M}_n} a_m b_n < \frac{\pi}{\sin(\pi/p)} (\Sigma a_m{}^p)^{1/p} (\Sigma b_n{}^{p'})^{1/p'}$$

unless (a) *or* (b) *is null.*

(2) Theorems **318** and **319** may be extended to multiple series or integrals of any order.

322.[d] *Suppose that the n numbers $p, q, \ldots, r$ satisfy*

$$p > 1, \quad q > 1, \quad \ldots, \quad r > 1, \quad \frac{1}{p} + \frac{1}{q} + \ldots + \frac{1}{r} = 1;$$

[a] For the case $p = 2$ see Schur (**1**).

[b] Cf. § 6.4; and see, for example, § 9.11.

[c] Owen (**1**) gives a more general but less precise result.

[d] For the case $p = q = \ldots = r$ see Schur (**1**).

that $K(x, y, \ldots, z)$ is a positive function of the n variables $x, y, \ldots, z$, homogeneous of degree $-n+1$; and that

$$(9.4.1)\quad \int_0^\infty \ldots \int_0^\infty K(1, y, \ldots, z)\, y^{-\frac{1}{q}} \ldots z^{-\frac{1}{r}}\, dy \ldots dz = k.$$

Then

$$\int_0^\infty \int_0^\infty \ldots \int_0^\infty K(x, y, \ldots, z) f(x) g(y) \ldots h(z)\, dx\, dy \ldots dz$$
$$\leqq k\left(\int_0^\infty f^p dx\right)^{1/p} \left(\int_0^\infty g^q dy\right)^{1/q} \ldots \left(\int_0^\infty h^r dz\right)^{1/r}.$$

If further

$$y^{-1/q} \ldots z^{-1/r} K(1, y, \ldots, z), \quad x^{-1/p} \ldots z^{-1/r} K(x, 1, \ldots, z), \ldots$$

are decreasing functions of all the variables which they involve, then

$$\Sigma\Sigma \ldots \Sigma K(m, n, \ldots, s)\, a_m b_n \ldots c_s \leqq k\, (\Sigma a_m{}^p)^{1/p} (\Sigma b_n{}^q)^{1/q} \ldots (\Sigma c_s{}^r)^{1/r}.$$

In virtue of the homogeneity of K, the convergence of (9.4.1) implies the convergence and equality of all the n integrals of the same type.

Theorem **322** may be proved by straightforward generalisations of the proofs of Theorems **318** and **319**.

9.5. Best possible constants: proof of Theorem 317. We have still to prove Theorem **317**, which asserts that the constant $\pi \operatorname{cosec}(\pi/p)$ of Theorems **315** and **316** is 'the best possible', that is to say that the inequalities asserted by the theorems would be false, for some a_m, b_n or $f(x)$, $g(y)$, if $\pi \operatorname{cosec}(\pi/p)$ were replaced by any smaller number. The method which we use illustrates an important general principle and may be used in the proof of many theorems of this 'negative' character.

We take $\qquad a_m = m^{-(1+\epsilon)/p}, \quad b_n = n^{-(1+\epsilon)/p'},$

where ϵ is small and positive; we may suppose that $\epsilon < p'/2p$. We denote by $O(1)$ a number which may depend upon p and ϵ, but is bounded when p is fixed and $\epsilon \to 0$; and by $o(1)$ a number which satisfies these conditions and tends to zero with ϵ. Then

$$\frac{1}{\epsilon} = \int_1^\infty x^{-1-\epsilon}\, dx < \sum_1^\infty m^{-1-\epsilon} < 1 + \int_1^\infty x^{-1-\epsilon}\, dx = 1 + \frac{1}{\epsilon},$$

and so

$$(9.5.1)\quad \Sigma a_m^p = \Sigma m^{-1-\epsilon} = \frac{1}{\epsilon} + O(1), \quad \Sigma b_n^{p'} = \frac{1}{\epsilon} + O(1).$$

Also

$$\Sigma\Sigma \frac{a_m b_n}{m+n} > \int_1^\infty \int_1^\infty x^{-(1+\epsilon)/p} y^{-(1+\epsilon)/p'} \frac{dx\,dy}{x+y}$$

$$= \int_1^\infty x^{-1-\epsilon}\,dx \int_{1/x}^\infty u^{-(1+\epsilon)/p'} \frac{du}{1+u}.$$

The error in replacing the lower limit in the inner integral by 0 is less than $x^{-\alpha}/\alpha$, where α is positive and independent of ϵ;[a] and

$$\frac{1}{\alpha}\int_1^\infty x^{-1-\alpha-\epsilon}\,dx < \frac{1}{\alpha^2}.$$

Hence

$$\Sigma\Sigma \frac{a_m b_n}{m+n} > \int_1^\infty x^{-1-\epsilon}\,dx \int_0^\infty u^{-(1+\epsilon)/p'} \frac{du}{1+u} + O(1) \tag{9.5.2}$$

$$= \frac{1}{\epsilon}\left\{\frac{\pi}{\sin(\pi/p)} + o(1)\right\} + O(1) = \frac{1}{\epsilon}\left\{\frac{\pi}{\sin(\pi/p)} + o(1)\right\}.$$

It is plain from (9.5.1) and (9.5.2) that, if k is any number less than $\pi \operatorname{cosec}(\pi/p)$, then

$$\Sigma\Sigma \frac{a_m b_n}{m+n} > k\,(\Sigma a_m^p)^{1/p}\,(\Sigma b_n^{p'})^{1/p'}$$

when ϵ is sufficiently small.

This proves that the constant in (9.1.1) is the best possible. Since (9.1.1) can be deduced from (9.1.2) as in § 9.4, it follows that the constant in (9.1.2) is also the best possible. We could of course also prove this directly.

An alternative method is to take

$$a_m = m^{-1/p}, \quad b_n = n^{-1/p'}$$

when $m \leqq \mu$, $n \leqq \mu$, and $a_m = 0$, $b_n = 0$ otherwise, and to make μ tend to infinity. The principle is the same in each case; we make a_m and b_n depend upon a parameter (ϵ or μ) in such a manner that the series involved tend to infinity when the parameter tends

[a] It is less than

$$\int_0^{1/x} u^{-(1+\epsilon)/p'}\,du = \frac{x^{-\beta}}{\beta},$$

where

$$\beta = 1 - \frac{1+\epsilon}{p'} = \frac{1}{p} - \frac{\epsilon}{p'};$$

and we may take $\alpha = 1/2p$, if $\epsilon < p'/2p$.

to a limit, and compare their values for values of the parameter near this limit. The method is effective in the proof of very many theorems of the type of Theorem **317**. The inequalities (9.1.1) and (9.1.2) assert unattained upper bounds; except when both sides vanish, equality cannot occur; and it is for this reason that the introduction of a parameter (ϵ or μ) is necessary in the proof of the complementary theorem.

9.6. Further remarks on Hilbert's theorems[a]. Theorems **315** and **316** have been proved in many different ways and have very varied applications. In this and the next section we collect a number of remarks which concern both proofs and applications, and are intended to illustrate the connections between the theorems and various parts of the theory of functions.

(1) Theorem **315** may be deduced from Theorem **316** by the process which led us to Theorem **321**. We define $f(x)$ and $g(y)$ by

$$f(x)=a_m \quad (m-1\leqq x<m), \qquad g(y)=b_n \quad (n-1\leqq x<n),$$

and observe that then

$$\int_{m-1}^{m}\int_{n-1}^{n}\frac{f(x)g(y)}{x+y}\,dx\,dy\geqq\frac{a_m b_n}{m+n}.$$

Here, however, we can go a little further, since

$$\frac{1}{m+n-1-\alpha}+\frac{1}{m+n-1+\alpha}>\frac{2}{m+n-1}$$

for $0<\alpha<1$, and so[b]

$$\int_{m-1}^{m}\int_{n-1}^{n}\frac{dx\,dy}{x+y}>\frac{1}{m+n-1}.$$

If now we replace m and n by $m+1$ and $n+1$, we obtain a slightly sharper form of Theorem **315**, viz.

323. *If the conditions of Theorem* **315** *are satisfied, then*

$$\sum_0^\infty\sum_0^\infty\frac{a_m b_n}{m+n+1}<\frac{\pi}{\sin(\pi/p)}\left(\sum_0^\infty a_m^p\right)^{1/p}\left(\sum_0^\infty b_n^{p'}\right)^{1/p'}.$$

[a] We describe Theorems **315** (together with the sharper Theorem **323**) and **316** as 'Hilbert's theorems'. Strictly, Hilbert's theorem is Theorem **315**, with $p=2$.

[b] Associate elements of the integral symmetrically situated about the centre of the square of integration.

Several other proofs of Hilbert's theorem, for example the proof of Mulholland (**1**), the proof of Schur given in § 8.12, the proof of Fejér and F. Riesz given below, and the proof of Pólya and Szegö (**1**, I, 290), also give the result in this form. The last three are limited to the case $p=2$.

(2) The proof of Fejér and F. Riesz is based upon the theory of analytic functions, and proceeds as follows. Suppose that $f(z)=\Sigma a_n z^n$ is a polynomial of degree N with non-negative coefficients, not identically zero. Then, by Cauchy's Theorem,

$$\int_{-1}^{1} f^2(x)\,dx = -i\int_0^{\pi} f^2(e^{i\theta})\,e^{i\theta}\,d\theta;$$

and so

$$\int_0^1 f^2(x)\,dx < \int_{-1}^{1} f^2(x)\,dx \leqq \tfrac{1}{2}\int_{-\pi}^{\pi} |f(e^{i\theta})|^2\,d\theta, \tag{9.6.1}$$

or

$$\sum_0^N \sum_0^N \frac{a_m a_n}{m+n+1} < \pi \sum_0^N a_n^2.$$

If we make $N\to\infty$ we obtain Hilbert's theorem, with $a_m=b_m$ and '$\leqq$' for '$<$'. The first restriction is unimportant, since, after § 8.8, a symmetric bilinear form in [2, 2] has a bound equal to that of the corresponding quadratic form. To replace '$\leqq$' by '$<$' requires a refinement of the argument which we shall not discuss here.

The second inequality in (9.6.1) may be written

$$\int_{-1}^{1} |f(x)|^2\,dx \leqq \tfrac{1}{2}\int_{-\pi}^{\pi} |f(e^{i\theta})|^2\,d\theta,$$

and in this form it is valid whether the coefficients a_n are real or complex, and has important function-theoretic applications[a].

(3) Hilbert's original proof depended upon the identity

$$\int_{-\pi}^{\pi} t\left\{\sum_1^n (-1)^r (a_r \cos rt - b_r \sin rt)\right\}^2 dt = 2\pi(S-T), \tag{9.6.2}$$

where

$$S=\sum_1^n\sum_1^n \frac{a_r b_s}{r+s},\quad T=\sum_1^n{\sum_1^n}' \frac{a_r b_s}{r-s}$$

(the dash implying that pairs r, s for which $r=s$ are omitted). From this it follows that

$$2\pi|S-T| \leqq \pi\int_{-\pi}^{\pi}\left\{\sum_1^n (-1)^r (a_r \cos rt - b_r \sin rt)\right\}^2 dt = \pi^2 \sum_1^n (a_r^2+b_r^2). \tag{9.6.3}$$

[a] See Fejér and F. Riesz (**1**). The inequality is actually true (and in the stricter form with '$<$') for any $f(z)$, except $f(z)=0$, for which $\Sigma|a_n|^2$ is convergent; and this is a corollary of Hilbert's theorem, if this theorem is proved in some other way.

If $a_r = b_r$, T disappears, and we obtain

$$\sum_1^n \sum_1^n \frac{a_r a_s}{r+s} \leqq \pi \sum_1^n a_r^2; \tag{9.6.4}$$

and from (9.6.4), and the remark of § 8.8 quoted under (2), we deduce

$$\sum_1^n \sum_1^n \frac{a_r b_s}{r+s} \leqq \pi \left(\sum_1^n a_r^2\right)^{\frac{1}{2}} \left(\sum_1^n b_r^2\right)^{\frac{1}{2}} \leqq \tfrac{1}{2}\pi \left(\sum_1^n a_r^2 + \sum_1^n b_r^2\right). \tag{9.6.5}$$

From (9.6.3) and (9.6.5) it follows that

$$|T| = \left|\sum_1^n \sum_1^n{}' \frac{a_r b_s}{r-s}\right| \leqq \pi \left(\sum_1^n a_r^2 + \sum_1^n b_r^2\right);$$

and hence, on grounds of homogeneity, that

$$\left|\sum_1^n \sum_1^n{}' \frac{a_r b_s}{r-s}\right| \leqq 2\pi \left(\sum_1^n a_r^2\right)^{\frac{1}{2}} \left(\sum_1^n b_r^2\right)^{\frac{1}{2}}.$$

This gives the second result of Theorem **294**, except that the constant 2π is not the best possible constant.

9.7. Applications of Hilbert's theorems. (1) As an application of Hilbert's theorem to the theory of analytic functions we select the following. Suppose that $f(z)$ is regular in $|z| < 1$ and belongs to the 'complex Lebesgue class L', i.e. that

$$\frac{1}{2\pi} \int_{-\pi}^{\pi} |f(re^{i\theta})|\, d\theta$$

is bounded for $r < 1$. If $f(z)$ is 'wurzelfrei', i.e. has no zeros in $|z| < 1$, then

$$f(z) = \Sigma c_n z^n = g^2(z) = (\Sigma a_n z^n)^2,$$

where $g(z)$ also is regular in $|z| < 1$. Since $\int |g(re^{i\theta})|^2 d\theta$ is bounded, $\Sigma |a_n|^2$ is convergent, and therefore, by Theorem **323**,

$$\Sigma \frac{|a_m|\,|a_n|}{m+n+1}$$

is convergent. *A fortiori*

$$\sum_\nu \frac{|c_\nu|}{\nu+1} = \sum_\nu \frac{1}{\nu+1} \left| \sum_{m+n=\nu} a_m a_n \right|$$

is convergent.

It is fairly easy, by a method which is familiar in this part of the theory of analytic functions, to extend the conclusion to general f (not necessarily 'wurzelfrei')[a]. We thus obtain the

[a] See F. Riesz (3), Hardy and Littlewood (2). We can express f as the sum of two 'wurzelfrei' functions of L.

theorem: *if $f(z)$ belongs to L in $|z|<1$, then its integrated power series is absolutely convergent for $|z|=1$.*[a]

(2) As an application of Hilbert's series theorem to the theory of functions of a real variable, we prove

324.[b] *If $f(x)$ is real, L^2, and not null, in $(0, 1)$, and*

$$a_n = \int_0^1 x^n f(x)\,dx \quad (n=0, 1, 2, \ldots),$$

then
$$\Sigma a_n^2 < \pi \int_0^1 f^2(x)\,dx.$$

The constant is the best possible.

We may plainly suppose $f(x) \geqq 0$. Then, if (b_n) is any non-negative and non-null sequence,

$$\Sigma a_n b_n = \Sigma b_n \int_0^1 x^n f(x)\,dx = \int_0^1 (\Sigma b_n x^n) f(x)\,dx,$$

$$(\Sigma a_n b_n)^2 \leqq \int_0^1 (\Sigma b_n x^n)^2\,dx \int_0^1 f^2(x)\,dx$$

$$= \Sigma\Sigma \frac{b_m b_n}{m+n+1} \cdot \int_0^1 f^2(x)\,dx < \pi\, \Sigma b_n^2 \cdot \int_0^1 f^2(x)\,dx,$$

by Hilbert's theorem. The result now follows from Theorem **15**.

To prove the constant π the best possible, consider

$$f(x) = (1-x)^{\epsilon - \frac{1}{2}},$$

and make ϵ tend to zero.

The integrals a_n are called the *moments* of $f(x)$ in $(0, 1)$ and are important in many theories.

Here we have deduced Theorem **324** from Theorem **323** (with $p=2$) and Theorem **15** (the converse of Hölder's inequality). We can, if we please, reverse the argument, deducing Theorem **323** (with $p=2$) from Theorem **324** and Theorem **191** (the integral analogue of Theorem **15**). Suppose that $g(x) = \Sigma b_n x^n$, where b_n is

[a] Hardy and Littlewood (**2**). The theorem may also be stated in the form 'if a power series $g(z) = \Sigma b_n z^n$ is of bounded variation in $|z|<1$, then $\Sigma |b_n|$ is convergent'. For this form of the theorem, and for more precise results, see Fejér (**1**).

[b] A much more general inequality, but without the best possible value of the constant, is proved by Hardy and Littlewood (**1**). See also Hardy (**10**).

non-negative and not always 0, and that $f(x)$ is any non-negative and non-null function. Then

$$\int_0^1 fg\,dx = \int_0^1 (\Sigma b_n x^n) f\,dx = \Sigma b_n \int_0^1 x^n f\,dx = \Sigma a_n b_n,$$

$$\left(\int_0^1 fg\,dx\right)^2 = (\Sigma a_n b_n)^2 \leqq \Sigma a_n{}^2 \Sigma b_n{}^2 < \pi \Sigma b_n{}^2 \int_0^1 f^2\,dx,$$

by Theorem **324**. Since this is true for all f, it follows, by Theorem **191**, that

$$\int_0^1 g^2\,dx < \pi \Sigma b_n{}^2;$$

and this is equivalent to Theorem **323**.

It is plain that when two inequalities, each involving a constant factor, are 'reciprocal' in this sense, each being deducible from the other in this way by the converse of Hölder's inequality, then one constant must be best possible if the other is. We shall meet with another application of this principle later (§ 9.10 (1)).

(3) As a corollary of Theorem **316** (with $p=2$), we prove

325.[a] *Suppose that* $a_n \geqq 0$ *and that the summations run from* 0 *to* ∞, *and that*

$$A(x) = \Sigma a_n x^n, \quad A^*(x) = \Sigma \frac{a_n x^n}{n!}. \tag{9.7.1}$$

Then

$$\Sigma\Sigma \frac{a_m a_n}{m+n+1} \leqq \pi\, \Sigma\Sigma \frac{(m+n)!}{m!\,n!} \frac{a_m a_n}{2^{m+n+1}}, \tag{9.7.2}$$

$$\int_0^1 A^2(x)\,dx \leqq \pi \int_0^\infty \{e^{-x} A^*(x)\}^2\,dx. \tag{9.7.3}$$

It may be verified at once by expansion and term-by-term integration that (9.7.2) and (9.7.3) are equivalent.

To prove (9.7.3) we observe that

$$A(x) = \int_0^\infty e^{-t} A^*(xt)\,dt = \frac{1}{x}\int_0^\infty e^{-u/x} A^*(u)\,du$$

and so

$$\int_0^1 A^2(x)\,dx = \int_0^1 \frac{dx}{x^2}\left(\int_0^\infty e^{-u/x} A^*(u)\,du\right)^2 = \int_1^\infty dy \left(\int_0^\infty e^{-uy} A^*(u)\,du\right)^2$$

$$= \int_0^\infty dw \left(\int_0^\infty e^{-uw} \alpha(u)\,du\right)^2,$$

where

$$\alpha(u) = e^{-u} A^*(u).$$

[a] Widder (1), Hardy (9).

This is

$$\int_0^\infty dw \int_0^\infty e^{-uw}\alpha(u)\,du \int_0^\infty e^{-vw}\alpha(v)\,dv = \int_0^\infty\int_0^\infty \frac{\alpha(u)\alpha(v)}{u+v}\,du\,dv$$

$$\leqq \pi\int_0^\infty \alpha^2(u)\,du = \pi\int_0^\infty \{e^{-u}A^*(u)\}^2\,du,$$

by Theorem **316**.

It is easy to see that the constant π is the best possible. The relations between the functions (9.7.1) are important in the theory of divergent series, particularly in connection with singularities of analytic functions.

9.8. Hardy's inequality. The two theorems which we discuss next were discovered in the course of attempts to simplify the proofs then known of Hilbert's theorems[a].

We might require only an imperfect form of Theorem **315**: *the double series is convergent whenever Σa_n^p and $\Sigma b_n^{p'}$ are convergent.* It would then be natural to argue as follows. We divide the double series into two parts S_1, S_2 by the diagonal $m=n$, and consider the part S_1 in which $m \leqq n$. Then

$$S_1 = \sum_{m\leqq n}\sum \frac{a_m b_n}{m+n} \leqq \sum_{m\leqq n}\sum \frac{a_m b_n}{n} = \Sigma \frac{A_n}{n} b_n,$$

where
$$A_n = a_1 + a_2 + \dots + a_n.$$

Since $\Sigma b_n^{p'}$ is convergent, the last series is convergent whenever $\Sigma n^{-p} A_n^p$ is convergent, and, to prove the convergence of S_1, it is enough to prove that the convergence of the last series is a consequence of that of Σa_n^p. The convergence of S_2 could then be proved in the same way.

This line of argument leads up to and is completed by the following theorem.

326. *If $p>1$, $a_n \geqq 0$, and $A_n = a_1 + a_2 + \dots + a_n$, then*

$$\Sigma\left(\frac{A_n}{n}\right)^p < \left(\frac{p}{p-1}\right)^p \Sigma a_n^p, \tag{9.8.1}$$

unless all the a are zero. The constant is the best possible.

[a] It was a considerable time before any really simple proof of Hilbert's double series theorem was found.

The corresponding theorem for integrals is

327.[a] *If $p>1, f(x)\geqq 0$, and $F(x)=\int_0^x f(t)\,dt$, then*

$$\int_0^\infty \left(\frac{F}{x}\right)^p dx < \left(\frac{p}{p-1}\right)^p \int_0^\infty f^p\,dx, \tag{9.8.2}$$

unless $f\equiv 0$. The constant is the best possible.

These theorems were first proved by Hardy (**2**), except that Hardy was unable to fix the constant in Theorem **326**. This imperfection was removed by Landau (**4**). A great many alternative proofs of the theorems have been given by various writers, for example by Broadbent (**1**), Elliott (**1**), Grandjot (**1**), Hardy (**4**), Kaluza and Szegö (**1**), Knopp (**1**). We begin by giving Elliott's proof of Theorem **326** and Hardy's proof of Theorem **327.**[b]

(i) In proving Theorem **326** we may suppose that $a_1>0$. For if we suppose that $a_1=0$, and replace a_{n+1} by b_n, (9.8.1) becomes

$$\left(\frac{b_1}{2}\right)^p + \left(\frac{b_1+b_2}{3}\right)^p + \ldots < \left(\frac{p}{p-1}\right)^p (b_1^p + b_2^p + \ldots),$$

an inequality weaker than (9.8.1) itself.

Let us write α_n for A_n/n, and agree that any number with suffix 0 is 0. We have then

$$\begin{aligned}
\alpha_n^p - \frac{p}{p-1}\alpha_n^{p-1}a_n &= \alpha_n^p - \frac{p}{p-1}\{n\alpha_n-(n-1)\alpha_{n-1}\}\alpha_n^{p-1}\\
&= \alpha_n^p\left(1-\frac{np}{p-1}\right) + \frac{(n-1)p}{p-1}\alpha_n^{p-1}\alpha_{n-1}\\
&\leqq \alpha_n^p\left(1-\frac{np}{p-1}\right) + \frac{n-1}{p-1}\{(p-1)\alpha_n^p + \alpha^p{}_{n-1}\}^{\text{c}}\\
&= \frac{1}{p-1}\{(n-1)\alpha^p{}_{n-1} - n\alpha_n^p\}.
\end{aligned}$$

Hence
$$\sum_1^N \alpha_n^p - \frac{p}{p-1}\sum_1^N \alpha_n^{p-1}a_n \leqq -\frac{N\alpha_N^p}{p-1} \leqq 0;$$

[a] We have already encountered this theorem in Ch. VII, but the proof which we gave there (in detail only when $p=2$) was intended primarily as an illustration of variational methods and has no particular pretensions to simplicity.

[b] We have expanded the proofs so as to deal with the question of equality.

[c] By Theorem **9**.

and therefore, by Theorem **13**,

$$(9.8.3)\quad \sum_1^N \alpha_n{}^p \leqq \frac{p}{p-1}\sum_1^N \alpha_n{}^{p-1}a_n \leqq \frac{p}{p-1}\left(\sum_1^N a_n{}^p\right)^{1/p}\left(\sum_1^N \alpha_n{}^p\right)^{1/p'}.$$

Dividing by the last factor on the right (which is certainly positive), and raising the result to the pth power, we obtain

$$(9.8.4)\qquad \sum_1^N \alpha_n{}^p \leqq \left(\frac{p}{p-1}\right)^p \sum_1^N a_n{}^p.$$

When we make N tend to infinity we obtain (9.8.1), except that we have '$\leqq$' in place of '$<$'. In particular we see that $\Sigma\alpha_n{}^p$ is finite.

Returning to (9.8.3), and replacing N by ∞, we obtain

$$(9.8.5)\quad \Sigma\alpha_n{}^p \leqq \frac{p}{p-1}\Sigma\alpha_n{}^{p-1}a_n \leqq \frac{p}{p-1}(\Sigma a_n{}^p)^{1/p}(\Sigma\alpha_n{}^p)^{1/p'}.$$

There is inequality in the second place unless $(a_n{}^p)$ and $(\alpha_n{}^p)$ are proportional, i.e. unless $a_n = C\alpha_n$, where C is independent of n. If this is so then (since $a_1 = \alpha_1 > 0$) C must be 1, and then $A_n = na_n$ for all n. This is only possible if all the a are equal, and this is inconsistent with the convergence of $\Sigma a_n{}^p$. Hence

$$(9.8.6)\qquad \Sigma\alpha_n{}^p < \frac{p}{p-1}(\Sigma a_n{}^p)^{1/p}(\Sigma\alpha_n{}^p)^{1/p'};$$

and (9.8.1) follows from (9.8.6) as (9.8.4) followed from (9.8.3).

To prove the constant factor the best possible, we take

$$a_n = n^{-1/p}\ (n \leqq N),\quad a_n = 0\ (n > N).$$

Then

$$\Sigma a_n{}^p = \sum_1^N \frac{1}{n},$$

$$A_n = \sum_1^n \nu^{-1/p} > \int_1^n x^{-1/p}\,dx = \frac{p}{p-1}\{n^{(p-1)/p} - 1\}\quad (n \leqq N),$$

$$\left(\frac{A_n}{n}\right)^p > \left(\frac{p}{p-1}\right)^p \frac{1-\epsilon_n}{n}\quad (n \leqq N),$$

where $\epsilon_n \to 0$ when $n \to \infty$. It follows that

$$\Sigma\left(\frac{A_n}{n}\right)^p > \sum_1^N \left(\frac{A_n}{n}\right)^p > \left(\frac{p}{p-1}\right)^p (1-\eta_N)\,\Sigma a_n{}^p,$$

where $\eta_N \to 0$ when $N \to \infty$. Hence any inequality of the type

$$\Sigma\left(\frac{A_n}{n}\right)^p < \left(\frac{p}{p-1}\right)^p (1-\epsilon)\,\Sigma a_n{}^p$$

is false if a_n is chosen as above and N is sufficiently large.

An alternative procedure is to take $a_n = n^{-(1/p)-\epsilon}$ for all n, and to make ϵ small. Compare § 9.5: it was this procedure which we followed there.

(ii) We may suppose f is not null. Let $n > 0$, $f_n = \mathrm{Min}\,(f, n)$, $F_n = \int_0^x f_n\,dx$, and let X_0 be so large that f, and so f_n, F_n, are not null in $(0, X)$ when $X > X_0$. We have

$$\int_0^X \left(\frac{F_n}{x}\right)^p dx = -\frac{1}{p-1}\int_0^X F_n^p \frac{d}{dx}(x^{1-p})\,dx$$

$$= \left[-\frac{x^{1-p}F_n^p(x)}{p-1}\right]_0^X + \frac{p}{p-1}\int_0^X \left(\frac{F_n}{x}\right)^{p-1} f_n\,dx$$

$$\leqq \frac{p}{p-1}\int_0^X \left(\frac{F_n}{x}\right)^{p-1} f_n\,dx,$$

since the integrated term vanishes at $x = 0$ in virtue of $F_n = o(x)$. Hence

$$\int_0^X \left(\frac{F_n}{x}\right)^p dx \leqq \left(\frac{p}{p-1}\right)\left(\int_0^X \left(\frac{F_n}{x}\right)^p dx\right)^{1/p'}\left(\int_0^X f_n^p\,dx\right)^{1/p}. \tag{9.8.7}$$

The left-hand side being positive (and finite), this gives

$$\int_0^X \left(\frac{F_n}{x}\right)^p dx \leqq \left(\frac{p}{p-1}\right)^p \int_0^X f_n^p\,dx.$$

We make $n \to \infty$ in this, the result being to suppress the two suffixes n; now making $X \to \infty$, we have

$$\int_0^\infty \left(\frac{F}{x}\right)^p dx \leqq \left(\frac{p}{p-1}\right)^p \int_0^\infty f^p\,dx,$$

the desired result with '$\leqq$' for '$<$'. Making $n \to \infty$ and then $X \to \infty$ in (9.8.7) we have

$$\int_0^\infty \left(\frac{F}{x}\right)^p dx \leqq \frac{p}{p-1}\left(\int_0^\infty \left(\frac{F}{x}\right)^p\right)^{1/p'}\left(\int_0^\infty f^p\,dx\right)^{1/p}. \tag{9.8.8}$$

The integrals in this being now known to be all finite and positive, (9.8.8) gives

$$\int_0^\infty \left(\frac{F}{x}\right)^p dx < \left(\frac{p}{p-1}\right)^p \int_0^\infty f^p\, dx,$$

unless $x^{-p}F^p$ and f^p are effectively proportional, which is impossible, since it would make f a power of x, and $\int f^p dx$ divergent.

The proof that the constant is the best possible follows the same lines as before: take $f(x) = 0$ for $x < 1$, $f(x) = x^{-(1/p)-\epsilon}$ for $x \geqq 1$.

Elliott's proof of Theorem **326** applies to Theorem **327** also, with the obvious modifications. The proof of Theorem **327** given in (ii) may be adapted to series, but does not give the best possible value of the constant.

(iii) The following proof of Theorem **327** (due to Ingham) is also interesting: we shall be content with proving the form with '$\leqq$'. We use Theorem **203**, supposing that the intervals of integration are each $(0, 1)$, that the weight-functions are 1, and that

$$r = 1, \quad s = p > 1, \quad f(x, y) = f(xy).$$

Then
$$\mathfrak{M}_1^{(x)} f(xy) = \int_0^1 f(xy)\, dx = \frac{F(y)}{y},$$

$$\mathfrak{M}_p^{(y)} f(xy) = \left\{\int_0^1 f^p(xy)\, dy\right\}^{1/p} = \left\{\frac{1}{x}\int_0^x f^p(t)\, dt\right\}^{1/p} \leqq \left\{\frac{1}{x}\int_0^1 f^p(t)\, dt\right\}^{1/p}$$

for $x \leqq 1$. Hence, by Theorem **203**,

$$\left\{\int_0^1 \left(\frac{F}{y}\right)^p dy\right\}^{1/p} \leqq \left(\int_0^1 f^p\, dt\right)^{1/p} \int_0^1 x^{-1/p}\, dx = \frac{p}{p-1}\left(\int_0^1 f^p\, dx\right)^{1/p}.$$

We then obtain the result by putting

$$x = X/c, \quad f(X/c) = g(X),$$

replacing X, g by x, f, and making $c \to \infty$.

9.9. Further integral inequalities. There are many analogues and extensions of Theorems **326** and **327**, which have been proved by different writers in different ways; and we give some of these theorems here. We consider integral inequalities first, since we can derive most of these in a simple and uniform manner from Theorem **319**, and the corresponding theorems for series sometimes involve slight additional complications.

(1) Take, in Theorem **319**,

$$K(x, y) = 1/y \ (x \leqq y), \quad K(x, y) = 0 \ (x > y).$$

Then, if $p>1$,

$$k=\int_0^\infty K(x,1)\,x^{-1/p}\,dx=\int_0^1 x^{-1/p}\,dx=\frac{p}{p-1},$$

and all the conditions on K are satisfied. Hence (b) and (c) of Theorem **319** give

$$\int_0^\infty dy\left(\frac{1}{y}\int_0^y f(x)\,dx\right)^p \leqq \left(\frac{p}{p-1}\right)^p \int_0^\infty f^p\,dx \tag{9.9.1}$$

and

$$\int_0^\infty dx\left(\int_x^\infty \frac{g(y)}{y}\,dy\right)^{p'} \leqq \left(\frac{p}{p-1}\right)^{p'} \int_0^\infty g^{p'}\,dy. \tag{9.9.2}$$

Of these inequalities, (9.9.1) is Theorem **327**, with '$\leqq$' for '$<$'; we cannot quote '$<$' from the general theorem because K is not always positive. But equality in (9.9.1), with a non-null f, involves

$$\iint K(x,y)\,f(x)\,g(y)\,dx\,dy=\frac{p}{p-1}\left(\int f^p\,dx\right)^{1/p}\left(\int g^{p'}\,dx\right)^{1/p'},$$

with non-null f and g. The argument of § 9.3 then shows that (9.3.1) is true for $x<y$, and that $f\equiv Cx^{-1/p}$ for small x, which is inconsistent with the convergence of $\int f^p\,dx$.

Similarly we can prove that there is inequality in (9.9.2) unless g is null. A trivial transformation then gives

328. *If* $p>1$ *and*

$$F(x)=\int_x^\infty f(t)\,dt,$$

then

$$\int_0^\infty F^p\,dx<p^p\int_0^\infty (xf)^p\,dx, \tag{9.9.3}$$

unless $f\equiv 0$. *The constant is the best possible.*

(2) More generally, take

$$K(x,y)=\frac{1}{\Gamma(r)}\frac{(y-x)^{r-1}}{y^r}\quad (x<y),\qquad K(x,y)=0\quad (x\geqq y)$$

with $r>0$. With $r=1$, we come back to (1). We now have

$$k=\frac{1}{\Gamma(r)}\int_0^1 x^{-1/p}\,(1-x)^{r-1}\,dx=\frac{\Gamma\left(1-\frac{1}{p}\right)}{\Gamma\left(r+1-\frac{1}{p}\right)}.$$

We are thus led to

329. *If $p>1$, $r>0$, and*

$$f_r(x)=\frac{1}{\Gamma(r)}\int_0^x (x-t)^{r-1}f(t)\,dt, \tag{9.9.4}$$

then

$$\int_0^\infty \left(\frac{f_r}{x^r}\right)^p dx < \left\{\frac{\Gamma\left(1-\frac{1}{p}\right)}{\Gamma\left(r+1-\frac{1}{p}\right)}\right\}^p \int_0^\infty f^p\,dx, \tag{9.9.5}$$

unless $f\equiv 0$. If

$$f_r(x)=\frac{1}{\Gamma(r)}\int_x^\infty (t-x)^{r-1}f(t)\,dt, \tag{9.9.6}$$

then

$$\int_0^\infty f_r^p\,dx < \left\{\frac{\Gamma\left(\frac{1}{p}\right)}{\Gamma\left(r+\frac{1}{p}\right)}\right\}^p \int_0^\infty (x^r f)^p\,dx, \tag{9.9.7}$$

unless $f\equiv 0$. In each case the constant is the best possible.

The function $f_r(x)$ of (9.9.4) is the 'Riemann-Liouville integral'[a] of $f(x)$ of order r, with 'origin' 0. The function (9.9.6) is the 'Weyl integral' of order r, which is in some ways more convenient, especially in the theory of Fourier series.

(3) Take

$$K(x,y)=\frac{y^{\alpha-1}}{x^\alpha}\quad (x\leqq y),\qquad K(x,y)=0\quad (x>y),$$

with $\alpha<1/p'$. Then

$$k=\int_0^1 x^{-\alpha-1/p}\,dx=\frac{p}{p-p\alpha-1},$$

and (*b*) and (*c*) of Theorem **319** give

$$\int_0^\infty y^{p(\alpha-1)}\left(\int_0^y x^{-\alpha}f(x)\,dx\right)^p dy<\left(\frac{p}{p-\alpha p-1}\right)^p\int_0^\infty f^p\,dx, \tag{9.9.8}$$

$$\int_0^\infty x^{-\alpha p'}\left(\int_x^\infty y^{\alpha-1}g(y)\,dy\right)^{p'} dx<\left(\frac{p'}{1-\alpha p'}\right)^{p'}\int_0^\infty g^{p'}\,dy. \tag{9.9.9}$$

Changing the notation, we obtain

330. *If $p>1$, $r\neq 1$, and $F(x)$ is defined by*

$$F(x)=\int_0^x f(t)\,dt\quad (r>1),\qquad F(x)=\int_x^\infty f(t)\,dt\quad (r<1),$$

[a] See § 10.17. Part of Theorem **329** is proved by Knopp (**3**).

then

$$\int_0^\infty x^{-r} F^p\,dx < \left(\frac{p}{|r-1|}\right)^p \int_0^\infty x^{-r}(xf)^p\,dx, \tag{9.9.10}$$

unless $f \equiv 0$.[a]

The constant is the best possible. It is also easy to verify that, when $p=1$, the two sides of (9.9.10) are equal.

9.10. Further theorems concerning series. Among the analogues and extensions of Theorem **326** we select the following.

(1) The theorem related to Theorem **326** as Theorem **328** is to Theorem **327** is

331.[b] *If* $p>1$, *then*

$$\Sigma(a_n + a_{n+1} + \ldots)^p < p^p \Sigma(na_n)^p,$$

unless (a_n) *is null. The constant is the best possible.*

This theorem is 'reciprocal' to Theorem **326** in the sense of § 9.7 (2), i.e. deducible from the latter theorem by the converse of Hölder's inequality. It may be instructive to set out the proof in detail, although what we say amounts to a repetition in a special case of what we have explained more generally before[c].

If $K(x,y)$ is defined as in § 9.9 (1), then

(9.10.1)

$$\Sigma\Sigma K(m,n)\,a_m b_n = \sum_{m\leqq n}\sum \frac{a_m b_n}{n} = \Sigma \frac{a_1+a_2+\ldots+a_n}{n} b_n = \Sigma \frac{A_n}{n} b_n$$

$$\leqq \left\{\Sigma\left(\frac{A_n}{n}\right)^p\right\}^{1/p} (\Sigma b_n^{p'})^{1/p'} < \frac{p}{p-1} (\Sigma a_m^{p})^{1/p} (\Sigma b_n^{p'})^{1/p'},$$

by Theorems **13** and **326**, unless (a) or (b) is nul.

On the other hand

$$\Sigma\Sigma K(m,n)\,a_m b_n = \Sigma a_m \left(\frac{b_m}{m} + \frac{b_{m+1}}{m+1} + \ldots\right);$$

and the maximum of this, for all (a) for which $\Sigma a_m^p = 1$, is, by Theorem **15**,

$$\left\{\Sigma\left(\frac{b_m}{m} + \frac{b_{m+1}}{m+1} + \ldots\right)^{p'}\right\}^{1/p'}.$$

[a] For a direct proof see Hardy (**5**).
[b] Copson (**1**); see also Hardy (**6**).
[c] See § 8.7.

Hence, by (9.10.1),

$$\left\{\Sigma\left(\frac{b_m}{m}+\frac{b_{m+1}}{m+1}+\dots\right)^{p'}\right\}^{1/p'}<\frac{p}{p-1}(\Sigma b_n{}^{p'})^{1/p'}=p'(\Sigma b_n{}^{p'})^{1/p'}.$$

Changing b_m into ma_m, and p' into p, we obtain Theorem **331**.

That the constant p^p is the best possible follows from the last remark of § 9.7 (2).

(2) **332.** *If* $p>1$, $a_n \geqq 0$, $\lambda_n>0$, *and*

$$\Lambda_n=\lambda_1+\lambda_2+\dots+\lambda_n,\quad A_n=\lambda_1 a_1+\lambda_2 a_2+\dots+\lambda_n a_n,$$

then
$$\Sigma\lambda_n\left(\frac{A_n}{\Lambda_n}\right)^p<\left(\frac{p}{p-1}\right)^p\Sigma\lambda_n a_n{}^p,$$

unless (a_n) *is null.*

This theorem, which is related to Theorem **326** as Theorem **321** is related to Theorem **315**, may be proved in various ways. In the first place, it may be deduced from Theorem **320** by a specialisation of K (as Theorem **327** was deduced from Theorem **319** in § 9.9); but the question of possible equality then needs a little attention. Perhaps the simplest proof is by a direct adaptation of Elliott's argument in § 9.8. If $\alpha_n=A_n/\Lambda_n$, we find that

$$\lambda_n\alpha_n{}^p-\frac{p}{p-1}\lambda_n\alpha_n{}^{p-1}a_n\leqq\frac{1}{p-1}(\Lambda_{n-1}\alpha^p{}_{n-1}-\Lambda_n\alpha_n{}^p);$$

and the proof may be completed as in § 9.8.[a]

The theorem may also be deduced from Theorem **327** by taking f to be an appropriate step-function (the process by which, in § 9.4, we deduced Theorem **320** from Theorem **319**). We shall not set this out in detail[b]; but the remark raises questions of which we say something more in the next section.

9.11. Deduction of theorems on series from theorems on integrals[c]. The process of deduction just referred to, and actually used in § 9.4, is very natural and often effective. It is however apt to lead to difficulties of detail, so that direct methods are usually preferable. We illustrate this by giving a deduction of

[a] For the details see Copson (**1**).
[b] For the details (which are rather troublesome) see Hardy (**4**).
[c] Compare §§ 6.4 and 9.4 (1).

Theorem **326** from Theorem **327**, which leads us incidentally to a remark of considerable intrinsic interest.

We observe first that *it is sufficient to prove Theorem* **326** *on the hypothesis that* a_n *decreases as* n *increases.* This follows from a theorem which is of sufficient interest to be stated separately.

333. *If the* a_n *are given except in arrangement, and* $\phi(u)$ *is a positive increasing function of* u, *then*

$$\Sigma\phi\left(\frac{A_n}{n}\right)$$

is greatest when the a_n *are arranged in decreasing order.*

To prove Theorem **333** we have only to observe that, if $\nu > \mu$ and $a_\nu > a_\mu$, the effect of exchanging a_μ and a_ν is to leave A_n unchanged when $n < \mu$ or $n \geqq \nu$, and to increase A_n when $\mu \leqq n < \nu$. The theorem is one of a type which we shall discuss in much greater detail in Ch. X.

Suppose now that a_n decreases, and that Theorem **327** has been proved. We define $f(x)$ by

$$f(x) = a_n \quad (n-1 \leqq x < n).$$

Then

$$\Sigma a_n{}^p = \int_0^\infty f^p\,dx. \tag{9.11.1}$$

If $n < x < n+1$, then

$$\frac{F(x)}{x} = \frac{a_1 + a_2 + \dots + a_n + (x-n)a_{n+1}}{x} = \frac{A_n - na_{n+1} + xa_{n+1}}{x},$$

and

$$A_n - na_{n+1} \geqq 0,$$

so that F/x decreases from A_n/n to $A_{n+1}/(n+1)$ when x increases from n to $n+1$. Hence

$$\frac{F}{x} \geqq \frac{A_{n+1}}{n+1} \quad (n < x < n+1)$$

and so

$$\int_0^\infty \left(\frac{F}{x}\right)^p dx \geqq \sum_1^\infty \left(\frac{A_n}{n}\right)^p. \tag{9.11.2}$$

Theorem **326** now follows from (9.11.1), (9.11.2), and Theorem **333**.

If the reader will try to deduce Theorem **331** from Theorem **328** similarly, he will find some difficulty. Something is lost in the passage from integrals to series, and it is by no means always that (as here) the passage can be made without damage to the final result.

9.12. Carleman's inequality. If in Theorem **326** we write a_n for $a_n{}^p$, we obtain

$$(9.12.1)\quad \Sigma\left(\frac{a_1{}^{1/p}+a_2{}^{1/p}+\ldots+a_n{}^{1/p}}{n}\right)^p < \left(\frac{p}{p-1}\right)^p \Sigma a_n.$$

If we make $p\to\infty$, and use Theorem **3**, we obtain

$$\Sigma\,(a_1 a_2 \ldots a_n)^{1/n} \leqq e\,\Sigma a_n;$$

and this suggests the more complete theorem which follows.

334.[a] $$\Sigma\,(a_1 a_2 \ldots a_n)^{1/n} < e\,\Sigma a_n,$$

unless (a_n) *is null. The constant is the best possible.*

It is natural to attempt to prove the complete theorem by means of Theorem **9**; but a direct application of Theorem **9** to the left-hand side of (9.12.1) is insufficient[b]. To remedy this, we apply Theorem 9 not to $a_1, a_2, \ldots, a_n$ but to $c_1 a_1, c_2 a_2, \ldots, c_n a_n$, and choose the c so that, when Σa_n is near the boundary of convergence, these numbers shall be 'roughly equal'. This requires that c_n shall be roughly of the order of n.

These considerations suggest the following proof. We have

$$\Sigma\,(a_1 a_2 \ldots a_n)^{1/n} = \Sigma\left(\frac{c_1 a_1 . c_2 a_2 \ldots c_n a_n}{c_1 c_2 \ldots c_n}\right)^{1/n}$$

$$\leqq \sum_n (c_1 c_2 \ldots c_n)^{-1/n} \frac{1}{n} \sum_{m\leqq n} c_m a_m$$

$$= \sum_m a_m c_m \sum_{n\geqq m} \frac{1}{n} (c_1 c_2 \ldots c_n)^{-1/n}.$$

[a] Carleman (**1**). The proof given here is due to Pólya (**2**). The less precise convergence theorem (without the constant e) was found independently by other writers, and there are a number of proofs of one form or the other of the theorem. See Collingwood (in Valiron, **1**, 186, where there is a proof due to Littlewood), Kaluza and Szegö (**1**), Knopp (**1**), Ostrowski (**2**, 201–204).

[b] $$\sum_n (a_1 a_2 \ldots a_n)^{1/n} \leqq \sum_n \frac{1}{n} \sum_{m\leqq n} a_m = \Sigma a_m \sum_{n\geqq m} \frac{1}{n};$$

but the right-hand side is generally divergent. The proof fails because the a in $a_1 a_2 \ldots a_n$ are 'too unequal', and too much is lost in replacing $\mathfrak{G}(a)$ by $\mathfrak{A}(a)$.

In order that the inner summation should be easily effected, we choose

$$(c_1 c_2 \dots c_n)^{1/n} = n+1,$$

when

$$c_m = \frac{(m+1)^m}{m^{m-1}}, \quad \sum_{n \geqq m} \frac{1}{n}(c_1 c_2 \dots c_n)^{-1/n} = \sum_{n \geqq m} \frac{1}{n(n+1)} = \frac{1}{m},$$

and then

$$\Sigma (a_1 a_2 \dots a_n)^{1/n} \leqq \Sigma \frac{a_m c_m}{m} = \Sigma a_m \left(1+\frac{1}{m}\right)^m < e \Sigma a_m,$$

by Theorem **140**, unless a_m is null.

We can prove the constant best possible as in § 9.5. We may, for example, take $a_n = 1/n$ for $n \leqq \mu$, $a_n = 0$ for $n > \mu$, and make μ tend to infinity.

The corresponding integral theorem is

335.[a] *If f is not null, then*

$$\int_0^\infty \exp\left\{\frac{1}{x}\int_0^x \log f(t)\,dt\right\} dx < e\int_0^\infty f(x)\,dx.$$

9.13. Theorems with $0 < p < 1$. We have supposed so far that the parameter p involved in our theorems is greater than 1. A good many of them, however, have analogues with a p less than 1, and we give a selection of them in this section. The characteristic difference between the two cases lies (as is to be expected after our experience with Hölder's and Minkowski's inequalities) in a reversal of the sign of inequality.

(1) **336.** *If* $K(x,y)$ *is non-negative and homogeneous of degree* -1, $0<p<1$, *and*

$$\int_0^\infty K(x,1)\,x^{-1/p}\,dx = \int_0^\infty K(1,y)\,y^{-1/p'}\,dy = k < \infty,$$

then

$$(a) \quad \int_0^\infty\int_0^\infty K(x,y) f(x) g(y)\,dx\,dy \geqq k\left(\int_0^\infty f^p\,dx\right)^{1/p}\left(\int_0^\infty g^{p'}\,dy\right)^{1/p'},$$

$$(b) \quad \int_0^\infty dy\left(\int_0^\infty K(x,y) f(x)\,dx\right)^p \geqq k^p\int_0^\infty f^p(x)\,dx.$$

Here, in accordance with the conventions of §§ 5.1 and 6.5,

[a] Knopp (1).

(*a*) means 'if the double integral, and the second integral on the right, are finite, then the first integral on the right is finite, and ...'; and (*b*) means 'if the integral on the left is finite, then the integral on the right is finite, and ...'.

If we use the second method of § 9.3, the proof of (*a*) is the same as that of (*a*) of Theorem **319**. The sign is reversed because we use Hölder's inequality with $p<1$. To deduce (*b*) from (*a*), we appeal to Theorem **234**. We leave it to the reader to frame the corresponding theorem for $p<0$, and to consider the question of equality.

We cannot take $K=1/(x+y)$, since then $k=\infty$. There is therefore no exact analogue of Hilbert's theorem.

(2) **337.** *If* $0<p<1$, $f(x)\geqq 0$,

$$\int_0^\infty f^p\,dx<\infty,$$

and

$$F(x)=\int_x^\infty f(t)\,dt,$$

then

$$\int_0^\infty \left(\frac{F}{x}\right)^p dx>\left(\frac{p}{1-p}\right)^p\int_0^\infty f^p\,dx,$$

unless $f\equiv 0$. *The constant is the best possible.*

We may deduce Theorem **337**, in an imperfect form, from Theorem **336**, by taking

$$K(x,y)=0\ \ (x<y),\quad K(x,y)=\frac{1}{y}\ \ (x\geqq y),$$

when

$$k=\int_1^\infty x^{-1/p}\,dx=\frac{p}{1-p}.$$

To prove the complete theorem in this way would involve a discussion of the sign of inequality in Theorem **336** (and so in Theorem **234**). We therefore follow a direct method analogous to that of § 9.8.

We may suppose

$$\int_0^\infty f(t)\,dt,\quad \int_0^\infty \left(\frac{F}{x}\right)^p dx$$

finite, since otherwise there would be nothing to prove.

We have

(9.13.1)

$$\int_\xi^X \left(\frac{F}{x}\right)^p dx = \frac{1}{1-p}\left[x^{1-p}F^p(x)\right]_\xi^X + \frac{p}{1-p}\int_\xi^X \left(\frac{F}{x}\right)^{p-1} f\,dx.$$

Since F decreases as x increases,

$$x^{1-p}F^p(x) = 2\left(\frac{F(x)}{x}\right)^p \tfrac{1}{2}x \leq 2\int_{\frac{1}{2}x}^{x} \left(\frac{F}{t}\right)^p dt$$

tends to 0 both when $x \to 0$ and when $x \to \infty$. Hence (9.13.1) gives in the limit

$$\int_0^\infty \left(\frac{F}{x}\right)^p dx = \frac{p}{1-p}\int_0^\infty \left(\frac{F}{x}\right)^{p-1} f\,dx.$$

We leave the completion of the proof to the reader.

For a more complete result, corresponding to Theorem **330**, see Theorem **347**.

(3) Finally we prove a theorem which is related to Theorem **326** roughly as Theorem **337** is related to Theorem **327**. The correspondence is not quite precise, and the theorem illustrates very instructively the slight additional complications which are sometimes inherent in a theorem concerning series.

338.[a] *If* $0 < p < 1$ *and* $\Sigma a_n^p < \infty$, *then*

$$\Sigma'\left(\frac{a_n + a_{n+1} + \dots}{n}\right)^p > \left(\frac{p}{1-p}\right)^p \Sigma a_n^p,$$

unless (a_n) *is null. The dash over the summation on the left-hand side implies that the term for which* $n = 1$ *is to be multiplied by*

$$1 + \frac{1}{1-p}.$$

The constant is the best possible.

In Theorem **337**, take

$$f(x) = 0 \quad (0 < x < 1), \quad f(x) = a_n \quad (0 < n \leqq x < n+1).$$

Then, if $0 < n \leqq x < n+1$,

$$\frac{F}{x} = \frac{(n+1-x)a_n + a_{n+1} + \dots}{x} \leqq \frac{a_n + a_{n+1} + \dots}{n}.$$

Hence
$$\int_1^\infty \left(\frac{F}{x}\right)^p dx \leqq \sum_1^\infty \left(\frac{a_n + a_{n+1} + \dots}{n}\right)^p,$$

while
$$\int_0^1 \left(\frac{F}{x}\right)^p dx = (a_1 + a_2 + \dots)^p \int_0^1 x^{-p}\,dx = \frac{(a_1 + a_2 + \dots)^p}{1-p};$$

and the result follows from Theorem **337**.

[a] The substance of this theorem was communicated to us by Prof. Elliott in 1927.

Some such gloss as that contained in the last clause of the theorem is necessary; the result is not necessarily true if the dash is omitted[a].

9.14. A theorem with two parameters p and q. We conclude this chapter with a theorem which, although again an extension of Hilbert's theorem, has peculiarities which do not occur in any of the earlier theorems of the chapter. It involves two independent indices p and q and an undetermined constant $K(p,q)$.

339. *If* $$p>1,\quad q>1,\quad \frac{1}{p}+\frac{1}{q}\geqq 1,$$

so that $$0<\lambda=2-\frac{1}{p}-\frac{1}{q}=\frac{1}{p'}+\frac{1}{q'}\leqq 1,$$

then $$\sum_1^\infty\sum_1^\infty \frac{a_m b_n}{(m+n)^\lambda}\leqq K\left(\sum_1^\infty a_m{}^p\right)^{1/p}\left(\sum_1^\infty b_n{}^q\right)^{1/q},$$

where $K=K(p,q)$ depends on p and q only.

This theorem reduces to Theorem **315** when $q=p'$, $\lambda=1$: in that case we know the best possible value of K. The best value has not been found in the general case, and the problem of determining it appears to be difficult. We shall prove later (§ 10.17) a deeper theorem in which $\lambda<1$ and $m+n$ is replaced by $|m-n|$ (equal values being then excluded from the summation).

It is sufficient to prove that, if $\Sigma a_m{}^p=A$, $\Sigma b_n{}^q=B$, then

$$\sum_m a_m \sum_{n\leqq m}\frac{b_n}{(m+n)^\lambda}\leqq KA^{1/p}B^{1/q}, \tag{9.14.1}$$

and for this, by Theorem **13**, that

$$\sum_m \beta_m{}^{p'}\leqq KB^{p'/q}, \tag{9.14.2}$$

where $$\beta_m=\sum_{n\leqq m}\frac{b_n}{(m+n)^\lambda}.$$

Now $$\beta_m\leqq m^{-\lambda}\sum_{n\leqq m} b_n=m^{-\lambda}B_m$$

and $p'\geqq q$. Hence

$$\Sigma\beta_m{}^{p'}\leqq\Sigma m^{-p'\lambda}B_m{}^{p'}=\Sigma\left(\frac{B_m}{m}\right)^q B_m{}^{p'-q}m^{q-p'\lambda}.$$

[a] Take $a_1=1$, $a_2=a_3=\ldots=0$. Then the result is false if $p>\frac{1}{2}$. For an alternative form of the result see Theorem **345**.

But
$$B_m = \sum_1^m b_n \leqq m^{1/q'} \left(\sum_1^m b_n{}^q\right)^{1/q} \leqq B^{1/q} m^{1/q'},$$
and
$$\frac{p'-q}{q'} + q - p'\lambda = 0.$$
Hence
$$\Sigma \beta_m{}^{p'} \leqq B^{(p'-q)/q} \Sigma \left(\frac{B_m}{m}\right)^q \leqq \left(\frac{q}{q-1}\right)^q B^{(p'-q)/q+1} = K B^{p'/q},$$
by Theorem **326**. This proves (9.14.2).

Similarly we prove

340. *Under the same conditions as those of Theorem* **339**
$$\int_0^\infty \int_0^\infty \frac{f(x)g(y)}{(x+y)^\lambda} dx\, dy \leqq K \left(\int_0^\infty f^p dx\right)^{1/p} \left(\int_0^\infty g^q dy\right)^{1/q}.$$

MISCELLANEOUS THEOREMS AND EXAMPLES

341. If (i) a_m, b_n, $f(x)$, $g(y)$ are non-negative, (ii) the summations go from 1 to ∞ and the integrations from 0 to ∞,

(iii) $(\Sigma a_m{}^p)^{1/p} = A$, $(\Sigma b_n{}^{p'})^{1/p'} = B$, $(\int f^p dx)^{1/p} = F$, $(\int g^{p'} dy)^{1/p'} = G$,

and (iv) $p > 1$, then

(1)
$$\Sigma\Sigma \frac{a_m b_n}{\text{Max}(m,n)} < pp'AB,$$

(2)
$$\iint \frac{f(x)g(y)}{\text{Max}(x,y)} dx\, dy < pp'FG,$$

unless (a_m) or (b_n) or $f(x)$ or $g(y)$ is null. The constants are the best possible.

[Cases of Theorems **318** and **319**, (a). In order to shorten the statements of the following theorems we agree that conditions (i), (ii), and (iii) are pre-supposed in all of them; and that, whenever the conclusion is expressed by an inequality
$$X < KY \quad (\text{or } X > KY),$$
with a *definite* K, then K has its best possible value (unless the contrary is stated explicitly) and equality is excluded unless a sequence or function involved in the theorem is null.

When, on the other hand, the conclusion is
$$X \leqq KY,$$
with an *unspecified* K, then K is a function of any parameters of the theorem.]

342. If $p>1$, then

(1) $$\Sigma\Sigma \frac{\log (m/n)}{m-n} a_m b_n < \pi^2 \operatorname{cosec}^2 \frac{\pi}{p} . AB,$$

(2) $$\iint \frac{\log (x/y)}{x-y} f(x) g(y)\, dx dy < \pi^2 \operatorname{cosec}^2 \frac{\pi}{p} . FG.$$

[Also cases of Theorems **318** and **319**, (*a*). Here

$$k = \int_0^\infty \frac{\log x}{x-1} x^{-1/p} dx = \pi^2 \operatorname{cosec}^2 \frac{\pi}{p}.]$$

343. If $p>1$, then

$$\sum_2^\infty \sum_2^\infty \frac{a_m b_n}{mn \log mn} < \frac{\pi}{\sin (\pi/p)} \left(\sum_2^\infty \frac{a_m{}^p}{m} \right)^{1/p} \sum_2^\infty \left(\frac{b_n{}^{p'}}{n} \right)^{1/p'}.$$

[Mulholland (**2**). Since

$$\log \frac{m+1}{m} < \frac{1}{m},$$

the result is slightly stronger than that obtained by taking $\Lambda_m = \log m$, $M_n = \log n$ in Theorem **321**.]

344. If $0<p<1$ then

$$\Sigma (a_n + a_{n+1} + \dots)^p > p^p \Sigma (na_n)^p.$$

[Copson (**2**). This theorem, with Theorems **326**, **331**, and **338**, forms a systematic set of four.]

345. If $0<p<1$ then

$$\Sigma \left(\frac{a_n + a_{n+1} + \dots}{n} \right)^p > p^p \Sigma a_n{}^p.$$

[Corollary of Theorem **344**. Compare Theorem **338**; here there is no gloss, but the constant is less favourable and is presumably not the best possible.]

346. If (*a*) $c>1$, $s_n = a_1 + a_2 + \dots + a_n$, or (*b*) $c<1$, $s_n = a_n + a_{n+1} + \dots$, then

(α) $$\Sigma n^{-c} s_n{}^p \leqq K \Sigma n^{-c} (na_n)^p \quad (p>1),$$

(β) $$\Sigma n^{-c} s_n{}^p \geqq K \Sigma n^{-c} (na_n)^p \quad (0<p<1).$$

[In each of the four cases $K = K(p, c)$, as laid down under Theorem **341**. See Hardy and Littlewood (**1**).

We prove (α) when $c>1$. If

$$\phi_n = n^{-c} + (n+1)^{-c} + \dots,$$

then $\phi_n < Kn^{1-c}$. Hence, if we agree that $s_0 = 0$, we have

$$\sum_1^m n^{-c} s_n{}^p = \sum_1^m (\phi_n - \phi_{n+1}) s_n{}^p \leqq \sum_1^m \phi_n (s_n{}^p - s_{n-1}{}^p)$$

$$\leqq K \sum_1^m n^{1-c} s_n{}^{p-1} a_n \leqq K \left(\sum_1^m n^{-c} (na_n)^p \right)^{1/p} \left(\sum_1^m n^{-c} s_n{}^p \right)^{1/p'};$$

and (α) follows. Hardy and Littlewood (**2**) give function-theoretic applications of (α) and (β). The important case is that in which $c=2$.]

347. If r and F satisfy the conditions of Theorem **330**, but $0<p<1$, then

$$\int x^{-r} F^p\, dx > \left(\frac{p}{|r-1|}\right)^p \int x^{-r} (xf)^p\, dx.$$

[Hardy (**5**).]

348. If
$$\sigma(y) = \Sigma a_m e^{-m/y}$$
and $p>1$, then
$$2^{-p} \Sigma \left\{\frac{\sigma(n)}{n}\right\}^p < \int_0^\infty \left\{\frac{\sigma(y)}{y}\right\}^p dy < \left\{\Gamma\left(\frac{1}{p'}\right)\right\}^p A^p.$$

[Take $K(x,y) = y^{-1} e^{-x/y}$ and apply Theorem **319**, (*b*). More general but less precise results are given by Hardy and Littlewood (**1**), and some function-theoretic applications by Hardy and Littlewood (**2**).]

349. If λ_n and Λ_n satisfy the conditions of Theorem **332**, then
$$\Sigma \lambda_n (a_1^{\lambda_1} a_2^{\lambda_2} \ldots a_n^{\lambda_n})^{1/\Lambda_n} < e \Sigma \lambda_n a_n.$$

[See Hardy (**4**).]

350. If $p>1$, $K(x)>0$, and
$$\int K(x) x^{s-1}\, dx = \phi(s),$$
then
$$\iint K(xy) f(x) g(y)\, dx\, dy < \phi\left(\frac{1}{p}\right) \left(\int x^{p-2} f^p\, dx\right)^{1/p} \left(\int g^{p'}\, dy\right)^{1/p'},$$

$$\int dx \left(\int K(xy) f(y)\, dy\right)^p < \phi^p\left(\frac{1}{p}\right) \int x^{p-2} f^p\, dx,$$

$$\int x^{p-2}\, dx \left(\int K(xy) f(y)\, dy\right)^p < \phi^p\left(\frac{1}{p'}\right) \int f^p\, dx.$$

In particular, when $K(x) = e^{-x}$, and $F(x) = \int K(xy) f(y)\, dy$ is the 'Laplace transform' of $f(x)$,

$$\int F^p\, dx < \Gamma^p\left(\frac{1}{p}\right) \int x^{p-2} f^p\, dx, \quad \int x^{p-2} F^p\, dx < \Gamma^p\left(\frac{1}{p'}\right) \int f^p\, dx.$$

351. If also $K(x)$ is a decreasing function of x, and
$$A(x) = \Sigma a_n K(nx), \quad A_n = \int a(x) K(nx)\, dx,$$
then
$$\int A^p(x)\, dx < \phi^p\left(\frac{1}{p}\right) \Sigma n^{p-2} a_n^p,$$

$$\Sigma A_n^p < \phi^p\left(\frac{1}{p}\right) \int x^{p-2} a^p(x)\, dx,$$

$$\int x^{p-2} A^p(x)\, dx < \phi^p\left(\frac{1}{p'}\right) \Sigma a_n^p,$$

$$\Sigma n^{p-2} A_n^p < \phi^p\left(\frac{1}{p'}\right) \int a^p(x)\, dx.$$

352. If $F(x)$ is the Laplace transform of $f(x)$, and $1 < p \leqq 2$, then

$$\int F^{p'} dx \leqq \frac{2\pi}{p'} \left(\int f^p dx \right)^{p'/p}.$$

[For the last three theorems see Hardy (**10**). Theorem **350** may be deduced from Theorem **319** by transformation. It is not asserted that the constant in Theorem **352** is the best possible.]

353. If
$$K_0(x) \geqq 0,$$
$$K_1(x, y) = \int K_0(xt) K_0(yt)\, dt, \quad K_2(x, y) = \int K_1(x, t) K_1(y, t)\, dt,$$
$$\int \frac{K_1(x, 1)}{\sqrt{x}} dx = k,$$
then
$$\Sigma\Sigma K_2(m, n) a_m a_n \leqq k\, \Sigma\Sigma K_1(m, n) a_m a_n.$$

[See Hardy (**9**). The theorem is one concerning quadratic, not bilinear, forms.]

354.
$$\Sigma\Sigma \frac{\log(m/n)}{m-n} a_m a_n \leqq \pi\, \Sigma\Sigma \frac{a_m a_n}{m+n}.$$

355.
$$\Sigma\Sigma \frac{|\log(m/n)|}{\operatorname{Max}(m, n)} a_m a_n \leqq 2\, \Sigma\Sigma \frac{a_m a_n}{\operatorname{Max}(m, n)}.$$

[Corollaries of Theorem **353**. Observe that Theorem **354**, when combined with Theorem **315**, gives

$$\Sigma\Sigma \frac{\log(m/n)}{m-n} a_m a_n \leqq \pi^2\, \Sigma a_m^2,$$

in agreement with Theorem **342**.]

356. If
$$c(x) = \int_0^x a(t)\, b(x-t)\, dt,$$
$$A^p = \int x^{-1} \left(x^\alpha a(x) \right)^p dx, \quad B^q = \int x^{-1} \left(x^\beta b(x) \right)^q dx, \quad C^r = \int x^{-1} \left(x^\gamma c(x) \right)^r dx.$$
$$p > 1, \quad q > 1, \quad \frac{1}{r} \leqq \frac{1}{p} + \frac{1}{q}, \quad \alpha < 1, \quad \beta < 1, \quad \gamma = \alpha + \beta - 1,$$
then
$$C < KAB,$$
where
$$K = \frac{\Gamma(1-\alpha)\,\Gamma(1-\beta)}{\Gamma(1-\gamma)}.$$

357. If $\quad a_0 = b_0 = 0, \quad c_n = a_0 b_n + a_1 b_{n-1} + \ldots + a_n b_0,$
$$A^p = \Sigma n^{-1} (n^\alpha a_n)^p, \quad B^q = \Sigma n^{-1} (n^\beta b_n)^q, \quad C^r = \Sigma n^{-1} (n^\gamma c_n)^r,$$
p, q, r, and γ satisfy the conditions of Theorem **356**, and $0 \leqq \alpha < 1$, $0 \leqq \beta < 1$, then $C < KAB$, with the K of Theorem **356**. If $\alpha < 0$ or $\beta < 0$, the inequality is true with *some* K.

358. If
$$a_0=b_0=\dots=c_0=0$$
and
$$u_n=\Sigma a_{r_1} b_{r_2} \dots c_{r_k} \quad (r_j \geqq 0,\ \Sigma r_j=n),$$
then
$$\Sigma u_n^2 < \frac{1}{\pi}\left\{\Gamma\left(\frac{1}{2k}\right)\right\}^{2k} (\Sigma n^{2k-2} a_n^{2k})^{1/k} \dots (\Sigma n^{2k-2} c_n^{2k})^{1/k}.$$

359. If $p>1$, $l>0$, $m>0$ and $c(x)$ is defined as in Theorem **356**, then
$$\int x^{(1-l-m)(p-1)} c^p(x)\,dx \leqq K \int x^{(1-l)(p-1)} a^p(x)\,dx \int x^{(1-m)(p-1)} b^p(x)\,dx,$$
where
$$K=\left\{\frac{\Gamma(l)\,\Gamma(m)}{\Gamma(l+m)}\right\}^{p-1}.$$
There is equality if and only if
$$a(x)\equiv Ax^{l-1}e^{-Cx}, \quad b(x)\equiv Bx^{m-1}e^{-Cx},$$
where A, B, C are non-negative constants and C is positive.

[For Theorems **356–359** see Hardy and Littlewood (**3**, **5**, and **12**).]

360. If $L(x)$ is the Laplace transform of $f(x)$, and $q \geqq p > 1$, then
$$\int x^{-(p+q-pq)/p} L^q(x)\,dx \leqq KF^q.$$

361. If $p>1$, $q>1$,
$$\mu=\frac{1}{p}+\frac{1}{q}-1 \geqq 0,$$
and L, M are the Laplace transforms of f, g, then
$$\int x^{-\mu} LM\,dx \leqq KFG.$$

362. If $p>1$, $0 \leqq \mu < 1/p$, and
$$\alpha_n=\sum_m \frac{a_m}{(m+n)^{1-\mu}},$$
then
$$\Sigma \alpha_n^{p/(1-\mu p)} \leqq KA^{p/(1-\mu p)}.$$

[This may be deduced from Theorem **339** by the converse of Hölder's inequality. Many further theorems of the same general character as Theorems **360–362** are given by Hardy and Littlewood (**1**).]

363. If λ_n is positive and
$$p>1, \quad A_n=a_1+a_2+\dots+a_n, \quad \lambda_1+\lambda_2+\dots+\lambda_n \leqq cn,$$
then
$$\Sigma\left(\frac{A_n}{n}\right)^p \lambda_n \leqq KA^p.$$

364. If λ_n is positive and
$$p>1, \quad r>1, \quad \lambda_1^r+\lambda_2^r+\dots+\lambda_n^r \leqq cn,$$
then
$$\Sigma\Sigma \frac{a_m b_n}{m+n} \lambda_{m+n} \leqq KAB.$$
The result is not necessarily true when $r=1$ (as it is when $\lambda_n=1$).

[For these two theorems, which are corresponding extensions of Theorems **326** and **315**, see Hardy and Littlewood (**11**).]

365. The inequalities in Theorems **326** and **334** are the special cases $\phi = x^t\ (0<t<1)$, $\phi = \log x$, of

(i) $$\Sigma\phi^{-1}\left(\frac{\phi(a_1)+\ldots+\phi(a_n)}{n}\right) < K(\phi)\,\Sigma a_n.$$

[Knopp (**2**). This remark has led Knopp to a systematic investigation of forms of ϕ for which (i) is true. See also Mulholland (**4**).]

366. Suppose that ϕ and ψ are continuous and strictly increasing for $x>0$, and have the limits 0 or $-\infty$ when $x\to 0$; and that ϕ is convex with respect to ψ (§ 3.9). Then (i), if true for ϕ, is also true for ψ, with $K(\psi) \leqq K(\phi)$.

[Knopp (**2**).]

367. $$\Sigma\left(\log\frac{e^{1/a_1}+e^{1/a_2}+\ldots+e^{1/a_n}}{n}\right)^{-1} < 2\,\Sigma a_n.$$

[Knopp (**2**).]

CHAPTER X

REARRANGEMENTS

10.1. Rearrangements of finite sets of variables. In what follows we are concerned with finite sets of non-negative numbers such as

$$a_1, a_2, \ldots, a_j, \ldots, a_n; \quad b_1, b_2, \ldots, b_j, \ldots, b_n;$$

$$a_{-n}, \ldots, a_0, \ldots, a_j, \ldots, a_n:$$

we denote such sets by (a), (b), ...

Taking for example the first set, in which j assumes the values 1, 2, ..., n, we define a *permutation function* $\phi(j)$ as a function which takes each of the values 1, 2, ..., n just once when j varies through the same aggregate of values. If

$$a_{\phi(j)} = a_j' \quad (j = 1, 2, \ldots, n)$$

then we describe (a') as a *rearrangement* of (a). Similar definitions apply to other cases in which the range of variation of j is different.

There are certain special rearrangements of (a) which are particularly important here. These rearrangements, which we denote by

$$(\bar{a}),\ (a^+),\ (^+a),\ (a^*),$$

are defined as follows.

The set $(\bar{a})$ is the set (a) rearranged in ascending order, so that, when the values of j are 1, 2, ..., n,

$$\bar{a}_1 \leqq \bar{a}_2 \leqq \ldots \leqq \bar{a}_n.$$

The set $(\bar{a})$ is defined unambiguously by the set (a) although, when the a are not all different, there are ambiguities in the definition of the permutation function by which we pass from (a) to $(\bar{a})$.

In defining the sets (a^+), (^+a), (a^*) we suppose that j varies from $-n$ to n. The set (a^+) is defined by

$$a_0^+ \geqq a_1^+ \geqq a_{-1}^+ \geqq a_2^+ \geqq a_{-2}^+ \geqq \ldots$$

and the set (^+a) by

$$^+a_0 \geqq {}^+a_{-1} \geqq {}^+a_1 \geqq {}^+a_{-2} \geqq {}^+a_2 \geqq \ldots.$$

There is one particularly important case, that in which every value of an a, except the largest, occurs an even number of times, while the largest value occurs an odd number of times. In this case we shall say that the set (a) is *symmetrical*. The sets (a^+) and (^+a) are then identical, and we write

$$a^+ = {}^+a = a^*,$$

so that a^* is defined by

$$a_0^* \geqq a_1^* = a_{-1}^* \geqq a_2^* = a_{-2}^* \geqq \ldots.$$

A set (a^*) may be said to be *symmetrically decreasing*. The sets (a^+) and (^+a) are sets arranged so as to be as nearly symmetrically decreasing as possible, but with the inevitable overweight of one side arranged systematically to the advantage of the right or the left respectively. All these sets are defined unambiguously by (a), though there may be ambiguities in the definitions of the corresponding permutation functions.

We note that

(10.1.1) $$a_j^+ = {}^+a_{-j}.$$

10.2. A theorem concerning the rearrangements of two sets. We begin by proving a very simple, but important, theorem concerning the set $(\bar{a})$.

368.[a] *If (a) and (b) are given except in arrangement, then*

$$\Sigma ab$$

is greatest when (a) and (b) are monotonic in the same sense and least when they are monotonic in opposite senses; that is to say

(10.2.1) $$\sum_{j=1}^{n} \bar{a}_j \bar{b}_{n+1-j} \leqq \sum_{1}^{n} a_j b_j \leqq \sum_{1}^{n} \bar{a}_j \bar{b}_j.$$

It will be observed that, since we can add up the sum Σab in any order, we may suppose *one* set, say (a), arranged from the beginning in any order we please (in particular in ascending order).

We may express the theorem equally well by saying that the maximum corresponds to 'similar ordering' of (a) and (b) in the

[a] This theorem and Theorem **369** are valid for all real, not necessarily positive, a and b.

sense of § 2.17, the minimum to 'opposite ordering'[a]. The theorem becomes 'intuitive' if we interpret the a as distances along a rod to hooks and the b as weights suspended from the hooks. To get the maximum statical moment with respect to an end of the rod, we hang the heaviest weights on the hooks farthest from that end.

To prove the theorem, suppose that the (a) are in ascending order, but not the (b). Then there are a j and a k such that $a_j \leqq a_k$ and $b_j > b_k$. Since

$$a_j b_k + a_k b_j - (a_j b_j + a_k b_k) = (a_k - a_j)(b_j - b_k) \geqq 0,$$

we do not diminish Σab by exchanging b_j and b_k. A finite number of such exchanges leads to an ascending order of the b, so that

$$\Sigma ab \leqq \Sigma \bar{a}\bar{b}.$$

The other half of the theorem is proved in the same way.

This argument establishes incidentally a variant of Theorem **368** which is sometimes useful.

369. *If*

$$\Sigma ab' \leqq \Sigma ab \tag{10.2.2}$$

for all rearrangements (b') *of* (b), *then* (a) *and* (b) *are similarly ordered.*

For, if $(a_j - a_k)(b_j - b_k) < 0$ for any j, k, we can falsify (10.2.2) by exchanging b_j and b_k.

10.3. A second proof of Theorem 368. We have to consider analogues of Theorem **368** for more than two sets of variables. These lie a good deal deeper and cannot be proved in so simple a manner. We therefore give a second proof of Theorem **368** which, though quite unnecessarily complicated for its immediate object, will serve to introduce the method which we use later. We confine ourselves to the second inequality (10.2.1), and divide the proof into three stages.

(1) Suppose first that the sets considered consist entirely of

[a] Theorem **43** (with $r=1$ and $p=1$) may be expressed, in our present notation, in the form

$$n\Sigma \bar{a}_j \bar{b}_{n+1-j} \leqq \Sigma a_j \, \Sigma b_j \leqq n \Sigma \bar{a}_j \bar{b}_j.$$

0's and 1's; we indicate such special sets by the use of German letters $\mathfrak{a}$, $\mathfrak{b}$, ... Then

$$\mathfrak{a}^2 = \mathfrak{a}, \quad \mathfrak{b}^2 = \mathfrak{b} \tag{10.3.1}$$

for all j. In this case

$$\Sigma\mathfrak{a}\mathfrak{b} \leqq \Sigma\mathfrak{a}, \quad \Sigma\mathfrak{a}\mathfrak{b} \leqq \Sigma\mathfrak{b},$$

and so $$\Sigma\mathfrak{a}\mathfrak{b} \leqq \text{Min}(\Sigma\mathfrak{a}, \Sigma\mathfrak{b}) = \Sigma\bar{\mathfrak{a}}\bar{\mathfrak{b}}.$$

(2) Any set (a) may be decomposed into a linear combination of sets

$$(\mathfrak{a}^1), \quad (\mathfrak{a}^2), \quad \dots, \quad (\mathfrak{a}^l)\text{[a]}$$

of the special type considered under (1), in such a way that

$$a_j = \alpha^1 \mathfrak{a}_j^{\,1} + \alpha^2 \mathfrak{a}_j^{\,2} + \dots + \alpha^l \mathfrak{a}_j^{\,l} \quad (j = 1, 2, \dots, n), \tag{10.3.2}$$

and

$$\bar{a}_j = \alpha^1 \bar{\mathfrak{a}}_j^{\,1} + \alpha^2 \bar{\mathfrak{a}}_j^{\,2} + \dots + \alpha^l \bar{\mathfrak{a}}_j^{\,l} \quad (j = 1, 2, \dots, n), \tag{10.3.3}$$

the coefficients α being non-negative.

The method of decomposition will become clear by considering a special case. Suppose that (a) contains (in some order) the three numbers A, B, C, where $0 \leqq A \leqq B \leqq C$, so that

$$\bar{a}_1 = A, \quad \bar{a}_2 = B, \quad \bar{a}_3 = C.$$

Then

$$\bar{a}_1 = A\,.\,1 + (B - A)\,0 + (C - B)\,0,$$
$$\bar{a}_2 = A\,.\,1 + (B - A)\,1 + (C - B)\,0,$$
$$\bar{a}_3 = A\,.\,1 + (B - A)\,1 + (C - B)\,1;$$

and we may write

$$\bar{a}_j = \alpha^1 \bar{\mathfrak{a}}_j^{\,1} + \alpha^2 \bar{\mathfrak{a}}_j^{\,2} + \alpha^3 \bar{\mathfrak{a}}_j^{\,3},$$

where $$\alpha^1 = A, \quad \alpha^2 = B - A, \quad \alpha^3 = C - B$$

and $(\bar{\mathfrak{a}}^1)$, $(\bar{\mathfrak{a}}^2)$, $(\bar{\mathfrak{a}}^3)$ are the three sets

$$(1, 1, 1), \quad (0, 1, 1), \quad (0, 0, 1).$$

If then we perform the permutation which changes $(\bar{a})$ into (a) and at the same time $(\bar{\mathfrak{a}}^1)$, ... into $(\mathfrak{a}^1)$, ...,[b] we obtain

$$a_j = \alpha^1 \mathfrak{a}_j^{\,1} + \alpha^2 \mathfrak{a}_j^{\,2} + \alpha^3 \mathfrak{a}_j^{\,3}.$$

[a] $\mathfrak{a}^i$ means $\mathfrak{a}^{(i)}$ and α^i means $\alpha^{(i)}$: in (10.3.1) above, $\mathfrak{a}^2$ is a power, but this use does not recur.

[b] $\mathfrak{a}^1$ is *defined* by this permutation.

In the general case we proceed in the same way, writing

$$\bar{a}_1 = \bar{a}_1 . 1 + (\bar{a}_2 - \bar{a}_1) . 0 + (\bar{a}_3 - \bar{a}_2) . 0 + \dots,$$
$$\bar{a}_2 = \bar{a}_1 . 1 + (\bar{a}_2 - \bar{a}_1) . 1 + (\bar{a}_3 - \bar{a}_2) . 0 + \dots,$$
$$\dots\dots\dots\dots$$

This secures (10.3.3), and (10.3.2) then follows by rearrangement, as in the special case.

(3) From (1) and (2) we can deduce the general theorem. For, decomposing (b) as in (2), we have

$$a_j = \sum_\rho \alpha^\rho \mathfrak{a}_j{}^\rho, \quad \bar{a}_j = \sum_\rho \alpha^\rho \bar{\mathfrak{a}}_j{}^\rho, \quad b_j = \sum_\sigma \beta^\sigma \mathfrak{b}_j{}^\sigma, \quad \bar{b}_j = \sum_\sigma \beta^\sigma \bar{\mathfrak{b}}_j{}^\sigma,$$

$$\sum_j a_j b_j = \sum_\rho \sum_\sigma \alpha^\rho \beta^\sigma \sum_j \mathfrak{a}_j{}^\rho \mathfrak{b}_j{}^\sigma$$
$$\leqq \sum_\rho \sum_\sigma \alpha^\rho \beta^\sigma \sum_j \bar{\mathfrak{a}}_j{}^\rho \bar{\mathfrak{b}}_j{}^\sigma = \sum_j \bar{a}_j \bar{b}_j.$$

10.4. Restatement of Theorem 368. It will also be useful to restate Theorem **368** in different language. We suppose now that, in the sets (a), (b), j runs from $-n$ to n. We write

$$f(x) = \sum a_j x^j, \quad g(x) = \sum b_j x^j,$$

and call

$$a_0 = \mathfrak{C}(f(x))$$

the *central coefficient* of f. Plainly

$$\mathfrak{C}(f(x^{-1})) = \mathfrak{C}(f(x)).$$

Also

$$\sum_{r+s=0} a_r b_s = \sum a_j b_{-j} = \mathfrak{C}(fg).$$

The sets $(a_j{}^+)$ and $({}^+b_{-j})$ are similarly ordered, and if we write

$$f^+(x) = \sum a_j{}^+ x^j, \quad {}^+f(x) = \sum {}^+a_j x^j,$$

so that, by (10.1.1),

$$f^+(x^{-1}) = {}^+f(x),$$

then Theorem **368** gives

$$\mathfrak{C}(fg) = \sum_{r+s=0} a_r b_s = \sum a_j b_{-j} \leqq \sum a_j{}^+ {}^+b_{-j} = \sum_{r+s=0} a_r{}^+ {}^+b_s = \mathfrak{C}(f^+ {}^+g).$$

Hence we deduce

370. *The central coefficient of*

$$\sum_{-n}^{n} a_j x^j \sum_{-n}^{n} b_j x^j$$

is greatest, for all rearrangements of the a and b, when (a_j) *and* (b_{-j}) *are similarly ordered, in particular when* (a) *is* (a^+) *and* (b) *is* $({}^+b)$ *or when* (a) *is* $({}^+a)$ *and* (b) *is* (b^+).

10.5. Theorems concerning the rearrangements of three sets. We pass now to theorems involving three sets of variables.

371.[a] *Suppose that the c, x, and y are non-negative, and the c symmetrically decreasing, so that*

$$c_0 \geqq c_1 = c_{-1} \geqq c_2 = c_{-2} \geqq \ldots \geqq c_{2k} = c_{-2k},$$

while the x and y are given except in arrangement. Then the bilinear form

$$S = \sum_{r=-k}^{k} \sum_{s=-k}^{k} c_{r-s} x_r y_s$$

attains its maximum when (x) *is* (x^+) *and y is* (y^+).

It is evident that, if this is so, then the maximum must also be attained when (x) is $({}^+x)$ and (y) is $({}^+y)$.

372.[b] *Suppose that* (a), (b), (c) *are three sets satisfying*

$$a_0 \geqq a_r = a_{-r}, \quad b_0 \geqq b_s = b_{-s}, \quad c_0 \geqq c_t = c_{-t}. \tag{10.5.1}$$

Then the maximum of

$$\sum_{r+s+t=0} a_r b_s c_t = \mathfrak{C}\,(\Sigma a_r x^r \,\Sigma b_s x^s \,\Sigma c_t x^t),$$

for rearrangements of the sets which leave a_0, b_0, c_0 *unaltered, is attained when* (a), (b), (c) *are* (a^*), (b^*), (c^*).

373.[c] *If* (a), (b), (c) *are three sets, of which* (c) *is symmetrical in the sense of* § 10.1, *then*

$$\sum_{r+s+t=0} a_r b_s c_t \leqq \sum_{r+s+t=0} a_r^{+}\,{}^{+}b_s c_t^{*} = \sum_{r+s+t=0} {}^{+}a_r b_s^{+} c_t^{*}.$$

It will be sufficient to prove Theorem **373**, since this includes the other two theorems. In the first place, Theorem **373** is Theorem **372** freed from the restrictions (10.5.1), wholly in regard to (a) and (b) and partly in regard to (c). To deduce Theorem **371** from Theorem **373**, we put $2k = n$, $x_r = a_{-r}$, $y_s = b_s$, and suppose that the a and b outside the range $(-k, k)$ are zero. We may observe

[a] Hardy, Littlewood, and Pólya (**1**).
[b] Hardy and Littlewood (**4**), Gabriel (**1**).
[c] Gabriel (**3**).

finally that Theorem **370** is the simple case of Theorem **373** in which $c_0 = 1$ and the remaining c are 0.

10.6. Reduction of Theorem 373 to a special case. We divide the proof of Theorem **373** into three stages, as in § 10.3. The whole difficulty of the proof lies in stage (1), in which (a), (b), (c) are of types $(\mathfrak{a})$, $(\mathfrak{b})$, $(\mathfrak{c})$: and we take this stage for granted for the moment and dispose of the easier stages (2) and (3).

First, we may decompose (a), (b), (c) into sums of sets $(\mathfrak{a}^\rho)$, $(\mathfrak{b}^\sigma)$, $(\mathfrak{c}^\tau)$, in such a way that

$$a_j = \sum_\rho \alpha^\rho \mathfrak{a}_j^\rho, \quad b_j = \sum_\sigma \beta^\sigma \mathfrak{b}_j^\sigma, \quad c_j = \sum_\tau \gamma^\tau \mathfrak{c}_j^\tau,$$

and
$$a_j^+ = \sum_\rho \alpha^\rho \mathfrak{a}_j^{\rho+}, \quad {}^+b_j = \sum_\sigma \beta^\sigma\, {}^+\mathfrak{b}_j^\sigma, \quad c_j^* = \sum_\tau \gamma^\tau \mathfrak{c}_j^{\tau *}.$$

Here the $\mathfrak{a}$, $\mathfrak{b}$, $\mathfrak{c}$ are all 0 or 1, the α, β, γ are non-negative, and (a point which does not arise in § 10.3) the sets $(\mathfrak{c}^\tau)$ are symmetrical. All this is proved by the method of § 10.3 (2).[a] When we have done this, and proved the theorem for sets of type $(\mathfrak{a})$, $(\mathfrak{b})$, $(\mathfrak{c})$, we have

$$\sum_{r+s+t=0} a_r b_s c_t = \sum_{\rho,\sigma,\tau} \alpha^\rho \beta^\sigma \gamma^\tau \sum_{r+s+t=0} \mathfrak{a}_r^\rho \mathfrak{b}_s^\sigma \mathfrak{c}_t^\tau$$

$$\leqq \sum_{\rho,\sigma,\tau} \alpha^\rho \beta^\sigma \gamma^\tau \sum_{r+s+t=0} \mathfrak{a}_r^{\rho+}\, {}^+\mathfrak{b}_s^\sigma \mathfrak{c}_t^{\tau *} = \sum_{r+s+t=0} a_r^+\, {}^+b_s c_t^*,$$

and the proof is completed.

It remains to prove the theorem in the special case in which all a, b, c are 0 or 1.[b] The set c, being symmetrical, contains an even number of 0's and an odd number of 1's. We write

$$f(x) = \sum a_r x^r, \quad g(x) = \sum b_s x^s, \quad h(x) = \sum c_t x^t.$$

Since we may add any number of 0's to the sets, we may suppose that all the summations run from $-n$ to n.

We have also

$$f^+(x) = \sum a_r^+ x^r = x^{-R} + \dots + 1 + \dots + x^{R'},$$
$${}^+g(x) = \sum {}^+b_s x^s = x^{-S'} + \dots + 1 + \dots + x^S,$$
$$h^*(x) = \sum c_t^* x^t = x^{-T} + \dots + 1 + \dots + x^T,$$

[a] In order that the sets $(\mathfrak{c}^\tau)$ obtained by the process of § 10.3 (2) should be symmetrical, in the sense of § 10.1, we drop those $\mathfrak{c}^\tau$ which correspond to zero γ^τ.

[b] So that, strictly, we should write $\mathfrak{a}$, $\mathfrak{b}$, $\mathfrak{c}$ for a, b, c. There is, however, no further necessity for this notation.

where R, R', S, S', T are non-negative integers and

$$(10.6.1) \qquad R \leqq R' \leqq R+1, \quad S \leqq S' \leqq S+1.$$

We have to prove that

$$(10.6.2) \qquad \mathfrak{C}(fgh) \leqq \mathfrak{C}(f^{+}{}^{+}g\,h^{*}).$$

The inequality (10.6.2) may be made 'intuitive' by a geometrical representation. Let x, y be rectangular coordinates in a plane, and represent each non-zero coefficient of f, g, h by a line, $x=r$ for $a_r=1$, $y=s$ for $b_s=1$, and $x+y=-t$ for $c_t=1$. If $a_r b_s c_t$ contributes a unit to $\mathfrak{C}(fgh)$, these three lines intersect. Each of the functions f, g, h is represented by a family of parallel lines,

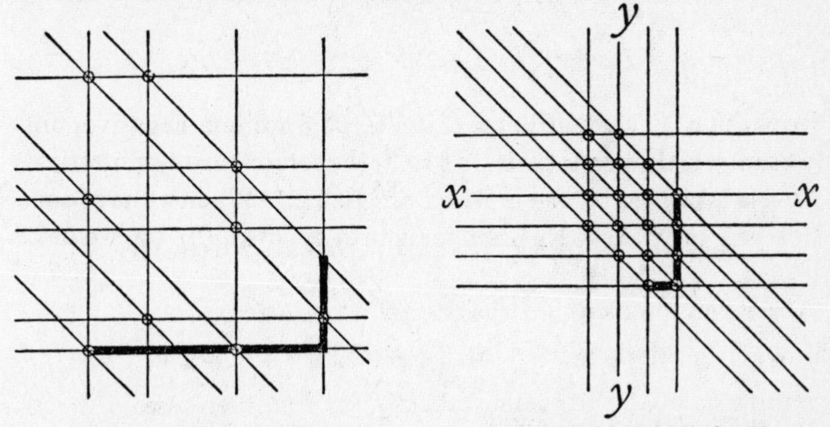

Fig. 1. Graph of f, g, h.

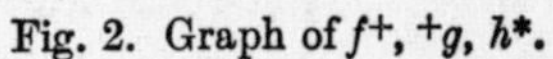

Fig. 2. Graph of f^{+}, ${}^{+}g$, h^{*}.

and $\mathfrak{C}(fgh)$ is the total number of triple intersections of these lines. We represent f^{+}, ${}^{+}g$, h^{*} similarly; f^{+} is also represented by $R+1+R'$ vertical lines, but now these lines are shifted as near as possible together. Typical figures are shown in Figs. 1 and 2: here (a), (b), (c) are the sets

$$1, 0, 1, 0, 0, 1, 0, 0, 1;$$

$$1, 1, 0, 0, 1, 1, 1, 0, 0, 1;$$

$$1, 0, 1, 0, 0, 0, 1, 0, 1, 0, 0, 1, 0$$

respectively; and

$$R=1, \quad R'=2, \quad S=2, \quad S'=3, \quad T=2.$$

It is intuitive that the number of intersections is greatest when, as in Fig. 2, the diagram is as condensed as possible.

Our proof of (10.6.2) may be presented geometrically[a] and followed on the figures. We reduce the actual case considered to a simpler one by taking away one horizontal and one vertical line from each figure, as is suggested by the thick lines in the figures. We prefer however to state the proof in a purely analytical form.

10.7. Completion of the proof. There are three subordinate cases in which the proof is easy.

(1) If $R'=0$, f^+ reduces to 1 and the result is included in Theorem **370**.

(2) If $S'=0$, ${}^+g$ reduces to 1 and again the result is included in Theorem **370**.

(3) Suppose that

$$R+S'\leqq T,\quad R'+S\leqq T. \tag{10.7.1}$$

We have in any case

(10.7.2)
$$\mathfrak{C}(fgh)=\sum_{r+s+t=0} a_r b_s c_t \leqq \Sigma a_r \Sigma b_s = (R+1+R')(S+1+S').$$

But, when the inequalities (10.7.1) are satisfied,

$$\mathfrak{C}(f^{+}{}^{+}gh^*)=\mathfrak{C}\{(x^{-(R+S')}+\dots+x^{R'+S})(x^{-T}+\dots+1+\dots+x^{T})\}$$

is the sum of all the coefficients of $f^{+}{}^{+}g$, and therefore

(10.7.3)
$$\mathfrak{C}(f^{+}{}^{+}gh^*)=\sum_{r,s} a_r b_s=\sum_r a_r \sum_s b_s=(R+1+R')(S+1+S').$$

The result follows from (10.7.2) and (10.7.3).

We now consider the general case in which

$$R'>0,\quad S'>0,\quad \operatorname{Max}(R+S',\ R'+S)=n>T.$$

We assume that the result has been proved for

$$\operatorname{Max}(R+S',\ R'+S)<n$$

and argue inductively.

Let x^ρ be the highest power in f, x^σ the lowest in g, and write

$$f-x^\rho=\phi,\quad g-x^\sigma=\psi,\quad f^+-x^{R'}=\bar{\phi},\quad {}^+g-x^{-S'}=\bar{\psi}.$$

Since $R'>0$, $S'>0$, none of these functions vanishes identically.

[a] See Gabriel (3).

Then $$fgh = (\phi + x^\rho)(x^\sigma + \psi)h = \phi\psi h + \chi h,$$
where $$\chi = x^\sigma\phi + x^{\rho+\sigma} + x^\rho\psi.$$
Since the highest power in $x^\sigma\phi$ is lower than $x^{\rho+\sigma}$, and the lowest power in $x^\rho\psi$ is higher than $x^{\rho+\sigma}$, there is no overlapping, and all coefficients in χ are 0 or 1. Since the sum of the coefficients in h is $2T+1$, it follows that

$$\mathfrak{C}(\chi h) \leqq 2T+1,$$

(10.7.4) $$\mathfrak{C}(fgh) \leqq \mathfrak{C}(\phi\psi h) + 2T + 1.$$

On the other hand

(10.7.5) $$f^{+}{}^{+}g h^* = (\bar\phi + x^{R'})(x^{-S'} + \bar\psi)h^* = \bar\phi\bar\psi h^* + \bar\chi h^*,$$

where
$$\bar\chi = x^{-S'}\bar\phi + x^{R'-S'} + x^{R'}\bar\psi$$
$$= x^{-R-S'} + \dots + x^{R'-S'-1} + x^{R'-S'} + x^{R'-S'+1} + \dots + x^{R'+S}.$$

The sequence of exponents in $\bar\chi$ is an unbroken one, extending from $-R-S'$ to $R'+S$. We know that either $R+S'$ or $R'+S$ is greater than T. If $R+S' > T$ then, by (10.6.1),

$$R'+S \geqq R+S'-1 \geqq T,$$

and so the unbroken sequence from $-T$ to T, of length $2T+1$, is part of the sequence of exponents of $\bar\chi$. The same conclusion follows when $R'+S > T$. Since h^* has an unbroken sequence of exponents of length $2T+1$, centred round the constant term, it follows that
$$\mathfrak{C}(\bar\chi h^*) = 2T+1,$$
and so, by (10.7.5), that

(10.7.6) $$\mathfrak{C}(f^{+}{}^{+}g h^*) = \mathfrak{C}(\bar\phi\bar\psi h^*) + 2T + 1.$$

Now $$\phi^+(x) = x^{-(R'-1)} + \dots + x^R = \bar\phi(x^{-1}),^{\mathrm{a}}$$
$${}^{+}\psi(x) = x^{-S} + \dots + x^{S'-1} = \bar\psi(x^{-1}).$$

Also

$$\mathrm{Max}(R'-1+S,\ R+S'-1) = \mathrm{Max}(R'+S,\ R+S') - 1 = n-1,$$

and so, by our hypothesis,

(10.7.7) $$\mathfrak{C}(\phi\psi h) \leqq \mathfrak{C}(\phi^{+}{}^{+}\psi h^*) = \mathfrak{C}\{\bar\phi(x^{-1})\bar\psi(x^{-1})h^*(x)\}$$
$$= \mathfrak{C}\{\bar\phi(x)\bar\psi(x)h^*(x^{-1})\} = \mathfrak{C}(\bar\phi\bar\psi h^*).$$

[a] It is not necessarily true that
$$\phi^+(x) = \bar\phi(x) = x^{-R} + \dots + x^{R'-1},$$
since, if $R' = R$, this polynomial is 'overweighted' at the wrong end. When $R' = R+1$, either formula is correct.

Finally, comparing (10.7.4), (10.7.6), and (10.7.7), we see that

$$\mathfrak{C}(fgh) \leqq \mathfrak{C}(f^{++}gh^{*}),$$

and the proof is completed.

10.8. Another proof of Theorem 371. There is another proof of Theorem **371** which is interesting in itself, although it cannot be extended to prove the more general Theorem **373.**

We have to prove that, among the arrangements of the x and y which make S a maximum, there is one in which

$$(10.8.1) \qquad x_r - x_{r'} \geqq 0, \quad y_s - y_{s'} \geqq 0$$

if $\qquad |r'| > |r|, \quad |s'| > |s|$

or if $\qquad r' = -r < 0, \quad s' = -s < 0.$

We may suppose on grounds of continuity that the x, y, and c are all positive, that the x and the y are all different, and that the c are different except in so far as they are restricted by $c_{-n} = c_n$, the condition of symmetry.

We shall denote an arrangement of the x and y generally by A. We say that A is 'correct' if it satisfies (10.8.1); there is just one correct arrangement C. We say that A is 'almost correct' if it satisfies (10.8.1) except perhaps when $r' = -r$ or $s' = -s$; there are, including C, 2^{2k} almost correct arrangements, and we denote the class of such arrangements by C'. Finally we denote by K the class of those A which give the maximum value of S. We have to prove that C is a K.

Given p, we can associate the x and y in pairs

$$(10.8.2) \quad (x_{p-i}, x_{p+i}), \quad (y_{p-j}, y_{p+j}) \quad (i, j = 1, 2, 3, \ldots),$$

or in pairs

$$(10.8.3) \quad (x_{p-i}, x_{p+i+1}), \quad (y_{p-j}, y_{p+j+1}) \quad (i, j = 0, 1, 2, \ldots).$$

If a suffix falls outside the interval $(-k, k)$ then the corresponding x or y is to be replaced by 0. In the first case the elements x_p and y_p are left unpaired, and the pairing may be made, by choice of appropriate p, i, j, to include any pair of elements the difference of whose ranks is positive and even. In the second case no elements are left unpaired, and the pairing may be made to include any pair the difference of whose ranks is odd. We use both pairings, and the arguments are essentially the same whichever is being used.

Consider, for example, the pairing (10.8.2), and suppose, to fix our ideas, that $p \geqq 0$, so that

$$|p-i| \leqq |p+i|, \quad |p-j| \leqq |p+j|.$$

We denote by I, J those values of i and j for which

$$x_{p-I} < x_{p+I}, \quad y_{p-J} < y_{p+J},$$

so that the pairs corresponding to I and J do not satisfy (10.8.1). Such pairs we call 'wrong', others 'right'. If $p+i$ falls outside $(-k, k)$, but $p-i$ inside it, then x_{p+i} is to be replaced by 0, and the corresponding pair of x is certainly right. Hence, except perhaps when $p=0$, there are i which are not I and j which are not J.

If, for a given p and a given pairing[a], there are no wrong pairs, we say that A is 'right with respect to p', and otherwise 'wrong with respect to p'. It is clear that C is right with respect to every p, and that any C' is right with respect to all p except perhaps $p=0$ and the pairing (10.8.2). Further, any A other than C is wrong with respect to some p and pairing, and any A which is not a C' is wrong either with respect to a p other than 0 or with respect to $p=0$ and the pairing (10.8.3).

We now (again envisaging the first pairing, and supposing $p \geqq 0$) consider the effect on S of the substitution

$$\Omega_p \quad (x_{p-I}, x_{p+I}; \, y_{p-J}, y_{p+J})$$

which interchanges each pair x_{p-I}, x_{p+I} and each pair y_{p-J}, y_{p+J}.

We divide S into nine partial sums defined as follows:

S_1: $r=p$; $s=p$;

S_2: $r=p$; $s=p-j, p+j$ $(j \neq J)$;

S_3: $r=p-i, p+i$ $(i \neq I)$; $s=p$;

S_4: $r=p$; $s=p-J, p+J$;

S_5: $r=p-I, p+I$; $s=p$;

S_6: $r=p-i, p+i$ $(i \neq I)$; $s=p-j, p+j$ $(j \neq J)$;

S_7: $r=p-i, p+i$ $(i \neq I)$; $s=p-J, p+J$;

S_8: $r=p-I, p+I$; $s=p-j, p+j$ $(j \neq J)$;

S_9: $r=p-I, p+I$; $s=p-J, p+J$.

It is plain, first, that S_1, S_2, S_3, and S_6 are not affected by Ω_p.

[a] Either (10.8.2) or (10.8.3). In what follows 'right (or wrong) with respect to p' means always 'right (or wrong) with respect to p and the pairing under consideration'.

Next $$S_4 = x_p \sum_J (c_J y_{p-J} + c_{-J} y_{p+J})$$
is not affected because $c_{-J} = c_J$. Similarly S_5 and
$$S_9 = \sum_{I,J} (c_{-I+J} x_{p-I} y_{p-J} + c_{-I-J} x_{p-I} y_{p+J} + c_{I+J} x_{p+I} y_{p-J} + c_{I-J} x_{p+I} y_{p+J})$$
are not affected. It remains to consider S_7 and S_8.

The contribution of the pair x_{p-i}, x_{p+i} to S_7 is
$$x_{p-i} \sum_J (c_{-i+J} y_{p-J} + c_{-i-J} y_{p+J}) + x_{p+i} \sum_J (c_{i+J} y_{p-J} + c_{i-J} y_{p+J}),$$
and the increment produced by Ω_p is
$$-(x_{p-i} - x_{p+i}) \sum_J (c_{i-J} - c_{i+J})(y_{p-J} - y_{p+J}).$$
The total change in S_7 is the sum of this increment over $i \neq I$, and is positive, provided that there are J and $i \neq I$, since the three differences written are respectively positive, positive, and negative. Hence S_7 is increased if there are J and $i \neq I$; and similarly S_8 is increased if there are I and $j \neq J$. Finally, S is increased if either of these conditions is satisfied.

If $p \neq 0$, there are $i \neq I$ and $j \neq J$; and then S is increased unless A is right with respect to p. In any case, whatever p, S is not diminished.

Suppose now that A is not a C'. Then A is wrong with respect either to some $p \neq 0$ or to $p = 0$ and the pairing (10.8.3). The argument above, or the similar argument based upon the pairing (10.8.3), then shows that S is increased by Ω_p (or the corresponding substitution based upon the other pairing), and that A is not a K. Hence the K are included among the C'. But if a C' is not C, then the substitution Ω_0 replaces it by C and does not diminish S; and therefore C is a K.[a]

It does not seem to be possible to prove Theorem **373** by any equally simple argument based upon a substitution defined directly.

10.9. Rearrangements of any number of sets. There are

[a] The argument is the same in principle as that used by Hardy, Littlewood, and Pólya (**1**), and substantially reproduced by Hardy and Littlewood (**6**). We have however expanded it considerably, Dr R. Rado having pointed out to us that the original form of the argument was not conclusive. Another form of the proof is indicated in Theorem **389** at the end of the chapter.

analogues of Theorem **373** for more than three sets (a), ..., which may be deduced from Theorem **373** itself.

374.[a] *If $(a), (b), (c), (d), \ldots$ are finite sets of non-negative numbers, and $(c), (d), \ldots$ are symmetrical, then*

$$(10.9.1)\qquad \sum_{r+s+t+u+\ldots=0} a_r b_s c_t d_u \ldots \leqq \sum_{r+s+t+u+\ldots=0} a_r^{++} b_s c_t^* d_u^* \ldots.$$

We assume the theorem to be true when there are $k-1$ symmetrical sets (c), (d), ... involved, and prove that it is true when there are k. We shall make use of the following theorem, which is of some interest in itself.

375. *If (c^*), (d^*), ... are symmetrically decreasing sets, then the set (Q) defined by*

$$(10.9.2)\qquad Q_n = \sum_{t+u+\ldots=n} c_t^* d_u^* \ldots$$

is symmetrically decreasing.

It is enough to prove the theorem for two sets (c^*), (d^*), since its truth in general then follows by repetition of the argument. We may agree that, when there is no indication to the contrary, sums involving several suffixes are extended over values of the suffixes whose sum vanishes.

It is plain that $Q_{-n} = Q_n$. Further, for any set (x), we have

$$\sum_m x_m Q_m = \Sigma x_m Q_n = \Sigma x_m c_t^* d_u^* \leqq \Sigma x_m^+ c_t^* d_u^* = \sum_m x_m^+ Q_m,$$

by Theorem **373**. It follows, by Theorem **369**, that the Q_m are similarly ordered to the x_m^+, and therefore, since Q_m is an even function of m, that the set is symmetrically decreasing.

This is the most elegant proof, but there is a simpler one which does not depend upon Theorem **373**.

We drop the asterisks for convenience and suppose $n \geqq 0$. Then

$$Q_n = \Sigma c_{n+r} d_{-r} + \Sigma c_{n+1-r} d_{r-1},$$

the summations extending over $r \geqq 1$. Similarly

$$Q_{n+1} = \Sigma c_{n+r} d_{1-r} + \Sigma c_{n+1-r} d_r.$$

Subtracting, and using the equations $d_{-r} = d_r$ and $d_{1-r} = d_{r-1}$, we obtain

$$\begin{aligned} Q_n - Q_{n+1} &= \Sigma \{c_{n+r}(d_{-r} - d_{1-r}) + c_{n+1-r}(d_{r-1} - d_r)\} \\ &= \Sigma (c_{n+1-r} - c_{n+r})(d_{r-1} - d_r). \end{aligned}$$

Since $|n+1-r| < n+r$ for $n \geqq 0$, $r \geqq 1$, each term here is non-negative.

[a] Gabriel (3). The case of the theorem in which all the sets are symmetrical was proved by Hardy and Littlewood (4).

Returning to the proof of Theorem **374**, we define Q_n as in (10.9.2), and P_m by

$$P_m = \sum_{r+s=m} a_r b_s.$$

Then $\quad \Sigma a_r b_s c_t d_u \ldots = \Sigma P_m c_t d_u \ldots \leqq \Sigma P_m^+ c_t^* d_u^* \ldots,$

by the case $k-1$ of the theorem[a]. That is to say,

$$\Sigma a_r b_s c_t d_u \ldots \leqq \sum_m P_m^+ Q_m = \sum_m P_m Q_{\phi(m)},$$

where $\phi(m)$ is a permutation function for which $P_m = P_{\phi(m)}^+$; i.e.

$$\Sigma a_r b_s c_t d_u e_v \ldots \leqq \Sigma a_r b_s Q_{\phi(m)} \leqq \Sigma a_r^{+} {}^{+}b_s Q_m^*$$
$$= \Sigma a_r^{+} {}^{+}b_s Q_m = \Sigma a_r^{+} {}^{+}b_s c_t^* d_u^* e_v^* \ldots,$$

which is (10.9.1).

From Theorem **374** we can deduce[b]

376. *Given any finite number of sets* (a), (b), ..., *we have*

$$\Sigma a_{r_1} a_{-r_2} b_{s_1} b_{-s_2} c_{t_1} c_{-t_2} \ldots \leqq \Sigma a_{r_1}^{+} {}^{+}a_{r_2} b_{s_1}^{+} {}^{+}b_{s_2} c_{t_1}^{+} {}^{+}c_{t_2} \ldots$$
$$= \Sigma a_{r_1}^{+} a_{-r_2}^{+} b_{s_1}^{+} b_{-s_2}^{+} c_{t_1}^{+} c_{-t_2}^{+} \ldots.$$

10.10. A further theorem on the rearrangement of any number of sets. In Theorems **373** and **374** two of the sets, (a) and (b), were arbitrary, but the remainder were subject to the condition of 'symmetry'. This restriction is essential; if (a), (b) and (c) are unrestricted, it is not possible to specify the maximal arrangement generally by means of the symbols a^+, ${}^+a$, ...[c]

There is however a less precise theorem which is often equally effective in applications.

377. *For any system of k sets* (a), (b), (c), ...

$$\sum_{r+s+t+\ldots=0} a_r b_s c_t \ldots \leqq K(k) \sum_{r+s+t+\ldots=0} a_r^+ b_s^+ c_t^+ \ldots,$$

where $K = K(k)$ *is a number depending only on* k.

We suppose $k=3$; the argument is essentially the same in the general case.

We define the sets (β^*), (γ^*) by

(10.10.1) $\qquad \beta_m^* = b_m^+, \quad \gamma_m^* = c_m^+ \quad (m \geqq 0),$

(10.10.2) $\qquad \beta_{-m}^* = \beta_m^*, \quad \gamma_{-m}^* = \gamma_m^* \quad (m \geqq 0);$

[a] With $P, c, d, \ldots$ for $a, b, c, \ldots$: since (c) is symmetrical, ${}^+c_t = c_t^*$.

[b] Gabriel (3). [c] See Theorem 388 at the end of the chapter.

and (β) and (γ) as the sets into which (β^*) and (γ^*) are changed by the permutations which change (b^+) into (b) and (c^+) into (c) respectively. Then (β) and (γ) are symmetrical sets. Further, since $b^+_{-m} \leqq b^+_m$ and $c^+_{-m} \leqq c^+_m$ when $m \geqq 0$, we have

$$b^+_n \leqq \beta^*_n, \quad c^+_n \leqq \gamma^*_n$$

for all n, and so

$$b_n \leqq \beta_n, \quad c_n \leqq \gamma_n \tag{10.10.3}$$

for all n.

We shall also require an inequality for β^*_m and γ^*_m with $m < 0$. We have $b^+_n \leqq b^+_{-n+1}$ and $c^+_n \leqq c^+_{-n+1}$ for $n \geqq 1$, and so, by (10.10.1) and (10.10.2),

$$\beta^*_m \leqq b^+_{m+1}, \quad \gamma^*_m \leqq c^+_{m+1} \quad (m < 0). \tag{10.10.4}$$

Using (10.10.3) and the symmetry of (β) and (γ), we find

$$S = \sum_{r+s+t=0} a_r b_s c_t \leqq \sum_{r+s+t=0} a_r \beta_s \gamma_t \leqq \sum_{r+s+t=0} a^+_r \beta^*_s \gamma^*_t,$$

by Theorem **373**. The last sum is

$$\Big(\sum_{s \geqq 0, t \geqq 0} + \sum_{s < 0, t \geqq 0} + \sum_{s \geqq 0, t < 0} + \sum_{s < 0, t < 0}\Big) a^+_r \beta^*_s \gamma^*_t;$$

and so, by (10.10.1) and (10.10.4),

(10.10.5)

$$S \leqq \sum_{s \geqq 0, t \geqq 0} a^+_r b^+_s c^+_t + \sum_{s < 0, t \geqq 0} a^+_r b^+_{s+1} c^+_t$$
$$+ \sum_{s \geqq 0, t < 0} a^+_r b^+_s c^+_{t+1} + \sum_{s < 0, t < 0} a^+_r b^+_{s+1} c^+_{t+1} = S_1 + S_2 + S_3 + S_4.$$

In S_2, $s < 0$ and $r + s + t = 0$, so that either $r > 0$ or $t > 0$. In the first case $a^+_r \leqq a^+_{r-1}$ and in the second $c^+_t \leqq c^+_{t-1}$. Hence in any case, in S_2,

$$a^+_r b^+_{s+1} c^+_t \leqq a^+_{r-1} b^+_{s+1} c^+_t + a^+_r b^+_{s+1} c^+_{t-1}. \tag{10.10.6}$$

Similarly, in S_3,

$$a^+_r b^+_s c^+_{t+1} \leqq a^+_{r-1} b^+_s c^+_{t+1} + a^+_r b^+_{s-1} c^+_{t+1}. \tag{10.10.7}$$

Finally, in S_4, $s < 0$, $t < 0$ and $r + s + t = 0$, so that $r \geqq 2$ and $a^+_r \leqq a^+_{r-2}$, and

$$a^+_r b^+_{s+1} c^+_{t+1} \leqq a^+_{r-2} b^+_{s+1} c^+_{t+1}. \tag{10.10.8}$$

If now we substitute into (10.10.5) the upper bounds for the typical terms given by (10.10.6), (10.10.7), and (10.10.8), and observe that, in these upper bounds, the sum of the suffixes is always 0, we obtain

$$S \leqq (1+2+2+1) \sum_{r+s+t=0} a_r^+ b_s^+ c_t^+ = 6 \sum_{r+s+t=0} a_r^+ b_s^+ c_t^+;$$

which proves the theorem.

10.11. Applications. These theorems have important applications to the theory of Fourier series. It is easy to deduce from Theorem **376**[a] that if

$$f(\theta) = \sum_{-R}^{R} a_r e^{r\theta i}, \quad f^+(\theta) = \sum_{-R}^{R} \alpha_r e^{r\theta i},$$

where $\alpha_r = |a_r|^+$, and k is a positive integer, then

$$\int_{-\pi}^{\pi} |f(\theta)|^{2k} d\theta \leqq \int_{-\pi}^{\pi} |f^+(\theta)|^{2k} d\theta;$$

and this relation between trigonometrical polynomials may be extended to functions represented by general Fourier series. Series of the type

$$\Sigma \alpha_r e^{r\theta i}$$

have particularly simple properties. They converge uniformly except at the origin and congruent points, where the function which they represent has in general an infinite peak; and the ratio

$$\int_{-\pi}^{\pi} |f^+(\theta)|^{2k} d\theta : \Sigma (|r|+1)^{2k-2} \alpha_r^{2k}$$

lies between positive bounds depending only on k. We thus find, for example, that

$$\int_{-\pi}^{\pi} |f(\theta)|^{2k} d\theta \leqq K(k) \Sigma (|r|+1)^{2k-2} \alpha_r^{2k}.$$

For fuller developments see Hardy and Littlewood (**9**), Paley (**3**).

10.12. The rearrangement of a function.

The theorems of §§ 10.1–10.10 have analogues for functions of a continuous variable.

Suppose that $\phi(x)$ is non-negative and integrable in (0, 1), so that it is measurable and finite almost everywhere. If $M(y)$ is the measure of the set in which $\phi(x) \geqq y$, $M(y)$ is a decreasing function of y. The inverse $\bar{\phi}$ of M is defined by

$$\bar{\phi}\{M(y)\} = y;$$

and $\bar{\phi}(x)$ is a decreasing function of x defined uniquely in (0, 1) except for at most an enumerable set of values of x, viz. those corre-

[a] See Gabriel (**3**). A less precise inequality was given by Hardy and Littlewood (**9**).

sponding to intervals of constancy of $M(y)$. We may complete the definition of $\bar{\phi}(x)$ by agreeing, for example, that

$$\bar{\phi}(x)=\tfrac{1}{2}\{\bar{\phi}(x-0)+\bar{\phi}(x+0)\}$$

at a point of discontinuity[a].

We call $\bar{\phi}(x)$ the rearrangement of $\phi(x)$ in decreasing order. It is a decreasing function of x which has, in general, an infinite peak at the origin.

The measure of the set in which $\bar{\phi}(x)\geqq y$ is $M(y)$.[b] It follows that the two (in general quite different) sets in which

$$y_1\leqq\phi(x)<y_2,\quad y_1\leqq\bar{\phi}(x)<y_2$$

have the same measure, and that the same is true of the sets in which

$$\phi(x)>y,\quad \bar{\phi}(x)>y.$$

We may say that the functions $\phi(x)$ and $\bar{\phi}(x)$ are 'equi-measurable'; they have equal integrals over (0, 1) and

$$\int_0^1 F(\bar{\phi})\,dx=\int_0^1 F(\phi)\,dx$$

for any measurable F for which the integrals exist.

We may define $\bar{\phi}(x)$ similarly for a $\phi(x)$ defined in any interval of x, provided that, if the interval is infinite, $M(y)$ is finite for every positive y.

If $\phi_1(x)\leqq\phi(x)$ then plainly $\bar{\phi}_1(x)\leqq\bar{\phi}(x)$. Suppose in particular that $\phi_1(x)$ is $\phi(x)$ in E and zero in CE. Then

$$\int_E\phi(x)\,dx=\int\phi_1(x)\,dx=\int_0^{mE}\bar{\phi}_1(x)\,dx\leqq\int_0^{mE}\bar{\phi}(x)\,dx. \tag{10.12.1}$$

We shall use this inequality in § 10.19. In particular

$$\int_0^x\phi(t)\,dt\leqq\int_0^x\bar{\phi}(t)\,dt \tag{10.12.2}$$

if $\phi(x)$ is defined in $(0, a)$ and $0\leqq x\leqq a$.

[a] Compare § 6.15.

[b] This becomes obvious on drawing a figure. It must be remembered that $\bar{\phi}(x)$ may have intervals of constancy, corresponding to discontinuities of $M(y)$. It is however easy to prove that $M(y-0)=M(y)$ for all y, and so that the assertion in the text is true even for these exceptional y. In fact

$$M\left(y-\frac{1}{n}\right)-M(y)=mS_n$$

where S_n is the set in which $y-n^{-1}\leqq\phi<y$, and the limit of mS_n is zero.

Another type of rearrangement of a function will be important in what follows. Suppose, for example, that $\phi(x)$ is defined for all real, or almost all real x, and that $M(y)$ is finite for all positive y. We may define an even function $\phi^*(x)$ by agreeing that

$$\phi^*\{\tfrac{1}{2}M(y)\} = y$$

and that $\phi^*(-x) = \phi^*(x)$; or, what is the same thing, that $\phi^*(x)$ is even and

$$\phi^*(x) = \bar{\phi}(2x)$$

for positive x. Then $\phi^*(x)$ decreases symmetrically on each side of the origin, where it has generally an infinite cuspidal peak. We call $\phi^*(x)$ the rearrangement of $\phi(x)$ in symmetrical decreasing order.

10.13. On the rearrangement of two functions. We begin by proving an integral inequality corresponding to Theorem **368**.

378. *Whether a is finite or infinite,*

$$\int_0^a \phi\psi\,dx \leqq \int_0^a \bar{\phi}\bar{\psi}\,dx.$$

We prove this by an argument similar to that of § 10.3. In the first place, the theorem is true for functions which assume only the values 0 and 1. For suppose that E and F are the sets in which $\phi = 1$ and $\psi = 1$ respectively, and $\bar{E}$, $\bar{F}$ the analogous sets corresponding to $\bar{\phi}$, $\bar{\psi}$. Then the first integral is $m(EF)$, the measure of the set EF which is the product of E and F, and

$$m(EF) \leqq \text{Min}(mE, mF) = \text{Min}(m\bar{E}, m\bar{F}) = m(\bar{E}\bar{F}).$$

Next, the theorem is true for functions which assume only a finite number of non-negative values. In fact, following the lines of § 10.3, we can represent such a function ϕ in the form

$$\phi = \alpha_1\phi_1 + \alpha_2\phi_2 + \ldots + \alpha_n\phi_n,$$

where the α are non-negative, the ϕ are always 0 or 1, and

$$\bar{\phi} = \alpha_1\bar{\phi}_1 + \alpha_2\bar{\phi}_2 + \ldots + \alpha_n\bar{\phi}_n.$$

The inequality then follows from a linear combination of inequalities already proved.

Finally, we prove the theorem in the general case by approximating to ϕ and ψ, by functions of the type just considered. We do not give the last two stages of the proof in detail, since the arguments will recur in the proof of the more difficult Theorem **379.**

10.14. On the rearrangement of three functions. We come now to what is our main object in these sections, the integral theorem corresponding to Theorems **372** and **373.**

379.[a] *If $f(x)$, $g(x)$, and $h(x)$ are non-negative, and $f^*(x)$, $g^*(x)$, and $h^*(x)$ are the equi-measurable symmetrically decreasing functions, then*

$$(10.14.1)\quad I=\int_{-\infty}^{\infty}\int_{-\infty}^{\infty} f(x)g(y)h(-x-y)\,dx\,dy$$
$$\leqq\int_{-\infty}^{\infty}\int_{-\infty}^{\infty} f^*(x)g^*(y)h^*(-x-y)\,dx\,dy=I^*.$$

We may plainly suppose that none of f, g, h is null. We may also replace $-x-y$ by $\pm x\pm y$ without changing the significance of the inequality.

We prove the inequality (1) for functions which are always 0 or 1, (2) for functions which take only a finite number of values, and (3) for general functions. As with Theorem **373**, the whole difficulty lies in stage (1). We take this stage for granted for the moment and begin by showing that, if the theorem is true in this special case, it is true generally.

A function which takes only a finite number of non-negative values $0, a_1, a_2, \ldots, a_n$ can be expressed in the form

$$f(x)=\alpha_1 f_1(x)+\alpha_2 f_2(x)+\ldots+\alpha_n f_n(x),$$

where the α are positive, the f_j take only the values 0 and 1, and

$$f_1\geqq f_2\geqq\ldots\geqq f_n.$$

For we may suppose $0<a_1<a_2<\ldots<a_n$, take

$$\alpha_1=a_1,\quad \alpha_2=a_2-a_1,\quad \ldots,\quad \alpha_n=a_n-a_{n-1},$$

and

$$f_1=1\ (f\geqq a_1),\quad =0\ (f<a_1),$$
$$f_2=1\ (f\geqq a_2),\quad =0\ (f<a_2),$$
$$\cdots\cdots$$

A moment's consideration shows that we then have also

$$f^*(x)=\alpha_1 f_1^*(x)+\alpha_2 f_2^*(x)+\ldots+\alpha_n f_n^*(x).$$

[a] F. Riesz (8).

If we suppose that each of f, g, h takes only a finite number of values, and decompose them in this way, then (10.14.1) follows from the combination of similar inequalities involving triads f_i, g_j, h_k.[a]

To pass from this case to the general case, we approximate to f, g, h by functions which take only a finite number of values. We can approximate to f, for example, by the function f_n defined by

$$f_n=\frac{k}{n}\ \left(\frac{k}{n}\leqq f<\frac{k+1}{n},\ k=0,\ 1,\ 2,\ \ldots,\ n^2-1\right),\quad f_n=n\ (f\geqq n);$$

and to g and h similarly. Then $f_n\leqq f$, $f_n^*\leqq f^*$, and similarly for g and h. Hence (assuming that the theorem has been proved for the special type of functions) we have

$$I_n=\int_{-\infty}^{\infty}\int_{-\infty}^{\infty} f_n(x)g_n(y)h_n(-x-y)\,dx\,dy\leqq I_n^*\leqq I^*,$$

and so $I=\lim I_n\leqq I^*$.

It remains to prove the theorem in the special case when f, g, h assume only the values 0 and 1. It is however convenient first to make a further reduction of the problem.

First, we may suppose that the sets F, G, and H in which f, g, and h assume the value 1 are finite. If *two* of these sets are infinite, then two of f^*, g^*, and h^* are 1 for all x, in which case $I^*=\infty$[b] and there is nothing to prove. Suppose then that just one of the sets, say F, is infinite. Let F_N be the part of F in $(-N, N)$, let N be the smallest number for which $mF_N\geqq 2n$, and define f_n as being f in F_N and 0 outside. Then (assuming the theorem to have been proved when the sets are finite)

$$I_n=\int_{-\infty}^{\infty}\int_{-\infty}^{\infty} f_n gh\,dx\,dy\leqq\int_{-\infty}^{\infty}\int_{-\infty}^{\infty} f_n^* g^* h^*\,dx\,dy$$

$$=\int_{-n}^{n}\int_{-\infty}^{\infty} f^* g^* h^*\,dx\,dy\leqq\int_{-\infty}^{\infty}\int_{-\infty}^{\infty} f^* g^* h^*\,dx\,dy=I^*,$$

and so
$$I=\lim I_n\leqq I^*.$$

Suppose then that $f(x)$ assumes the value 1 in a set E of finite measure. We can represent E in the form $\mathcal{E}+e-e'$, where $\mathcal{E}$ is a

[a] Compare the similar argument in §10.6.

[b] Unless the third function is null.

finite set of non-overlapping intervals, and e and e' are sets of arbitrarily small measure[a]; and the sets in which g and h assume the value 1 can be represented similarly. It is also plain that, since f, g, and h do not exceed 1, small sets e, ... make a small difference in the integrals I and I^*. We may therefore suppose that the sets in which $f=1$, $g=1$, and $h=1$ are finite sets of intervals; if the theorem has been proved in this case, its truth in the more general case follows by approximation.

Next we may suppose, on similar grounds, that the ends of all the intervals are *rational*; and then, by a change of variable, that they are *integral*. The theorem is thus reduced to dependence upon the case in which each of the sets in which $f=1$, $g=1$, or $h=1$ consists of a finite number of intervals $(m, m+1)$, where m is an integer.

Finally we may suppose, if we please, that the number of intervals in any or all of the sets is *even*, since we can replace each interval by two by bisecting it and effecting another change of variable.

10.15. Completion of the proof of Theorem 379. It is convenient to replace $f(x)$ by $f(-x)$, as plainly we may without affecting the result. If we do this, write s, t for x, y and then make the substitution $s=x-t$, we obtain

(10.15.1)

$$I=\int_{-\infty}^{\infty}\int_{-\infty}^{\infty} f(t-x)\,g(t)\,h(-x)\,dx\,dt=\int_{-\infty}^{\infty} h(-x)\,\chi(x)\,dx,$$

(10.15.2)

$$I^*=\int_{-\infty}^{\infty}\int_{-\infty}^{\infty} f^*(t-x)\,g^*(t)\,h^*(-x)\,dx\,dt=\int_{-\infty}^{\infty} h^*(-x)\,{}^*\chi(x)\,dx,$$

where

(10.15.3)

$$\chi(x)=\int_{-\infty}^{\infty} f(t-x)\,g(t)\,dt,\qquad {}^*\chi(x)=\int_{-\infty}^{\infty} f^*(t-x)\,g^*(t)\,dt.\text{[b]}$$

We suppose for the moment merely that f, g, h are characteristic

[a] See for example de la Vallée Poussin (**2**, 20–23).

[b] ${}^*\chi(x)$ is naturally not to be confused with $\chi^*(x)$.

functions of sets (functions assuming the values 0 and 1 only), without using the further simplifications shown to be permissible at the end of § 10.14. We denote the sets in which $f, \ldots, f^*, \ldots$ assume the value 1 by $F, \ldots, F^*, \ldots$; each function vanishes outside the corresponding set, and $F^*, \ldots$ are intervals symmetric about the origin. We suppose that

$$mF = mF^* = 2R, \quad mG = mG^* = 2S, \quad mH = mH^* = 2T.$$

With this notation[a], we have

(10.15.4)

$$I \leqq \int_{-\infty}^{\infty} \int_{-\infty}^{\infty} f(t-x) g(t) \, dx \, dt = \int_{-\infty}^{\infty} f(-s) \, ds \int_{-\infty}^{\infty} g(t) \, dt = 4RS,$$

(10.15.5) $$I^* = \int_{-T}^{T} {}^*\chi(x) \, dx.$$

If x is fixed, and $t-x$ describes the set F, then t describes a set F_x obtained by translating F through a distance x. If we define $F^*_x = (F^*)_x$ similarly, then the functions (10.15.3) may be written in the form

(10.15.6) $$\chi(x) = m(F_x G), \quad {}^*\chi(x) = m(F^*_x G^*).$$

From this formula we can calculate ${}^*\chi(x)$. Let us suppose, as we may, that

(10.15.7) $$R \leqq S.$$

Then ${}^*\chi(x)$ is continuous,

(10.15.8) $${}^*\chi(x) = 0 \quad (|x| \geqq R+S), \quad {}^*\chi(x) = 2R \quad (|x| \leqq S-R),$$

and ${}^*\chi(x)$ is linear in the intervals $(-R-S, -S+R)$ and $(S-R, R+S)$. The graph of ${}^*\chi(x)$ is shown in Fig. 3.

Suppose now that

(10.15.9) $$R + S \leqq T.$$

Then it follows from (10.15.8) that

$$I^* = \int_{-T}^{T} {}^*\chi(x) \, dx = \int_{-R-S}^{R+S} {}^*\chi(x) \, dx = 4RS,$$

and the result of the theorem follows from (10.15.4). We have thus proved the theorem under the restriction (10.15.9). It is also plainly true if $R = 0$ or $S = 0$ (when F or G is null).

[a] Chosen to emphasize the parallelism of the argument with that of §§ 10.6–7.

So far F, G, H have been arbitrary sets of finite measure. We now make the further specialisation explained at the end of § 10.14, supposing that F, G, H are sets of intervals $(m, m+1)$, the numbers of intervals being $2R$, $2S$, $2T$ respectively. We may if we please suppose these numbers even, but we shall argue inductively, and it is more convenient to adopt a slightly more general hypothesis and to suppose only that $2R+2S+2T$ is even. In these circumstances R, S, and T are not necessarily integral, but $2R$, $2S$, $2T$ and

(10.15.10) $$\mu = R+S-T = R+S+T-2T$$

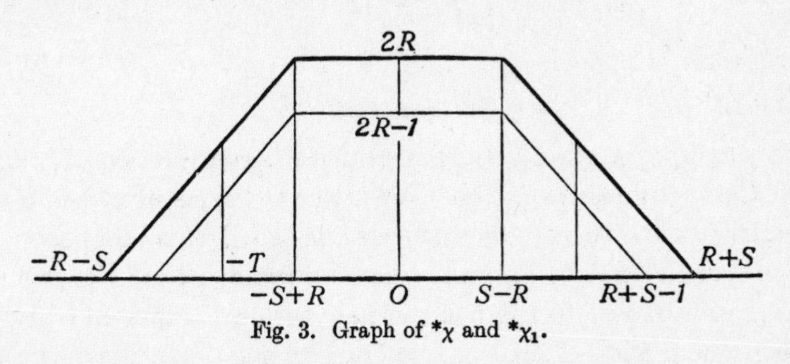

Fig. 3. Graph of $^*\chi$ and $^*\chi_1$.

are integral. We have already proved that the theorem is true if $\mu \leqq 0$, and it is also true if $R=0$ or $S=0$. It is therefore enough to establish its truth when

(10.15.11) $$\mu = n > 0, \quad R > 0, \quad S > 0,$$

on the assumption that it is true when $\mu = n-1$.

We denote by F_1 the set derived from F by omitting the last interval of F on the right; similarly G_1 is G less the last interval on the right of G. Generally, sets, functions, or numbers with suffix 1 are derived from F_1 and G_1 as the corresponding sets, functions, or numbers without suffixes are derived from F and G; thus f_1^* is the rearrangement of f_1, the characteristic function of F_1, and $^*\chi_1(x)$ is the 'Faltung' of f_1^* and g_1^*. F_1^* is the interval

$$(-R+\tfrac{1}{2}, \quad R-\tfrac{1}{2}),$$

and generally, R and S are replaced by $R-\frac{1}{2}$ and $S-\frac{1}{2}$ when we pass from F, G to F_1, G_1. By the inductive hypothesis

(10.15.12) $$I_1 \leqq I_1^*.$$

The function ${}^*\chi_1(x)$ vanishes for $|x| \geqq R+S-1$, is equal to $2R-1$ for $|x| \leqq S-R$, and is linear in the intervals remaining; and $T \leqq R+S-1$, by (10.15.10) and (10.15.11). Hence[a]

$${}^*\chi(x) - {}^*\chi_1(x) = 1$$

for $-T \leqq x \leqq T$, and, by (10.15.5),

$$I^* - I_1^* = \int_{-T}^{T} \{{}^*\chi(t) - {}^*\chi_1(t)\}\,dt = 2T. \tag{10.15.13}$$

We have now to consider

$$I - I_1 = \int_{-\infty}^{\infty} h(-x)\{\chi(x) - \chi_1(x)\}\,dx. \tag{10.15.14}$$

Here, after (10.15.6),

$$\chi(x) - \chi_1(x) = m(F_x G) - m(F_{1x} G_1). \tag{10.15.15}$$

This function is plainly linear in any interval $(m, m+1)$, and therefore assumes its extrema for integral values of x. Suppose then that x is integral. In this case the set $F_x G$ is composed of full intervals $(m, m+1)$, and, when we remove the intervals of F_x and G furthest to the right, either *one* or *no* interval of $F_x G$ is lost, one if the extreme interval of either set coincides with an interval of the other set, and none if there is no such coincidence[b]. Hence $\chi(x) - \chi_1(x)$ is 1 or 0 for integral x, and therefore

$$0 \leqq \chi(x) - \chi_1(x) \leqq 1 \tag{10.15.16}$$

for all x.

From (10.15.14) and (10.15.16) it follows that

$$0 \leqq I - I_1 = \int_H \{\chi(-x) - \chi_1(-x)\}\,dx \leqq \int_H dx = 2T, \tag{10.15.17}$$

and from (10.15.13) and (10.15.17) that

$$I - I_1 \leqq I^* - I_1^*. \tag{10.15.18}$$

Finally (10.15.12) and (10.15.18) give $I \leqq I^*$; and this completes the proof[c].

[a] See Fig. 3.

[b] We cannot lose *two* intervals because the intervals removed from F_x and G are the furthest to the right of their respective sets. This is the essential point of the proof.

[c] The proof follows the line indicated by Zygmund (1). It is, however, considerably longer, and necessarily so, since Zygmund's proof is not, as it stands, conclusive.

We proved that (10.15.16) is true when the intervals removed from F and G are

10.16. An alternative proof. The proof of Theorem **379** given by Riesz is also very interesting. We can simplify it by reducing the theorem, as in §10.14, to the case in which f, g, h are each equal to 1 in a finite set of intervals and to 0 elsewhere. We represent the variables x, y, z on the sides of an equilateral triangle, taking the middle point of each side as the origin and the positive directions on the sides cyclically. Then $x+y+z=0$ is the condition that the points x, y, z on the sides should be the three orthogonal projections of a point of the plane[a].

The functions $f(x)$, $g(y)$, $h(z)$ are the characteristic functions of three sets E_1, E_2, E_3, each consisting of a finite number of non-overlapping intervals, and $f^*(x)$, $g^*(y)$, $h^*(z)$ are the characteristic functions of the three intervals E_1^*, E_2^*, E_3^* of lengths E_1, E_2, E_3,[b] symmetrically disposed about the three origins. If E_{123} is the set of those points of the plane whose three projections belong to E_1, E_2, and E_3, and E^*_{123} is defined similarly, then[c]

$$I = \sin \tfrac{1}{3}\pi E_{123}, \quad I^* = \sin \tfrac{1}{3}\pi E^*_{123},$$

and what we have to prove is that

$$E_{123} \leqq E^*_{123}. \tag{10.16.1}$$

The figure E^*_{123} is defined by drawing six lines perpendicular to the sides, and is a hexagon unless one of E_1, E_2, E_3 is greater than the sum of the other two, in which case it reduces to a parallelogram. We begin by proving (10.16.1) in the latter case. Suppose for example that $E_3 \geqq E_1 + E_2$. Then E^*_{123} reduces to

the extreme intervals on the right. It would not have been true if we had removed two *arbitrary* intervals. Suppose, for example, that each of F and G is the interval $(-4, 4)$, that F_4 consists of the two intervals $(-4, -2)$ and $(2, 4)$, and G_4 of the interval $(-2, 2)$. We can pass from F, G to F_4, G_4 in four steps, taking away one unit interval from each set at each step; but

$$\chi(0) - \chi_4(0) = 8,$$

instead of being less than or equal to 4. The same example shows that Zygmund's assertion (**1**, 176) 'those [the values] of $\phi(x)$ in $(-\infty, \infty)$ increase at most by 2' is untrue unless his construction is restricted in a way which he does not state explicitly. It is essential to go closely into detail at this point, since it is the kernel of the proof.

[a] If P is the point in question and G is the centre of the triangle, then

$$x+y+z = PG\{\cos\alpha + \cos(\alpha + \tfrac{2}{3}\pi) + \cos(\alpha + \tfrac{4}{3}\pi)\} = 0.$$

[b] We use E_1 both for the set E_1 and for its measure.

[c] E_{123}, when used as a measure, is of course a plane measure.

E^*_{12}, the set of points projecting into E_1^* and E_2^* on two of the sides, while E_{123} is included in the set E_{12} defined similarly. Hence

$$E_{123} \leqq E_{12} = \operatorname{cosec} \tfrac{1}{3}\pi . E_1 E_2 = \operatorname{cosec} \tfrac{1}{3}\pi . E_1^* E_2^* = E^*_{12} = E^*_{123}.$$

This proves the theorem when E^*_{123} is a parallelogram.

Passing to the case of the hexagon, suppose for example that

$$E_3 > E_1 \geqq E_2, \quad E_3 < E_1 + E_2.$$

We define sets $\quad E_1(t), \quad E_2(t), \quad E_3(t),$

and corresponding intervals $E_1^*(t)$, $E_2^*(t)$, $E_3^*(t)$, by subtracting from each E_j a set of measure t at each end[a]. If t increases from 0 to

$$t_0 = \tfrac{1}{2}(E_1 + E_2 - E_3),$$

$E_1(t)$, $E_2(t)$, $E_3(t)$ decrease from E_1, E_2, E_3 to sets $E_1(t_0)$, $E_2(t_0)$, $E_3(t_0)$ whose measures satisfy

$$E_1(t_0) + E_2(t_0) = E_3(t_0).$$

The hexagon then reduces to a parallelogram, so that

$$E_{123}(t_0) \leqq E^*_{123}(t_0). \tag{10.16.2}$$

If we can prove also that

$$E_{123} - E_{123}(t_0) \leqq E^*_{123} - E^*_{123}(t_0), \tag{10.16.3}$$

our conclusion will follow by addition.

We prove (10.16.3) by comparing the derivatives of

$$\phi(t) = -E_{123}(t), \quad \phi^*(t) = -E^*_{123}(t).$$

In the first place, the difference between $E^*_{123}(t)$ and $E^*_{123}(t+h)$ is a hexagonal ring whose area is $hP(t) + O(h^2)$, where $P(t)$ is the perimeter of the hexagon corresponding to the value t, and so

$$\frac{d\phi^*}{dt} = P(t) = \operatorname{cosec} \tfrac{1}{3}\pi \{E_1(t) + E_2(t) + E_3(t)\}.$$

On the other hand the three sets

$$E_1(t) - E_1(t+h), \quad E_2(t) - E_2(t+h), \quad E_3(t) - E_3(t+h)$$

[a] That is to say
$$E_1 = E_1'(t) + E_1(t) + E_1''(t),$$
where $E_1'(t)$ lies to the left, and $E_1''(t)$ to the right, of $E_1(t)$, and
$$mE_1'(t) = mE_1''(t) = t.$$

consist in all of six intervals, each of length h, for small h. The twelve perpendiculars to the sides of the triangle drawn through the ends of the six intervals define a hexagonal ring[a] which includes the whole of $E_{123}(t)-E_{123}(t+h)$. The derivative $\phi'(t)$ is the total length of those parts of the outer boundary of this ring which also belong to $E_{123}(t)$. Projecting these parts of the

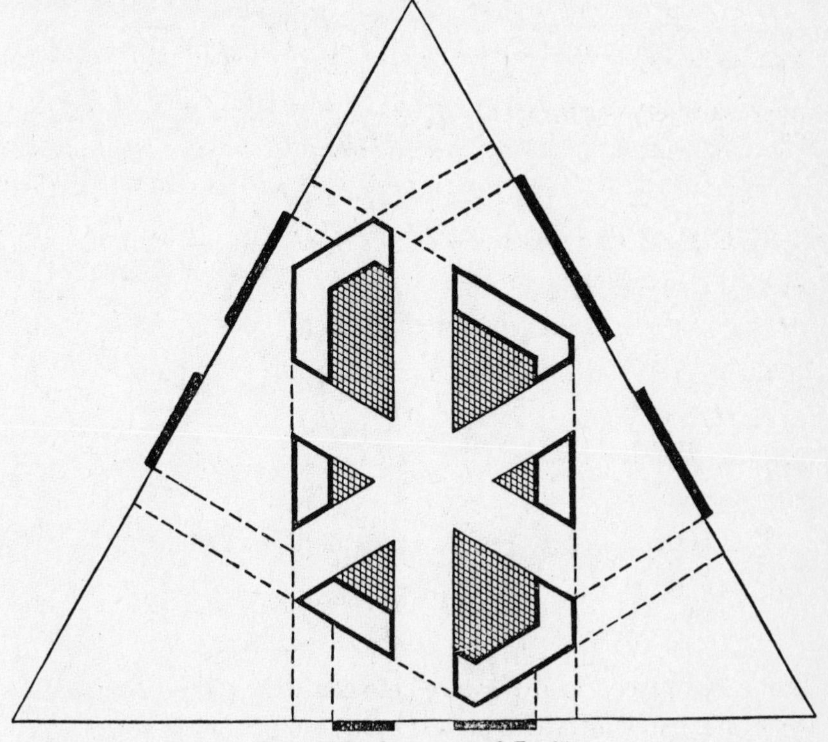

Fig. 4. The decrement of $E_{123}(t)$.

boundary on to the sides of the triangle, as indicated in the figure, we see that

$$\frac{d\phi}{dt} \leq \operatorname{cosec} \tfrac{1}{3}\pi \{E_1(t) + E_2(t) + E_3(t)\} = \frac{d\phi^*}{dt}.$$

From this (10.16.3) follows by integration, and this completes the proof of the theorem.

[a] See Fig. 4. In the figure the sets $E_1(t+h)$, ... are blackened on the sides of the triangle, the set $E_{123}(t+h)$ is shaded, the twelve perpendiculars are dotted, and the boundary of $E_{123}(t)-E_{123}(t+h)$ is indicated by a thick line.

10.17. Applications. The special case of Theorem **379** corresponding to Theorem **371** is

380. *If $h(x)$ is symmetrically decreasing, then*

$$I=\int_{-\infty}^{\infty}\int_{-\infty}^{\infty} f(x)g(y)h(x-y)\,dx\,dy$$
$$\leqq \int_{-\infty}^{\infty}\int_{-\infty}^{\infty} f^*(x)g^*(y)h(x-y)\,dx\,dy=I^*.$$

We shall now apply Theorems **371** and **380** to the special cases

$$c_{r-s}=|r-s|^{-\lambda}$$

and

$$h(x-y)=|x-y|^{-\lambda}.$$

381. *If* $$a_r\geqq 0,\quad b_s\geqq 0,$$

$$\text{(10.17.1)}\quad p>1,\quad q>1,\quad \frac{1}{p}+\frac{1}{q}>1,\quad \lambda=2-\frac{1}{p}-\frac{1}{q}$$

(so that $0<\lambda<1$*), and*

$$\Sigma a_r^p=A,\quad \Sigma b_s^q=B,$$

then

$$T=\Sigma\Sigma'\frac{a_r b_s}{|r-s|^{\lambda}}\leqq K A^{1/p}B^{1/q},$$

where the dash implies that $r\neq s$, and $K=K(p,q)$ depends on p and q only.

382. *If $f(x)\geqq 0$, $g(y)\geqq 0$, p and q satisfy* (10.17.1), *and*

$$\int_{-\infty}^{\infty} f^p(x)\,dx=F,\quad \int_{-\infty}^{\infty} g^q(y)\,dy=G,$$

then

$$I=\int_{-\infty}^{\infty}\int_{-\infty}^{\infty}\frac{f(x)g(y)}{|x-y|^{\lambda}}\,dx\,dy\leqq K F^{1/p}G^{1/q}.$$

The proofs of the two theorems are practically the same. We give that of Theorem **382**.[a]

It is plain, after Theorem **380**, that we may replace f and g by f^* and g^*. We then divide I into four parts corresponding to the four quadrants of integration. The north-east and south-west parts are equal, and so are the north-west and south-east parts, and the two latter do not exceed the two former[b]. We need

[a] For that of Theorem **381** see Hardy, Littlewood, and Pólya (**1**); for a deduction of Theorem **382** from Theorem **381** see Hardy and Littlewood (6).

[b] The north-west and south-east parts could be accounted for by the easier argument of § 9.14.

therefore only consider the north-east part. Hence, changing our notation again, it is sufficient to prove that

$$J=\int_0^\infty\int_0^\infty \frac{f(x)g(y)}{|x-y|^\lambda}\,dx\,dy \leqq K F^{1/p}G^{1/q},$$

where now f and g are positive and decreasing, and F and G are defined by integrals over $(0, \infty)$. We write

$$J=J_1+J_2,$$

where J_1 and J_2 are integrals over the octants $y \leqq x$ and $x \leqq y$ respectively.

We have $$J_1=\int_0^\infty f(x)\,dx\int_0^x \frac{g(y)}{(x-y)^\lambda}\,dy.$$

Since $g(y)$ decreases, and $(x-y)^{-\lambda}$ increases, in $(0, x)$,

$$x\int_0^x \frac{g(y)}{(x-y)^\lambda}\,dy \leqq \int_0^x g(y)\,dy\int_0^x \frac{dy}{(x-y)^\lambda}=\frac{x^{1-\lambda}}{1-\lambda}g_1(x),$$

say, by Theorem **236**. Hence

$$J_1 \leqq \frac{1}{1-\lambda}\int_0^\infty f(x)g_1(x)x^{-\lambda}\,dx.$$

By Theorem **189**

$$J_1 \leqq \frac{1}{1-\lambda}F^{1/p}\left(\int_0^\infty g_1^{p'}(x)x^{-p'\lambda}\,dx\right)^{1/p'}. \tag{10.17.2}$$

But $p' > q$, by (10.17.1), and

$$g_1(x)=\int_0^x g(y)\,dy \leqq G^{1/q}x^{1/q'},$$

again by Theorem **189**. Hence

$$g_1^{p'}(x)x^{-p'\lambda} \leqq g_1^{q}(x)\,(G^{1/q}x^{1/q'})^{p'-q}x^{-p'\lambda}=G^{(p'-q)/q}\left\{\frac{g_1(x)}{x}\right\}^q \tag{10.17.3}$$

(since

$$\frac{p'-q}{q'}-\lambda p'=\frac{p'-q}{q'}-\left(\frac{1}{p'}+\frac{1}{q'}\right)p'=-\frac{q}{q'}-1=-q).$$

From (10.17.2) and (10.17.3) it follows that

$$J_1 \leqq K F^{1/p}G^{(p'-q)/p'q}\left\{\int_0^\infty\left(\frac{g_1(x)}{x}\right)^q dx\right\}^{1/p'} \leqq K F^{1/p}G^{1/q},$$

by Theorem **327**.

The discussion of J_2 is similar, and the theorem follows.

383. *Suppose that $f(x)$ is non-negative and L^p, where $p>1$, in* $(0, \infty)$, *that*

$$0<\alpha<\frac{1}{p}, \quad q=\frac{p}{1-\alpha p}, \tag{10.17.4}$$

and that

$$f_\alpha(x)=\frac{1}{\Gamma(\alpha)}\int_0^x f(y)(x-y)^{\alpha-1}\,dy. \tag{10.17.5}$$

Then $f_\alpha(x)$ is L^q in $(0, \infty)$ and

$$\int_0^\infty f_\alpha^q\,dx \leqq K\left(\int_0^\infty f^p\,dx\right)^{q/p}, \tag{10.17.6}$$

where $K=K(p,\alpha)=K(p,q)$.

Suppose that $g(x)$ is any function of $L^{q'}$, and that

$$\lambda=1-\alpha=1-\frac{1}{p}+\frac{1}{q}=2-\frac{1}{p}-\frac{1}{q'}.$$

By Theorem **382**,

$$\int_0^\infty\int_0^\infty \frac{g(x)f(y)}{|x-y|^\lambda}\,dx\,dy \leqq K\left(\int_0^\infty f^p\,dx\right)^{1/p}\left(\int_0^\infty g^{q'}\,dx\right)^{1/q'},$$

and *a fortiori*

$$\int_0^\infty f_\alpha(x)g(x)\,dx=\frac{1}{\Gamma(\alpha)}\int_0^\infty g(x)\,dx\int_0^x \frac{f(y)}{(x-y)^\lambda}\,dy$$

$$\leqq K\left(\int_0^\infty f^p\,dx\right)^{1/p}\left(\int_0^\infty g^{q'}\,dx\right)^{1/q'}.$$

Since this is true for all g, it follows, by Theorem **191**, that

$$\left(\int_0^\infty f_\alpha^q\,dx\right)^{1/q} \leqq K\left(\int_0^\infty f^p\,dx\right)^{1/p},$$

which is (10.17.6).

The proof shows that the result is also true when $f_\alpha(x)$ is defined by

$$f_\alpha(x)=\frac{1}{\Gamma(\alpha)}\int_x^\infty f(y)(y-x)^{\alpha-1}\,dy.$$

Theorem **383** embodies a result in the theory of 'fractional integration'. Liouville (**1**) and Riemann (**1**, 331–344) defined the integral $f_\alpha(x)$ of $f(x)$, of order α, as

$$f_\alpha(x)=\frac{1}{\Gamma(\alpha)}\int_a^x f(y)(x-y)^{\alpha-1}\,dy. \tag{10.17.7}$$

The lower limit a is the 'origin of integration'; a change of origin changes f_α in a manner which is not trivial formally, though unimportant for theorems of the type considered here. It is easily deduced from Theorem 383[a] that, *if f is L^p in (a, b), where $-\infty<a<b\leqq\infty$, $\alpha<1/p$, and f_α is the integral of f, of order α and with origin a, then f_α is L^q in (a, b).* When $\alpha>1/p$, f_α is continuous, and indeed belongs to the 'Lipschitz class' of order $\alpha-1/p$.

In applications of the theory, f is usually periodic. It was observed by Weyl (**3**) that the reference to an origin a is in this case inappropriate: Weyl accordingly modified the definition as follows. If we suppose that the mean value of f over a period is zero (a condition which we can always satisfy by subtracting an appropriate constant from f), then

$$\int_{-\infty}^{x} f(y)(x-y)^{\alpha-1}\,dy$$

converges at the lower limit, and we may take $a=-\infty$ in (10.17.7). Our theorem concerning the Lebesgue classes may be extended to this case also.

10.18. Another theorem concerning the rearrangement of a function in decreasing order. The theorem with which we end is important primarily for its function-theoretic applications, but the proof which we give[b] is interesting independently.

The theorem may be stated in two forms.

384. *Suppose that $f(x)$ is non-negative and integrable in a finite interval $(0, a)$, that $\bar{f}(x)$ is the rearrangement of $f(x)$ in decreasing order, that*

$$\Theta(x)=\Theta(x,f)=\operatorname*{Max}_{0\leqq\xi<x}\frac{1}{x-\xi}\int_{\xi}^{x} f(t)\,dt, \tag{10.18.1}$$

and that $\bar{\Theta}(x)$ is the rearrangement of $\Theta(x)$ in decreasing order. Then

$$\bar{\Theta}(x)\leqq\frac{1}{x}\int_0^x \bar{f}(t)\,dt \tag{10.18.2}$$

for $0<x\leqq a$.

385. *Suppose that $f(x)$ satisfies the conditions of Theorem* **384**, *and that $s(y)$ is any increasing function of y defined for $y\geqq 0$. Then*

$$\int_0^a s\{\Theta(x)\}\,dx\leqq\int_0^a s\left\{\frac{1}{x}\int_0^x \bar{f}(t)\,dt\right\}dx. \tag{10.18.3}$$

We begin with two preliminary remarks.

[a] See Hardy and Littlewood (**6**).
[b] Due to F. Riesz (**10**).

(1) We shall prove Theorem **384** first and deduce Theorem **385**. Since $\Theta(x)$ and $\overline{\Theta}(x)$ are equimeasurable,

$$\int_0^a s\{\Theta(x)\}\,dx = \int_0^a s\{\overline{\Theta}(x)\}\,dx.$$

Hence (10.18.3) follows from (10.18.2).

That (10.18.2) follows from (10.18.3), so that the two forms of the theorem are equivalent, is a little less obvious, but is proved in Theorem **392**.[a] The first implication is sufficient for our purpose here, since it is in the second form that the theorem is used in the applications.

(2) If $$\Theta_0(x) = \Theta_0(x,f) = \frac{1}{x}\int_0^x f(t)\,dt$$

then $$\Theta_0(x,f) \leqq \Theta_0(x,\bar{f}) = \Theta(x,\bar{f}) = \frac{1}{x}\int_0^x \bar{f}(t)\,dt,$$

by (10.12.2), and

$$\int_0^a s\{\Theta_0(x)\}\,dx \leqq \int_0^a s\left\{\frac{1}{x}\int_0^x \bar{f}(t)\,dt\right\}dx. \tag{10.18.4}$$

This, a much more trivial inequality than (10.18.3), is the analogue for integrals of Theorem **333**.

10.19. Proof of Theorem 384. We suppose, as we may, that $a=1$.

We consider a point x_0 for which

$$x_0 > 0, \quad \overline{\Theta}(x_0) > 0,$$

write

$$\overline{\Theta}(x_0) = p + \epsilon \quad (p > 0,\ \epsilon > 0), \tag{10.19.1}$$

and consider the set E defined by

$$0 \leqq x \leqq 1, \quad \Theta(x) > p. \tag{10.19.2}$$

Since $\Theta(x)$ and $\overline{\Theta}(x)$ are equimeasurable, E has the same measure as the set in which $\overline{\Theta}(x) > p$. This set is at least as large as the set

[a] See the Miscellaneous Theorems at the end of the chapter.

Theorem **385** was proved by Hardy and Littlewood (8), who deduced it by a limiting process from the analogous theorem for finite sums (Theorem **394**). Their proof of Theorem **394** was elementary but long, and a much shorter proof was found by Gabriel (2). Riesz 'en combinant ce qui me paraît être l'idée essentielle de M. Gabriel avec un théorème appartenant aux éléments de l'analyse' (Lemma A below) was able to prove the theorem directly and without limiting processes.

in which $\overline{\Theta}(x) \geqq p+\epsilon$, and the measure of this last set is, after (10.19.1), at least x_0. Hence

$$x_0 \leqq mE. \tag{10.19.3}$$

Now the set E is composed of those points x for which

$$\frac{1}{x-\xi}\int_{\xi}^{x} f(t)\,dt > p \tag{10.19.4}$$

for some $\xi=\xi(x)<x$. We can write (10.19.4) in the form

$$\int_0^x f(t)\,dt - px > \int_0^{\xi} f(t)\,dt - p\xi \tag{10.19.5}$$

or

$$g(x) > g(\xi), \tag{10.19.6}$$

say. Thus E is the set of points in which a certain continuous function $g(x)$ assumes a value greater than some at any rate of the values which it has assumed before. This property enables us to characterise the structure of E.

Lemma A. *The set E is composed of a finite or enumerable system of non-overlapping intervals (α_k, β_k). All of these intervals are open, and*

$$g(\alpha_k)=g(\beta_k);$$

except possibly when $x=1$ is a point of E, in which case there is one interval $(\alpha_k, 1)$ closed on the right, and $g(\alpha_k) \leqq g(1)$, though $g(\alpha_k)$ is not necessarily equal to $g(1)$.[a]

In the first place, since $g(x)$ is continuous, E is an open set (except possibly for the point $x=1$). Hence E is a set of intervals (α_k, β_k), open if $\beta_k<1$.

If $\beta_k<1$ then β_k is not a point of E, and

$$g(\alpha_k) \geqq g(\beta_k) \tag{10.19.7}$$

by the definition of E.

Next, suppose that $\alpha_k<x_1<\beta_k$, and consider the minimum of $g(x)$ in the interval $0 \leqq x \leqq x_1$. This minimum cannot be attained for $\alpha_k<x \leqq x_1$, since all such x belong to E, and so $g(x)>g(\xi)$ for some $\xi<x$. Hence it is attained for $x \leqq \alpha_k$. But α_k is not a point of E, and therefore $g(\alpha_k) \leqq g(x)$ for all these x. Hence the mini-

[a] All that we need is that $g(\alpha_k) \leqq g(\beta_k)$; but the argument will probably be clearer if we make the lemma complete.

mum is attained at α_k, and $g(\alpha_k) \leqq g(x_1)$. Making $x_1 \to \beta_k$, we obtain

(10.19.8) $$g(\alpha_k) \leqq g(\beta_k);$$

and this, with (10.19.7), proves the lemma[a].

We can now prove Theorem **384**. We may write (10.19.8) in the form

$$p(\beta_k - \alpha_k) \leqq \int_{\alpha_k}^{\beta_k} f(x)\,dx,$$

and from this it follows that

$$p.mE = p\Sigma(\beta_k - \alpha_k) \leqq \Sigma \int_{\alpha_k}^{\beta_k} f(x)\,dx = \int_E f(x)\,dx.$$

Hence, by (10.12.1),

(10.19.9) $$p.mE \leqq \int_E f(x)\,dx \leqq \int_0^{mE} \bar{f}(x)\,dx;$$

and hence, by (10.19.1),

(10.19.10) $$\bar{\Theta}(x_0) - \epsilon = p \leqq \frac{1}{mE}\int_0^{mE} \bar{f}(x)\,dx.$$

Finally, since $\bar{f}(x)$ decreases, it follows from (10.19.10) and (10.19.3) that

$$\bar{\Theta}(x_0) - \epsilon \leqq \frac{1}{x_0}\int_0^{x_0} \bar{f}(x)\,dx.$$

Since ϵ is arbitrary, this gives (10.18.1), with x_0 for x.

The function-theoretic applications of Theorems **384** and **385** arise as follows. Suppose that $f(\theta)$ is integrable and has the period 2π, that

$$M(\theta) = M(\theta, f) = \operatorname*{Max}_{0<|t|\leqq\pi} \frac{1}{t}\int_0^t f(\theta+u)\,du,$$

and that $N(\theta)$ is the similar function formed with $|f(\theta+u)|$. These functions are of the same type as the $\Theta(x)$ of Theorem **384**. but are generated by means taken to either side of θ.

Consider now the integral

(i) $$h(\theta, p) = \frac{1}{2\pi}\int_{-\pi}^{\pi} f(\theta+t)\chi(t, p)\,dt,$$

where χ is a kernel which involves a parameter p and satisfies the conditions

(ii) $$\chi(t, p) \geqq 0, \quad \frac{1}{2\pi}\int_{-\pi}^{\pi} \chi(t, p)\,dt = 1.$$

The standard examples of such kernels are the 'Poisson kernel'

$$\chi = \frac{1-r^2}{1-2r\cos t + r^2},$$

[a] The argument here is due to M. Riesz (see F. Riesz, **10**).

in which $p=r$ is positive and less than 1, and the 'Fejér kernel'

$$\chi=\frac{\sin^2 \frac{1}{2}nt}{n\sin^2 \frac{1}{2}t},$$

in which $p=n$ is a positive integer. The corresponding values of h are $u(r, \theta)$, the harmonic function defined by the 'Poisson integral' of $f(\theta)$, and $\sigma_n(\theta)$, the Cesàro mean, of order 1, of the Fourier series of $f(\theta)$.

Suppose now (a) that $f(\theta)$ belongs to L^k, where $k>1$, and (b) that χ satisfies the additional condition

$$\text{(iii)} \qquad \frac{1}{2\pi}\int_{-\pi}^{\pi}\left|t\frac{\partial\chi}{\partial t}\right|dt \leqq A,$$

where A is independent of p. It follows from Theorems **385**, with $s(y)=y^k$, and **327**, that $M(\theta)$ also belongs to L^k.[a] And it is easy to deduce from (i), (ii), and (iii) that

$$|h(\theta, p)| \leqq AM(\theta),$$

where A is again independent of p. Hence *h has a majorant* (independent of p) *of the class L^k*.

It is easily verified that the Poisson kernel satisfies (iii). Hence $u(r, \theta)$ possesses a majorant $U(\theta)$ of the class L^k. The same is true of $\sigma_n(\theta)$, but in this case the proof is not quite so simple, since the Fejér kernel does not satisfy (iii). We can however prove that $|\sigma_n(\theta)| \leqq AN(\theta)$, and similar conclusions follow. All this is set out in detail by Hardy and Littlewood (**8**).

MISCELLANEOUS THEOREMS AND EXAMPLES

386. If $c_2 \geqq c_3 \geqq \dots \geqq c_{2n} \geqq 0$ and the sets (a), (b) are non-negative and given except in arrangement, then

$$\sum_{r=1}^{n}\sum_{s=1}^{n} c_{r+s}a_r b_s$$

is a maximum when (a) and (b) are both in decreasing order.

[F. Wiener (**1**).]

387. It is not true that

$$\sum_{r+s+t=0} a_r b_s c_t \leqq \sum_{r+s+t=0} a_r^+ b_s^+ c_t^+.$$

[Trivial: take (a), (b), (c) to be $(0, 2, 1)$, $(1, 2, 0)$, $(1, 2, 1)$. Then

$$\Sigma a_r b_s c_t = 14, \quad \Sigma a_r^+ b_s^+ c_t^+ = 12.]$$

388. There are sets (a), (b), (c) such that none of the eight sums

$$\Sigma a^+ b^+ c^+, \quad \Sigma^+ a b^+ c^+, \quad \Sigma a^{++} b c^+, \quad \dots, \quad \Sigma^+ a^+ b^+ c$$

gives the maximal sum Σabc.

[Suppose $0<h<1$ and ϵ positive and sufficiently small; and take (a) to be $0, 0, 0, 1, 2$, (b) to be $h-\epsilon, h, h+\epsilon, 1, 1$, and (c) to be formed of any five different elements.]

[a] See Theorem 398 below.

389. If $$M(x)=\Sigma|r|x_r,\quad M(y)=\Sigma|s|y_s,$$
and $p \neq 0$, then the substitution Ω_p of § 10.8 decreases $\mu = M(x)+M(y)$.

[The theorem is trivial, but may be used to construct another proof of Theorem **371**, which follows the general lines of that in § 10.8 but is free from an appeal to 'continuity'.

We use A, C, C', K as in § 10.8; there may now be more than one arrangement C. We define L as the sub-class of K formed by those members of K for which μ is least. If $p \neq 0$, and A is wrong with respect to p, then Ω_p decreases μ and does not diminish S. Hence any A of L is a C'; and we can then show as in § 10.8 that L includes a C.]

390. In the notation of Theorem **373**
$$\sum_{r+s+t=n} a_r b_s c_t \leqq \sum_{r+s+t=0} a_r^{+\,+} b_s c_t^*$$
for every n.

[Corollary of Theorem **373**.]

391. If (a), (a'), (b), (b'), (c), and (c') are six sets of positive numbers subject to (10.5.1), then
$$\sum_{r+s+t=0} a_r a_r' b_s b_s' c_t c_t' \leqq \sum_{r+s+t=0} a_r^* a_r'^* b_s^* b_s'^* c_t^* c_t'^*.$$

[Corollary of Theorem **372** if first reduced, by the method of § 10.3, to the special case in which every number is 0 or 1.]

392. If f and g are non-negative, and

(i) $$\int_0^a s\{f(x)\}\,dx \leqq \int_0^a s\{g(x)\}\,dx$$

for every positive and increasing $s(y)$, then

(ii) $$\bar{f} \leqq \bar{g}$$

except perhaps for an enumerable set of values of x.

[This is the theorem referred to in § 10.18, as proving the equivalence of Theorems **384** and **385**. It is an analogue of Theorem **107**.

Since the integrals (i) are unaltered when we replace f and g by $\bar{f}$ and $\bar{g}$, we may suppose f and g themselves decreasing, so that $f=\bar{f}$, $g=\bar{g}$ (except perhaps in an enumerable set of points).

If (ii) is not true for almost all x, we can find a b and a c such that

(iii) $$b<c,\quad f(c)>g(b).$$

For, if this were not so, we should have $f(b+0) \leqq g(b)$ for all b, and $f(b) \leqq g(b)$ at all points of continuity of the functions, and therefore except in an enumerable set.

Supposing then that b and c satisfy (iii), we choose r so that
$$g(b)<r<f(c)$$
and define $s(y)$ by
$$s(y)=0\ (y<r),\quad s(y)=1\ (y \geqq r).$$

Then $$\int_0^a s\{f(x)\}\,dx = \int_{f \geqq r} dx \geqq c > b \geqq \int_{g \geqq r} dx = \int_0^a s\{g(x)\}\,dx,$$
in contradiction with (i).]

393. If $a_1, a_2, \ldots, a_N$ are non-negative,

$$\Theta(n) = \Theta(n, a) = \operatorname*{Max}_{1 \leqq \nu \leqq n} \frac{a_\nu + a_{\nu+1} + \ldots + a_n}{n - \nu + 1},$$

and a bar denotes a rearrangement in decreasing order (a notation opposite to that of § 10.1), then

$$\overline{\Theta}(n) \leqq \frac{\bar{a}_1 + \bar{a}_2 + \ldots + \bar{a}_n}{n} \quad (1 \leqq n \leqq N).$$

394. If the conditions of Theorem **393** are satisfied, and $s(y)$ is a positive increasing function of y, then

$$\sum_1^N s\{\Theta(n)\} \leqq \sum_1^N s\left(\frac{\bar{a}_1 + \bar{a}_2 + \ldots + \bar{a}_n}{n}\right).$$

[The last two theorems are the analogues for finite sums of Theorems **384** and **385**, and the reader will find it instructive to prove them by an adaptation of the argument of §§ 10.18–19. The earlier proofs of Hardy and Littlewood and of Gabriel are referred to in § 10.18.]

395. If $\quad c_1 \geqq c_2 \geqq \ldots \geqq c_p > 0, \quad d_1 \geqq d_2 \geqq \ldots \geqq d_q > 0;$ $e_1, e_2, \ldots, e_{p+q}$ is the aggregate of the c and d rearranged in decreasing order;

$$C_n = c_1 + c_2 + \ldots + c_n,$$

and D_n and E_n are defined similarly; and $s(y)$ is positive and increasing; then

$$s(C_1) + s\left(\frac{C_2}{2}\right) + \ldots + s\left(\frac{C_p}{p}\right) + s(D_1) + s\left(\frac{D_2}{2}\right) + \ldots + s\left(\frac{D_q}{q}\right)$$
$$\leqq s(E_1) + s\left(\frac{E_2}{2}\right) + \ldots + s\left(\frac{E_{p+q}}{p+q}\right).$$

[This is a special case of Theorem **394**. For a direct proof by induction, due to Chaundy, see Hardy and Littlewood (8): the theorem is one of the lemmas on which they based their proof of Theorem **394**.]

396. If p, q, P, Q are positive integers and $s(y)$ is positive and increasing, then

(i) $$\sum_1^p s\left(\frac{p}{n}\right) + \sum_1^q s\left(\frac{q}{n}\right) \leqq \sum_1^{p+q} s\left(\frac{p+q}{n}\right),$$

(ii) $$\sum_{p+1}^{p+P} s\left(\frac{p}{n}\right) + \sum_{q+1}^{q+Q} s\left(\frac{q}{n}\right) \leqq \sum_{p+q+1}^{p+q+P+Q} s\left(\frac{p+q}{n}\right),$$

(iii) $$\sum_1^\infty s\left(\frac{p}{n}\right) + \sum_1^\infty s\left(\frac{q}{n}\right) \leqq \sum_1^\infty s\left(\frac{p+q}{n}\right).$$

[(i) and (ii) follow from Theorem **395** by appropriate specialisation, and (iii), which is true whether p and q are integers or not, is a corollary. A case of (iii) is

$$\frac{x^{1/a}}{1 - x^{1/a}} + \frac{x^{1/b}}{1 - x^{1/b}} \leqq \frac{x^{1/(a+b)}}{1 - x^{1/(a+b)}} \quad (a > 0,\ b > 0,\ 0 < x < 1):$$

this may naturally be proved independently (and with '$<$'), for example as an application of Theorem **103**.]

397. If a, b, α, β are positive and s positive and increasing, then

$$\int_a^{a+\alpha} s\left(\frac{a}{x}\right) dx + \int_b^{b+\beta} s\left(\frac{b}{x}\right) dx \leqq \int_{a+b}^{a+b+\alpha+\beta} s\left(\frac{a+b}{x}\right) dx.$$

398. If $k>1$, and $\Theta(x)$ is defined as in Theorem **384**, then

$$\int_0^a \Theta^k(x)\,dx \leqq \left(\frac{k}{k-1}\right)^k \int_0^a f^k(x)\,dx.$$

[From Theorems **385** and **327**. There is of course a corresponding theorem for finite sums. This theorem has particularly important applications.]

399. In order that an integrable function $\phi(x)$ should have the property

$$\int_0^1 s(x)\phi(x)\,dx \geqq 0,$$

for all positive, increasing, and bounded $s(x)$, it is necessary and sufficient that

$$\int_x^1 \phi(t)\,dt \geqq 0 \quad (0 \leqq x \leqq 1).$$

To prove the condition necessary, specialise $s(x)$ appropriately; to prove it sufficient, integrate partially or use the second mean value theorem. The condition is certainly satisfied if there is a ξ between 0 and 1 such that $\phi(x) \geqq 0$ for $x>\xi$, $\phi(x) \leqq 0$ for $x<\xi$, and

$$\int_0^1 \phi(x)\,dx = 0.$$

Theorem **397** is a special case of this theorem (after a simple transformation).]

400. If E and ξ are functions of x subject to

$$0 \leqq dE \leqq dx, \quad 0 \leqq \xi < x,$$

then

$$\int_0^1 \left\{\frac{E(x)-E(\xi)}{x-\xi}\right\}^k dx \leqq \frac{kE(1)-E^k(1)}{k-1} \quad (k>1),$$

$$\int_0^1 \frac{E(x)-E(\xi)}{x-\xi}\,dx \leqq E(1)\left\{1+\log\frac{1}{E(1)}\right\}.$$

[Suppose that $f(x)$ is always 0 or 1, and that $E(x)$ is the measure of the part of $(0,x)$ in which $f(x)=1$, and apply Theorem 385.]

401. If

$$p>1, \quad q>1, \quad \frac{1}{p}+\frac{1}{q} \geqq 1, \quad \lambda = 2-\frac{1}{p}-\frac{1}{q},$$

$$h<1-\frac{1}{p}, \quad k<1-\frac{1}{q}, \quad h+k \geqq 0,$$

and $h+k>0$ if $\dfrac{1}{p}+\dfrac{1}{q}=1$, then

$$\int_0^\infty \int_0^\infty \frac{f(x)g(y)}{x^h y^k |x-y|^{\lambda-h-k}}\,dx\,dy \leqq K\left(\int_0^\infty f^p\,dx\right)^{1/p}\left(\int_0^\infty g^q\,dx\right)^{1/q}.$$

[Here, and in Theorems **402** and **403**, K denotes a positive number depending only on the parameters of the theorem (here p, q, h, k).]

402. If

$$p>1, \quad 0 \leqq \alpha < \frac{1}{p}, \quad p \leqq q \leqq \frac{p}{1-\alpha p},$$

then

$$\int_0^\infty x^{-(p-q+pq\alpha)/p} f_\alpha{}^q\,dx \leqq K\left(\int_0^\infty f^p\,dx\right)^{q/p},$$

where f_α is defined as in (10.17.5). The result is still true if $\alpha \geqq 1/p$, when the second condition on q may be omitted.

[For the last two theorems see Hardy and Littlewood (**6**). The case $q=p$ gives
$$\int_0^\infty (x^{-\alpha} f_\alpha)^p\,dx \leqq K \int_0^\infty f^p\,dx:$$
compare Theorem **329**.]

403. The result of Theorem **383** is not necessarily true when $p=1$.

[Define $f(x)$ by
$$f(x)=\frac{1}{x}\left(\log\frac{1}{x}\right)^{-\beta} \quad (0<x\leqq \tfrac{1}{2}), \quad =0 \quad (x>\tfrac{1}{2}),$$
where $\beta>1$. Then
$$f_\alpha(x)=K\int_0^x \frac{1}{y}\left(\log\frac{1}{y}\right)^{-\beta}(x-y)^{\alpha-1}\,dy$$
$$>Kx^{\alpha-1}\int_0^x \frac{1}{y}\left(\log\frac{1}{y}\right)^{-\beta}dy = Kx^{\alpha-1}\left(\log\frac{1}{x}\right)^{1-\beta}.$$
Here
$$p=1, \quad q=\frac{1}{1-\alpha};$$
f is L, but f_α is L^q only if
$$\frac{\beta-1}{1-\alpha}>1, \quad \beta>2-\alpha.]$$

404. Suppose that $f(x)$ is defined in $(-1, 1)$ and has a continuous derivative $f'(x)$ which vanishes only at a finite number of points, and that
$$f(x)\geqq 0, \quad f(-1)=f(1)=0.$$
Then the length of the curve $y=f(x)$ is greater than that of $y=f^*(x)$, unless $f(x)=f^*(x)$.

[See Steiner (**1**, II, 265). If $0<y<Y=\operatorname{Max} f$ then (except perhaps for a finite number of values of y) the equation $y=f(x)$ has an even number $2n$ (depending on y) of roots. If we denote these roots, in ascending order, by $x_1, x_2, \ldots, x_{2n}$, and the derivative of x_ν with respect to y by x_ν', then, by Theorem **25**,
$$2\int_0^Y \{1+[\tfrac{1}{2}\Sigma(-1)^\nu x_\nu']^2\}^{\frac{1}{2}}\,dy \leqq \int_0^Y \Sigma(1+x_\nu'^2)^{\frac{1}{2}}\,dy.$$
There is equality only if $n=1$ for all y, and $x_1=-x_2$.]

405. Suppose that $f(x, y)\geqq 0$ for all x, y, and that the measure $M(z)$ of the set in which $f(x, y)\geqq z$ is finite for all positive z; define $\rho(z)$ by
$$M(z)=\pi\rho^2;$$
and write
$$f^*(x, y)=\rho^{-1}\{\surd(x^2+y^2)\},$$
where ρ^{-1} is the inverse of ρ. Then (under appropriate conditions of regularity) the area of the surface $z=f(x, y)$ is greater than that of
$$z=f^*(x, y).$$

[See Schwarz (**1**). The theorem is important in itself and interesting because it involves a two-dimensional analogue of the notion of $f^*(x)$.]

APPENDIX I

ON STRICTLY POSITIVE FORMS

11.1. Our aim in Appendix I is to present an elementary proof due to W. Habicht[a] for an important particular case of the theorem of Hilbert and Artin stated in § 2.23.

The term 'form' will be used as an abbreviation for 'a homogeneous polynomial with real coefficients'. A form

$$F = F(x_1, x_2, \ldots, x_m)$$

is said to be strictly positive if $F > 0$ for real values $x_1, x_2, \ldots, x_m$ of the variables unless $x_1 = x_2 = \ldots = x_m = 0$; we assume further that F is not a constant. We wish to prove the following theorem.

406. *Any strictly positive form F can be expressed as*

$$F = \frac{\Sigma_i M_i^2}{\Sigma_j N_j^2}, \tag{11.1.1}$$

where M_i and N_j are suitably chosen forms.

The proof is based on Theorem **56**. In that theorem the term 'a form with positive coefficients' is used implicitly with the following meaning (the reader should verify that with this meaning the proof in § 2.24 does apply): if $G = G(x_1, x_2, \ldots, x_m)$ is a form with positive coefficients and its degree is g, then

$$G = \Sigma a x_1^{\alpha_1} x_2^{\alpha_2} \ldots x_m^{\alpha_m},$$

where

$$a = a_{\alpha_1 \alpha_2 \ldots \alpha_m} > 0$$

for all integers $\alpha_1, \alpha_2, \ldots, \alpha_m$ satisfying the conditions

$$\alpha_1 \geqq 0, \quad \alpha_2 \geqq 0, \quad \ldots, \quad \alpha_m \geqq 0, \quad \alpha_1 + \alpha_2 + \ldots + \alpha_m = g.$$

We arrive at **406** in three steps from **56**.

[a] 'Über die Zerlegung strikte definiter Formen in Quadrate', *Commentarii Math. Helvetici*, 12 (1940), 317–322.

11.2. A particular case. Let the form $K(x_1, x_2, \dots, x_m)$ be even in each of its variables. That is,

$$K(-x_1, x_2, \dots, x_m) = K(x_1, -x_2, \dots, x_m) = \dots$$
$$= K(x_1, \dots, x_{m-1}, -x_m) = K(x_1, x_2, \dots, x_m).$$

Since only even powers of the variables occur in such a form K we have

$$K(x_1, x_2, \dots, x_m) = L(x_1^2, x_2^2, \dots, x_m^2), \tag{11.2.1}$$

where $L(y_1, y_2, \dots, y_m)$ is a form. If the degrees of K and L are k and l respectively we have of course $k = 2l$.

Assume now that $K(x_1, x_2, \dots, x_m)$ is strictly positive. Then

$$L(y_1, y_2, \dots, y_m) > 0$$

for $y \geqq 0$, $\Sigma y > 0$, and so, by Theorem **56**,

$$L = \frac{G}{(y_1 + y_2 + \dots + y_m)^p}, \tag{11.2.2}$$

where G is a form with positive coefficients. Any positive number is the square of some positive number, and we can write G as

$$G = \Sigma q^2 y_1^{\alpha_1} y_2^{\alpha_2} \dots y_m^{\alpha_m}, \tag{11.2.3}$$

where the q are positive numbers. It follows from (11.2.1), (11.2.2), and (11.2.3) that

$$K(x_1, x_2, \dots, x_m) = \frac{\Sigma(q x_1^{\alpha_1} x_2^{\alpha_2} \dots x_m^{\alpha_m})^2}{(x_1^2 + x_2^2 + \dots + x_m^2)^p}. \tag{11.2.4}$$

K is thus expressible in the desired form (11.1.1). In fact, the forms M_i and N_j reduce to monomials.[a] According to the remarks at the end of § 11.1 none of the monomials $q x_1^{\alpha_1} x_2^{\alpha_2} \dots x_m^{\alpha_m}$ compatible with the degree of K vanishes.

11.3. An intermediate representation. We consider now any strictly positive form $F(x_1, \dots, x_m)$ and the product of 2^m factors

$$\prod_{y_1=0}^{1} \prod_{y_2=0}^{1} \dots \prod_{y_m=0}^{1} F((-1)^{y_1} x_1, (-1)^{y_2} x_2, \dots, (-1)^{y_m} x_m) = K(x_1, x_2, \dots, x_m). \tag{11.3.1}$$

[a] Pólya (3).

The form $K(x_1, x_2, \ldots, x_m)$ is obviously even with respect to each of its variables, and strictly positive as a product of strictly positive factors. (11.2.4) accordingly applies to K, and combining this with (11.3.1) we arrive at the conclusion: *given any strictly positive form F we can find a form P such that the product FP is a sum of squares of monomials*:

$$(11.3.2) \qquad FP = \Sigma (q x_1^{\alpha_1} x_2^{\alpha_2} \ldots x_m^{\alpha_m})^2.$$

There is as yet no reason to suppose that P is a sum of squares and our goal (11.1.1) is not yet attained.

11.4. Proof of Theorem 406. Consider again the strictly positive form F. Let $2n$ be its degree. We introduce (this is the key-idea) an extra variable u and consider the form

$$u^{2n} + F(x_1, x_2, \ldots, x_m)$$

of $m+1$ variables; this is obviously strictly positive. After § 11.3 we can find a form $P^*(x_1, \ldots, x_m, u)$ such that

$$(11.4.1) \qquad (u^{2n} + F(x_1, \ldots, x_m)) P^*(x_1, \ldots, x_m, u) = \Sigma (q x_1^{\alpha_1} \ldots x_m^{\alpha_m} u^{\alpha})^2.$$

We attain our goal by 'reducing mod $(u^{2n}+F)$' the right-hand side of (11.4.1), and proceed to explain this process in detail.

We consider the general term of the sum of the right-hand side of (11.4.1). Dividing α by $2n$ we have two non-negative integers β and γ such that

$$(11.4.2) \qquad \alpha = 2n\beta + \gamma$$

and

$$(11.4.3) \qquad 0 \leqq \gamma \leqq 2n-1.$$

Then

$$(11.4.4) \qquad u^{\alpha} = u^{2n\beta+\gamma} = [(u^{2n})^{\beta} - (-F)^{\beta}] u^{\gamma} + (-F)^{\beta} u^{\gamma}.$$

The expression in square brackets is divisible by $u^{2n} - (-F)$. Hence, from (11.4.4),

$$(11.4.5) \qquad q x_1^{\alpha_1} \ldots x_m^{\alpha_m} u^{\alpha} = (u^{2n} + F) Q + R u^{\gamma},$$

where $\qquad R = R(x_1, \ldots, x_m) = q x_1^{\alpha_1} \ldots x_m^{\alpha_m} (-F)^{\beta}$

depends on m variables only and does not vanish identically. (The form $Q=Q(x_1, \ldots, x_m, u)$ may depend on $m+1$ variables and may vanish identically; Q and Ru^γ are the quotient and the remainder, respectively, in the division of the monomial on the left-hand side of (11.4.5) by $u^{2n}+F$, considered as a polynomial in u alone with coefficients depending on $x_1, \ldots, x_m$.) Introducing (11.4.5) on the right-hand side of (11.4.1) and transporting the terms divisible by $u^{2n}+F$ to the left-hand side, we obtain

$$(u^{2n}+F)P^{**}=\sum_{\gamma=0}^{2n-1} u^{2\gamma}\sum_{\delta} R_{\gamma\delta}^2. \tag{11.4.6}$$

Observe that γ satisfies (11.4.3), and also that, in accordance with the remarks at the end of § 11.1 and § 11.2, the sum on the right-hand side of (11.4.1) has some terms with $\alpha=0$ and, therefore, in view of (11.4.2), with $\gamma=0$. The sum

$$\sum_{\delta} R_{0\delta}^2$$

is therefore not empty and does not vanish identically.

From (11.4.6) it follows that P^{**} as a polynomial in u is of degree $\leqq 2n-2$. We can therefore write (11.4.6) as

$$(u^{2n}+F)\sum_{\gamma=0}^{2n-2} T_\gamma(x_1, \ldots, x_m)\, u^\gamma=\sum_{\gamma=0}^{2n-1} u^{2\gamma}\sum_{\delta} R_{\gamma\delta}^2.$$

Comparing the coefficients of u^0 and u^{2n} we obtain

$$FT_0=\sum_{\delta} R_{0\delta}^2,$$

$$T_0=\sum_{\delta} R_{n\delta}^2.$$

From the first equation we conclude that T_0 does not vanish identically; the two equations then give

$$F=\sum_{\delta} R_{0\delta}^2 \Big/ \sum_{\delta} R_{n\delta}^2,$$

which proves the desired (11.1.1) and Theorem **406**.

The proof actually yields more than we have stated. We have in fact

407. *If the coefficients of the strictly positive form F are rational numbers, the representation* (11.1.1) *can be chosen so that the coefficients of the forms M_i and N_j are also rational numbers.*

We need only one modification. In deriving (11.2.3) we used the obvious remark: any positive number is the square of a certain positive number. We have now to use instead the equally obvious remark: any positive rational number is a sum of squares of positive rational numbers. (In fact, if r and s are positive integers,

$$\frac{r}{s}=\frac{rs}{s^2}=\left(\frac{1}{s}\right)^2+\left(\frac{1}{s}\right)^2+\dots+\left(\frac{1}{s}\right)^2,$$

a sum of rs terms.)

APPENDIX II

THORIN'S PROOF AND EXTENSION OF THEOREM 295[a]

12.1. This is both brilliant and easy to follow: the reader who wishes to avoid all difficulties may substitute the present appendix for § 8.15. The proof, however, depends essentially on complex function-theory, which we have systematically excluded from our main account, and the argument of § 8.15, in spite of its greater difficulty and more limited range, retains its interest and importance. The extended theorem is

408. *Suppose that* $M_{\alpha,\beta}$ *is the maximum of* $|A|$, *where*

$$A = A(x, y) = \sum_{i=1}^{m} \sum_{j=1}^{n} a_{ij} x_i y_j$$

for

$$\sum_{i=1}^{m} |x_i|^{1/\alpha} \leqq 1, \quad \sum_{j=1}^{n} |y_j|^{1/\beta} \leqq 1, \tag{12.1.1}$$

it being understood that if $\alpha = 0$ *or* $\beta = 0$ (*or both*), *these inequalities are to be replaced by* $|x_i| \leqq 1$ *and* $|y_j| \leqq 1$, *respectively. Then* $\log M_{\alpha,\beta}$ *is convex in the quadrant* $\alpha \geqq 0$, $\beta \geqq 0$.

The proof turns on the following principles, all of them obvious. (i) The upper bound[b] of a family (possibly infinite in number) of convex functions is convex; (ii) the limit of a convergent sequence of convex functions is convex; (iii) in finding an upper bound taken with respect to a number of independent conditions we may take *successive* upper bounds, and in any order; for example,

$$\operatorname*{Max}_{(x,y)} f(x, y) = \operatorname*{Max}_{(x)} (\operatorname*{Max}_{(y)} f(x, y)).$$

[a] G. O. Thorin, 'Convexity theorems generating those of M. Riesz and Hadamard with some applications', *Seminar Math. Lund*, 9 (1948). The author proves still more general forms.

[b] 'Upper bound' throughout means 'least upper bound'.

We need the following

Lemma. Let $\Sigma = \Sigma a e^{bs}$ *be a finite sum with real* b *and* $s = \sigma + it$ *with real* σ *and* t, *and let*

$$m(\sigma) = \operatorname*{Max}_{(t)} |\Sigma|.$$

Then $\log m(\sigma)$ *is a convex function of* σ.

$m(\sigma)$ is continuous in the a, b; it is therefore, by principle (ii), enough to establish the case in which all b are rational. Σ is then a polynomial in $z = e^{s/Q}$, where Q is a common denominator of the b's, and the result is then a consequence of Hadamard's 'three circles theorem'.

[*If* $f(z)$ *is regular in an annulus* $r_1 \leqq |z| \leqq r_2$, *and* $M(r)$ *is the maximum of* $|f(z)|$ *for* $|z| = r$, *then* $\log M(r)$ *is a convex function of* $\log r$.

We give the proof for completeness. Choose μ so that

$$r_1^{\mu} M(r_1) = r_2^{\mu} M(r_2) = T.$$

Then $z^{\mu} f(z)$ is regular at every point of the annulus and has a modulus one-valued in the annulus. Its maximum modulus for the boundary is T; it follows that $r^{\mu} M(r) \leqq T$, whence we have

$$\log M(r) \leqq t \log M(r_1) + (1-t) \log M(r_2)$$

for $\log r = t \log r_1 + (1-t) \log r_2$.]

Coming now to Theorem **408**, it is enough, by continuity (as was observed in § 8.13), to prove $\log M_{\alpha,\beta}$ convex in σ on any segment $\alpha = \alpha_0 + \lambda_1 \sigma$, $\beta = \beta_0 + \lambda_2 \sigma$ lying in the open quadrant $\alpha > 0$, $\beta > 0$. Now for an internal (α, β) we may write

$$x_k = \xi_k^{\alpha} e^{i\phi_k}, \quad y_j = \eta_j^{\beta} e^{i\psi_j}, \quad \xi, \eta \geqq 0,$$

and then, for varying (real) ϕ, ψ, and non-negative ξ, η varying subject to

$$\Sigma \xi \leqq 1, \quad \Sigma \eta \leqq 1 \tag{12.1.2}$$

(a condition independent of α, β),

$$M_{\alpha,\beta} = \operatorname*{Max}_{\phi,\psi,\xi,\eta} |\Sigma\Sigma a \xi^{\alpha_0 + \lambda_1 \sigma} \eta^{\beta_0 + \lambda_2 \sigma} e^{i(\phi+\psi)}|.$$

If in this we replace σ by $s=\sigma+it$, then for any given real t the upper bound is unaltered ($\phi+\psi$ is merely 'translated'). We can now add an operation Max for varying t; and by principle (iii) we can make this the inner operation:

$$M_{\alpha,\beta}=\underset{\phi,\psi,\xi,\eta}{\text{Max}}\,(\underset{(t)}{\text{Max}}\,|\,\Sigma\Sigma a\xi^{\alpha_0+\lambda_1 s}\eta^{\beta_0+\lambda_2 s}\,e^{i(\phi+\psi)}\,|).$$

Now for fixed ϕ, ψ, ξ, η we may suppress in $\Sigma\Sigma$ any term for which either the ξ or the η is 0 (the indices are positive), obtaining a modified double sum $\Sigma\Sigma$, $F_{\xi,\eta,\phi,\psi}(s)$, say. Then

$$M_{\alpha,\beta}=\underset{\theta,\psi,\xi,\eta}{\text{Max}}\,m_{\phi,\psi,\xi,\eta}(\sigma),\quad m(\sigma)=\underset{(t)}{\text{Max}}\,|\,F(s)\,|.$$

But F is a Σ of the Lemma, so that its $\log m(\sigma)$ is convex (for *all* σ, and in particular on the segment). The upper bound of the family of $\log m(\sigma)$ is now convex, by principle (i), and our proof is completed.

APPENDIX III

ON HILBERT'S INEQUALITY

13.1. After the remarks in § 4.6 about the limitations of the theory of maxima and minima of functions of several variables it is of interest that some of the more sophisticated inequalities are capable of simple proof by this method. We take for illustration Hilbert's inequality, but for simplicity with the b's identical with the a's:[a]

$$\sum_{m,n=0}^{N} \frac{a_m a_n}{m+n+1} < \pi \sum_{0}^{N} a_n^2,$$

unless all the a's are 0.

We may suppose more than one a different from 0, the excepted cases being trivial. Consider

$$F(a) = F(a_0, a_1, \ldots, a_N) = \sum_{m,n=0}^{N} \frac{a_m a_n}{m+n+1}$$

subject to

$$G(a) = \sum_{0}^{N} a_n^2 = t, \tag{13.1.1}$$

where t is a positive constant. If any a_n is 0, a small increment δ in this a_n produces an increase of δ^2 in G and one of order δ in F, and so increases the ratio F/G. It follows, since F is continuous, that subject to (13.1.1) F attains its maximum $F^* = F^*(t)$ for some set of a_n, none of which is 0.

For this set we have

$$\frac{\partial F}{\partial a_n} - \lambda \frac{\partial G}{\partial a_n} = 0 \quad (n \leqq N)$$

for some λ, independent of n. This gives

$$\sum_{m=0}^{N} \frac{a_m}{m+n+1} = \lambda a_n \quad (n \leqq N), \tag{13.1.2}$$

[a] J. W. S. Cassels, 'An elementary proof of some inequalities', *Journ. L.M.S.* 23 (1948), 285–290. There are similar proofs for the general case, and for Hardy's and Carleman's inequalities.

and multiplying by a_n and adding we have

$$F^*(t)=\lambda t.$$

Let $(m+\frac{1}{2})^{\frac{1}{2}}a_m$ attain its maximum at $m=\mu$. Then (13.1.2) with $n=\mu$ gives

$$\lambda a_\mu=\sum_{m=0}^{N}\frac{a_m}{m+\mu+1}\leqq a_\mu(\mu+\tfrac{1}{2})^{\frac{1}{2}}\sum_{m=0}^{N}\frac{1}{(m+\mu+1)(m+\frac{1}{2})^{\frac{1}{2}}}.$$

Since $\{(x+\mu+1)(x+\frac{1}{2})^{\frac{1}{2}}\}^{-1}$ is strictly convex,

$$\sum_{m=0}^{N}\frac{1}{(m+\mu+1)(m+\frac{1}{2})^{\frac{1}{2}}}<\int_{-\frac{1}{2}}^{N+\frac{1}{2}}\frac{dx}{(x+\mu+1)(x+\frac{1}{2})^{\frac{1}{2}}}$$

$$=\int_0^{(N+1)^{\frac{1}{2}}}\frac{2dy}{y^2+\mu+\frac{1}{2}}<\int_0^\infty=(\mu+\tfrac{1}{2})^{-\frac{1}{2}}\pi,$$

and it follows, since $a_\mu\neq 0$, that $\lambda<\pi$.

For any non-null set (a_n) we have now

$$F(a)\leqq F^*(G)=\lambda G<\pi G=\pi\sum_{0}^{N}a_n{}^2.$$

BIBLIOGRAPHY

N. H. Abel

1. Sur les séries, *Œuvres complètes*, II (2nd ed., Christiania, 1881), 197–201.

E. Artin

1. Über die Zerlegung definiter Funktionen in Quadrate, *Abhandl. a. d. math. Seminar Hamburg*, 5 (1927), 100–115.

E. Artin and O. Schreier

1. Algebraische Konstruktion reeller Körper, *Abhandl. a. d. math. Seminar Hamburg*, 5 (1927), 85–99.

G. Aumann

1. Konvexe Funktionen und die Induktion bei Ungleichungen zwischen Mittelwerten, *Münchner Sitzungsber.* 1933, 403–415.

S. Banach

1. *Opérations linéaires* (Warsaw, 1932).

J. Bernoulli

1. *Unendliche Reihen* (Ostwald's Klassiker der exakten Wissenschaften Nr. 171, Leipzig, 1909).

F. Bernstein

1. Über das Gauss'sche Fehlergesetz, *Math. Annalen*, 64 (1907), 417–447.

F. Bernstein and G. Doetsch

1. Zur Theorie der konvexen Funktionen, *Math. Annalen*, 76 (1915), 514–526.

A. S. Besicovitch

1. On mean values of functions of a complex and of a real variable, *Proc. L.M.S.* (2), 27 (1928), 373–388.

Z. W. Birnbaum and W. Orlicz

1. Über die Verallgemeinerung des Begriffes der zueinander konjugierten Potenzen, *Studia Math.* 3 (1931), 1–67.

W. Blaschke

1. *Kreis und Kugel* (Leipzig, 1916).

G. A. Bliss

1. *Calculus of variations* (Chicago, 1927).
2. The transformation of Clebsch in the calculus of variations, *Proc. International Math. Congress* (Toronto, 1924), I, 589–603.
3. An integral inequality, *Journ. L.M.S.* 5 (1930), 40–46.

H. Blumberg

1. On convex functions, *Trans. Amer. Math. Soc.* 20 (1919), 40–44.

M. Bôcher

1. *Introduction to higher algebra* (New York, 1907).

H. Bohr

1. Zur Theorie der fastperiodischen Funktionen (I), *Acta Math.* 45 (1924), 29–127.

O. Bolza

1. *Vorlesungen über Variationsrechnung* (Leipzig, 1909).

L. S. Bosanquet

1. Generalisations of Minkowski's inequality, *Journ. L.M.S.* 3 (1928), 51–56.

W. Briggs and **G. H. Bryan**

1. *The tutorial algebra* (4th ed., London, 1928).

T. A. A. Broadbent

1. A proof of Hardy's convergence theorem, *Journ. L.M.S.* 3 (1928), 242–243.

G. H. Bryan and **W. Briggs.** *See* **W. Briggs** and **G. H. Bryan**

V. Buniakowsky

1. Sur quelques inégalités concernant les intégrales ordinaires et les intégrales aux différences finies, *Mémoires de l'Acad. de St-Pétersbourg* (VII), 1 (1859), No. 9.

T. Carleman

1. Sur les fonctions quasi-analytiques, *Conférences faites au cinquième congrès des mathématiciens scandinaves* (Helsingfors, 1923), 181–196.

A. L. Cauchy

1. *Cours d'analyse de l'École Royale Polytechnique.* Ire partie. Analyse algébrique (Paris, 1821). [*Œuvres complètes*, IIe série, III.]
2. *Exercices de mathématiques*, II (Paris, 1827). [*Œuvres complètes*, IIe série, VII.]

G. Chrystal

1. *Algebra*, II (2nd ed., London, 1900).

R. Cooper

1. Notes on certain inequalities (I): generalisation of an inequality of W. H. Young, *Journ. L.M.S.* 2 (1927), 17–21.
2. Notes on certain inequalities (II), *Journ. L.M.S.* 2 (1927), 159–163.
3. The converses of the Cauchy-Hölder inequality and the solutions of the inequality $g(x+y) \leqq g(x)+g(y)$, *Proc. L.M.S.* (2), 26 (1927), 415–432.
4. Note on the Cauchy-Hölder inequality, *Journ. L.M.S.* 3 (1928), 8–9.

E. T. Copson

1. Note on series of positive terms, *Journ. L.M.S.* 2 (1927), 9–12.
2. Note on series of positive terms, *Journ. L.M.S.* 3 (1928), 49–51.

G. E. Crawford

1. Elementary proof that the arithmetic mean of any number of positive quantities is greater than the geometric mean, *Proc. Edinburgh Math. Soc.* 18 (1900), 2–4.

G. Darboux

1. Sur la composition des forces en statique, *Bull. des sciences math.* 9 (1875), 281–288.

L. L. Dines

1. A theorem on orthogonal functions with an application to integral inequalities, *Trans. Amer. Math. Soc.* 30 (1928), 425–438.

A. L. Dixon

1. A proof of Hadamard's theorem as to the maximum value of the modulus of a determinant, *Quart. Journ. of Math.* (2), 3 (1932), 224–225.

G. Doetsch and **F. Bernstein.** *See* **F. Bernstein** and **G. Doetsch**

J. Dougall

1. Quantitative proofs of certain algebraic inequalities, *Proc. Edinburgh Math. Soc.* 24 (1906), 61–77.

J. M. C. Duhamel and **A. A. L. Reynaud**

1. *Problèmes et développemens sur diverses parties des mathématiques* (Paris, 1823).

E. B. Elliott

1. A simple exposition of some recently proved facts as to convergency, *Journ. L.M.S.* 1 (1926), 93–96.
2. A further note on sums of positive terms, *Journ. L.M.S.* 4 (1929), 21–23.

Euclid

1. *The thirteen books of Euclid's Elements* (translated by Sir Thomas Heath, Cambridge, 1908).

L. Fejér

1. Über gewisse Minimumprobleme der Funktionentheorie, *Math. Annalen*, 97 (1927), 104–123.

L. Fejér and F. Riesz

1. Über einige funktionentheoretische Ungleichungen, *Math. Zeitschr.* 11 (1921), 305–314.

B. de Finetti

1. Sul concetto di media, *Giornale dell' Istituto Italiano degli Attuari*, 2 (1931), 369–396.

E. Fischer

1. Über den Hadamardschen Determinantensatz, *Archiv d. Math. u. Physik* (3), 13 (1908), 32–40.

E. C. Francis and J. E. Littlewood

1. *Examples in infinite series with solutions* (Cambridge, 1928).

F. Franklin

1. Proof of a theorem of Tschebyscheff's on definite integrals, *American Journ. of Math.* 7 (1885), 377–379.

M. Fréchet

1. Pri la funkcia equacio $f(x+y)=f(x)+f(y)$, *L'enseignement math.* 15 (1913), 390–393.
2. A propos d'un article sur l'équation fonctionnelle $f(x+y)=f(x)+f(y)$, *L'enseignement math.* 16 (1914), 136.

G. Frobenius

1. Über Matrizen aus positiven Elementen (II), *Berliner Sitzungsber.* 1909, 514–518.

R. M. Gabriel

1. An additional proof of a theorem upon rearrangements, *Journ. L.M.S.* 3 (1928), 134–136.
2. An additional proof of a maximal theorem of Hardy and Littlewood, *Journ. L.M.S.* 6 (1931), 163–166.
3. The rearrangement of positive Fourier coefficients, *Proc. L.M.S.* (2), 33 (1932), 32–51.

C. F. Gauss

1. *Werke* (Göttingen, 1863–1929).

J. A. Gmeiner and O. Stolz

1. *Theoretische Arithmetik*, II Abteilung (Leipzig, 1902).

J. P. Gram

1. Über die Entwicklung reeller Funktionen in Reihen, mittelst der Methode der kleinsten Quadrate, *Journal f. Math.* 94 (1881), 41–73.

K. Grandjot

1. On some identities relating to Hardy's convergence theorem, *Journ. L.M.S.* 3 (1928), 114–117.

Grebe

1. Über die Vergleichung zwischen dem arithmetischen, dem geometrischen und dem harmonischen Mittel, *Zeitschr. f. Math. u. Physik*, 3 (1858), 297–298.

A. Haar

1. Über lineare Ungleichungen, *Acta Litt. ac Scient. Univ. Hung.* 2 (1924), 1–14.

J. Hadamard

1. Résolution d'une question relative aux déterminants, *Bull. des sciences math.* (2), 17 (1893), 240–248.

H. Hahn

1. *Theorie der reellen Funktionen*, I (Berlin, 1921).

G. Hamel

1. Eine Basis aller Zahlen und die unstetigen Lösungen der Funktionalgleichung $f(x+y)=f(x)+f(y)$, *Math. Annalen*, 60 (1905), 459–462.

G. H. Hardy

1. *A course of pure mathematics* (6th ed., Cambridge, 1928).

2. Note on a theorem of Hilbert, *Math. Zeitschr.* 6 (1920), 314–317.

3. Note on a theorem of Hilbert concerning series of positive terms, *Proc. L.M.S.* (2), 23 (1925), Records of Proc. xlv–xlvi.

4. Notes on some points in the integral calculus (LX), *Messenger of Math.* 54 (1925), 150–156.

5. Notes on some points in the integral calculus (LXIV), *Messenger of Math.* 57 (1928), 12–16.

6. Remarks on three recent notes in the *Journal*, *Journ. L.M.S.* 3 (1928), 166–169.

7. Notes on some points in the integral calculus (LXVIII), *Messenger of Math.* 58 (1929), 115–120.

8. Prolegomena to a chapter on inequalities, *Journ. L.M.S.* 4 (1929), 61–78 and 5 (1930), 80.

9. Remarks in addition to Dr Widder's note on inequalities, *Journ. L.M.S.* 4 (1929), 199–202.

10. The constants of certain inequalities, *Journ. L.M.S.* 8 (1933), 114–119.

G. H. Hardy and **J. E. Littlewood**

1. Elementary theorems concerning power series with positive coefficients and moment constants of positive functions, *Journal f. Math.* 157 (1927), 141–158.

2. Some new properties of Fourier constants, *Math. Annalen*, 97 (1927), 159–209 (199).

3. Notes on the theory of series (VI): two inequalities, *Journ. L.M.S.* 2 (1917), 196–201.

4. Notes on the theory of series (VIII): an inequality, *Journ. L.M.S.* 3 (1928), 105–110.

5. Notes on the theory of series (X): some more inequalities, *Journ. L.M.S.* 3 (1928), 294–299.

6. Some properties of fractional integrals (1), *Math. Zeitschr.* 27 (1928), 565–606.

7. Notes on the theory of series (XII): on certain inequalities connected with the calculus of variations, *Journ. L.M.S.* 5 (1930), 283–290.

8. A maximal theorem with function-theoretic applications, *Acta Math.* 54 (1930), 81–116.

9. Notes on the theory of series (XIII): some new properties of Fourier constants, *Journ. L.M.S.* 6 (1931), 3–9.

10. Some integral inequalities connected with the calculus of variations, *Quart. Journ. of Math.* (2), 3 (1932), 241–252.

11. Some new cases of Parseval's theorem, *Math. Zeitschr.* 34 (1932), 620–633.

12. Some more integral inequalities, *Tôhoku Math. Journal*, 37 (1933), 151–159.

13. Bilinear forms bounded in space $[p, q]$, *Quart. Journ. of Math.* (2), 5 (1934), 241–254.

G. H. Hardy, J. E. Littlewood and **E. Landau**

1. Some inequalities satisfied by the integrals or derivatives of real or analytic functions, *Math. Zeitschr.* 39 (1935), 677–695.

G. H. Hardy, J. E. Littlewood and **G. Pólya**

1. The maximum of a certain bilinear form, *Proc. L.M.S.* (2), 25 (1926), 265–282.

2. Some simple inequalities satisfied by convex functions, *Messenger of Math.* 58 (1929), 145–152.

F. Hausdorff

1. Summationsmethoden und Momentfolgen (I), *Math. Zeitschr.* 9 (1921), 74–109.

F. Hausdorff (*cont.*)

2. Eine Ausdehnung des Parsevalschen Satzes über Fourierreihen, *Math. Zeitschr.* 16 (1923), 163–169.

E. Hellinger and O. Toeplitz

1. Grundlagen für eine Theorie der unendlichen Matrizen, *Math. Annalen*, 69 (1910), 289–330.

C. Hermite

1. *Cours de la Faculté des Sciences de Paris* (4th lithographed ed., Paris, 1888).

D. Hilbert

1. Über die Darstellung definiter Formen als Summe von Formenquadraten, *Math. Annalen*, 32 (1888), 342–350. [*Werke*, II, 154–161.]
2. Über ternäre definite Formen, *Acta Math.* 17 (1893), 169–197. [*Werke*. II, 345–366.]

E. W. Hobson

1. *The theory of functions of a real variable and the theory of Fourier, series*, I, II (2nd ed., Cambridge, 1921, 1926).

O. Holder

1. Über einen Mittelwertsatz, *Göttinger Nachrichten*, 1889, 38–47.

A. Hurwitz

1. Über den Vergleich des arithmetischen und des geometrischen Mittels, *Journal f. Math.* 108 (1891), 266–268. [*Werke*, II, 505–507.]
2. Sur le problème des isopérimètres, *Comptes rendus*, 132 (1901), 401–403. [*Werke*, I, 490–491.]

J. L. W. V. Jensen

1. Sur une généralisation d'une formule de Tchebycheff, *Bull. des sciences math.* (2), 12 (1888), 134–135.
2. Sur les fonctions convexes et les inégalités entre les valeurs moyennes, *Acta Math.* 30 (1906), 175–193.

B. Jessen

1. Om Uligheder imellem Potensmiddelvaerdier, *Mat. Tidsskrift*, B (1931), No. 1.
2. Bemaerkinger om konvekse Funktiner og Uligheder imellem Middelvaerdier (I), *Mat. Tidsskrift*, B (1931), No. 2.

B. Jessen (*cont.*)

3. Bemaerkinger om konveksef Funktioner og Uligheder imellem Middelvaerdier (II), *Mat. Tidsskrift*, B (1931), Nos. 3–4.
4. Über die Verallgemeinerungen des arithmetischen Mittels, *Acta Litt. ac Scient. Univ. Hung.* 5 (1931), 108–116.

A. E. Jolliffe

1. An identity connected with a polynomial algebraic equation, *Journ. L.M.S.* 8 (1933), 82–85.

Th. Kaluza and G. Szegö

1. Über Reihen mit lauter positiven Gliedern, *Journ. L.M.S.* 2 (1927), 266–272.

J. Karamata

1. Sur une inégalité relative aux fonctions convexes, *Publ. math. Univ. Belgrade*, 1 (1932), 145–148.

K. Knopp

1. Über Reihen mit positiven Gliedern, *Journ. L.M.S.* 3 (1928), 205–211.
2. Neuere Sätze über Reihen mit positiven Gliedern, *Math. Zeitschr.* 30 (1929), 387–413.
3. Über Reihen mit positiven Gliedern (2te Mitteilung), *Journ. L.M.S.* 5 (1930), 13–21.

A. Kolmogoroff

1. Sur la notion de la moyenne, *Rend. Accad. dei Lincei* (6), 12 (1930), 388–391.

N. Kritikos

1. Sur une extension de l'inégalité entre la moyenne arithmétique et la moyenne géométrique, *Bull. soc. math. Grèce*, 9 (1928), 43–46.

E. Landau

1. Über einen Konvergenzsatz, *Göttinger Nachrichten*, 1907, 25–27.
2. Einige Ungleichungen für zweimal differentiierbare Funktionen, *Proc. L.M.S.* (2), 13 (1913), 43–49.
3. Die Ungleichungen für zweimal differentiierbare Funktionen, *Meddelelser København*, 6 (1925), Nr. 10.
4. A note on a theorem concerning series of positive terms, *Journ. L.M.S.* 1 (1926), 38–39.

H. Lebesgue

1. *Leçons sur l'intégration et la recherche des fonctions primitives* (2nd ed., Paris, 1928).

S. Lhuilier

1. *Polygonométrie, ou de la mesure des figures rectilignes. Et abrégé d'isopérimétrie élémentaire* (Genève and Paris, 1789).

A. Liapounoff

1. Nouvelle forme du théorème sur la limite de probabilité, *Memoires de l'Acad. de St-Pétersbourg* (VIII), 12 (1901), No. 5.

J. Liouville

1. Sur le calcul des différentielles à indices quelconques, *Journal de l'École Polytechnique*, 13 (1832), 1–69.

J. E. Littlewood

1. Note on the convergence of series of positive terms, *Messenger of Math.* 39 (1910), 191–192.
2. On bounded bilinear forms in an infinite number of variables, *Quart. Journ. of Math.* (2), 2 (1930), 164–174.

J. E. Littlewood and **E. C. Francis**. *See* **E. C. Francis** and **J. E. Littlewood**

J. E. Littlewood and **G. H. Hardy**. *See* **G. H. Hardy** and **J. E. Littlewood**

J. E. Littlewood, G. H. Hardy and **G. Pólya**. *See* **G. H. Hardy, J. E. Littlewood** and **G. Pólya**

C. Maclaurin

1. *A treatise of fluxions* (Edinburgh, 1742).
2. A second letter to Martin Folkes, Esq.; concerning the roots of equations, with the demonstration of other rules in algebra, *Phil. Transactions*, 36 (1729), 59–96.

E. Meissner

1. Über positive Darstellung von Polynomen, *Math. Annalen*, 70 (1911), 223–235.

E. A. Milne

1. Note on Rosseland's integral for the stellar absorption coefficient, *Monthly Notices R.A.S.* 85 (1925), 979–984.

H. Minkowski

1. *Geometrie der Zahlen*, I (Leipzig, 1896).
2. Discontinuitätsbereich für arithmetische Äquivalenz, *Journal f. Math.* 129 (1905), 220–274.

P. Montel

1. Sur les fonctions convexes et les fonctions sousharmoniques, *Journal de math.* (9), 7 (1928), 29–60.

(Sir) T. Muir

1. Solution of the question 14792 [*Educational Times*, 54 (1901), 83], *Math. from Educ. Times* (2), 1 (1902), 52–53.

R. F. Muirhead

1. Inequalities relating to some algebraic means, *Proc. Edinburgh Math. Soc.* 19 (1901), 36–45.
2. Some methods applicable to identities and inequalities of symmetric algebraic functions of n letters, *Proc. Edinburgh Math. Soc.* 21 (1903), 144–157.
3. Proofs that the arithmetic mean is greater than the geometric mean, *Math. Gazette*, 2 (1904), 283–287.
4. Proofs of an inequality, *Proc. Edinburgh Math. Soc.* 24 (1906), 45–50.

H. P. Mulholland

1. Note on Hilbert's double series theorem, *Journ. L.M.S.* 3 (1928), 197–199.
2. Some theorems on Dirichlet series with positive coefficients and related integrals, *Proc. L.M.S.* (2), 29 (1929), 281–292.
3. A further generalisation of Hilbert's double series theorem, *Journ. L.M.S.* 6 (1931), 100–106.
4. On the generalisation of Hardy's inequality, *Journ. L.M.S.* 7 (1932), 208–214.

M. Nagumo

1. Über eine Klasse der Mittelwerte, *Jap. Journ. of Math.* 7 (1930), 71–79.

E. J. Nanson

1. An inequality, *Messenger of Math.* 33 (1904), 89–90.

(Sir) I. Newton

1. *Arithmetica universalis: sive de compositione et resolutione arithmetica liber.* [*Opera*, I.]

A. Oppenheim

1. Note on Mr Cooper's generalisation of Young's inequality, *Journ. L.M.S.* 2 (1927), 21–23.
2. Inequalities connected with definite hermitian forms, *Journ. L.M.S.* 5 (1930), 114–119.

W. Orlicz and **Z. W. Birnbaum.** *See* **Z. W. Birnbaum** and **W. Orlicz**

A. Ostrowski

1. Zur Theorie der konvexen Funktionen, *Comm. Math. Helvetici*, 1 (1929), 157–159.
2. Über quasi-analytische Funktionen und Bestimmtheit asymptotischer Entwicklungen, *Acta. Math.* 53 (1929), 181–266.

P. M. Owen

1. A generalisation of Hilbert's double series theorem, *Journ. L.M.S.* 5 (1930), 270–272.

R. E. A. C. Paley

1. A proof of a theorem on averages, *Proc. L.M.S.* (2), 31 (1930), 289–300.
2. A proof of a theorem on bilinear forms, *Journ. L.M.S.* 6 (1931), 226–230.
3. Some theorems on orthogonal functions, *Studia Math*, 3 (1931), 226–238.
4. A note on bilinear forms, *Bull. Amer. Math. Soc.* (2), 39 (1933), 259–260.

H. R. Pitt

1. On an inequality of Hardy and Littlewood, *Journ. L.M.S.* 13 (1938), 95–101.

H. Poincaré

1. Sur les équations algébriques, *Comptes rendus*, 97 (1888), 1418–1419.

S. Pollard

1. The Stieltjes integral and its generalisations, *Quart. Journ. of Math.* 49 (1923), 73–138.

G. Pólya

1. On the mean-value theorem corresponding to a given linear homogeneous differential equation, *Trans. Amer. Math. Soc.* 24 (1922), 312–324.
2. Proof of an inequality, *Proc. L.M.S.* (2), 24 (1926), Records of Proc. lvii.

G. Pólya (*cont.*)

3. Über positive Darstellung von Polynomen, *Vierteljahrsschrift d. naturforschenden Gesellsch. Zürich*, 73 (1928), 141–145.
4. Untersuchungen über Lücken und Singularitäten von Potenzreihen, *Math. Zeitschr.* 29 (1929), 549–640.

G. Pólya, G. H. Hardy and **J. E. Littlewood.** *See* **G. H. Hardy, J. E. Littlewood** and **G. Pólya**

G. Pólya and **G. Szegö**

1. *Aufgaben und Lehrsätze aus der Analysis*, I, II (Berlin, 1925).

A. Pringsheim

1. Zur Theorie der ganzen transzendenten Funktionen, *Münchner Sitzungber.* 32 (1902), 163–192, 295–304.

J. Radon

1. Über die absolut additiven Mengenfunktionen, *Wiener Sitzungsber.* (IIa), 122 (1913), 1295–1438.

A. A. L. Reynaud and **J. M. C. Duhamel.** *See* **J. M. C. Duhamel** and **A. A. L. Reynaud**

B. Riemann

1. *Gesammelte math. Werke* (Leipzig, 1876).

F. Riesz

1. *Les systèmes d'équations linéaires à une infinité d'inconnues* (Paris, 1913).
2. Untersuchungen über Systeme integrierbarer Funktionen, *Math. Annalen*, 69 (1910), 449–497.
3. Über die Randwerte einer analytischen Funktion, *Math. Zeitschr.* 18 (1923), 87–95.
4. Über eine Verallgemeinerung der Parsevalschen Formel, *Math. Zeitschr.* 18 (1923), 117–124.
5. Sur les fonctions subharmoniques et leur rapport à la théorie du potentiel, *Acta Math.* 48 (1926), 329–343.
6. Su alcune disuguaglianze, *Boll. dell' Unione Mat. Italiana*, Anno 7 (1928), No. 2.
7. Sur les valeurs moyennes des fonctions, *Journ. LM.S.* 5 (1930), 120–121.
8. Sur une inégalité intégrale, *Journ, L.M.S.* 5 (1930), 162–168.
9. Sur les fonctions subharmoniques et leur rapport à la théorie du potentiel (seconde partie), *Acta Math.* 54 (1930), 162–168.

F. Riesz (*cont.*)

10. Sur un théorème de maximum de MM. Hardy et Littlewood, *Journ. L.M.S.* 7 (1932), 10–13.

F. Riesz and **L. Fejér**. *See* **L. Fejér** and **F. Riesz**

M. Riesz

1. Sur les maxima des formes bilinéaires et sur les fonctionnelles linéaires, *Acta Math.* 49 (1927), 465–497.
2. Sur les fonctions conjuguées, *Math. Zeitschr.* 27 (1928), 218–244.

L. J. Rogers

1. An extension of a certain theorem in inequalities, *Messenger of Math.* 17 (1888), 145–150.

S. Saks

1. Sur un théorème de M. Montel, *Comptes rendus*, 187 (1928), 276–277.

O. Schlömilch

1. Über Mittelgrössen verschiedener Ordnungen, *Zeitschr. f. Math. u. Physik*, 3 (1858), 310–308.

O. Schreier and **E. Artin**. *See* **E. Artin** and **O. Schreier**

I. Schur

1. Bemerkungen zur Theorie der beschränkten Bilinearformen mit unendlich vielen Veränderlichen, *Journal f. Math.* 140 (1911), 1–28.
2. Über eine Klasse von Mittelbildungen mit Anwendungen auf die Determinantentheorie, *Sitzungsber. d. Berl. Math. Gesellsch.* 22 (1923), 9–20.

H. A. Schwarz

1. Beweis des Satzes dass die Kugel kleinere Oberfläche besitzt, als jeder andere Körper gleichen Volumens, *Göttinger Nachrichten*, 1884, 1–13. [*Werke*, II, 327–340.]
2. Über ein die Flächen kleinsten Flächeninhalts betreffendes Problem der Variationsrechnung, *Acta soc. scient. Fenn.* 15 (1885), 315–362 [*Werke*, I, 224–269.]

W. Sierpinski

1. Sur l'équation fonctionnelle $f(x+y)=f(x)+f(y)$, *Fundamenta Math.* 1 (1920), 116–122.
2. Sur les fonctions convexes mesurables, *Fundamenta Math.* 1 (1920), 125–129.

H. Simon

1. Über einige Ungleichungen, *Zeitschr. f. Math. u. Physik*, 33 (1888), 56–61.

Ch. Smith

1. *A treatise on algebra* (London, 1888).

J. F. Steffensen

1. Et Bevis for Saetningen om, at det geometriske Middletal at positive Størrelser ikke større end det aritmetiske, *Mat. Tidsskrift*, A (1930), 115–116.
2. The geometrical mean, *Journ. of the Institute of Actuaries*, 62 (1931), 117–118.

J. Steiner

1. *Gesammelte Werke* (Berlin, 1881–2.)

E. Stiemke

1. Über positive Lösungen homogener linearer Gleichungen, *Math. Annalen*, 76 (1915), 340–342.

O. Stolz and **J. A. Gmeiner**. *See* **J. A. Gmeiner** and **O. Stolz**

R. Sturm

1. *Maxima und Minima in der elementaren Geometrie* (Leipzig, 1910).

G. Szegö and **Th. Kaluza**. *See* **Th. Kaluza** and **G. Szegö**

G. Szegö and **G. Pólya** *See* **G. Pólya** and **G. Szegö**

E. C. Titchmarsh

1. *The theory of functions* (Oxford, 1932).
2. Reciprocal formulae involving series and integrals, *Math. Zeitschr.* 25 (1926), 321–341.
3. An inequality in the theory of series, *Journ. L.M.S.* 3 (1928), 81–83.

O. Toeplitz

1. Zur Theorie der quadratischen Formen von unendlich vielen Veränderlichen, *Göttinger Nachrichten* (1910), 489–506.

O. Toeplitz and **E. Hellinger**. *See* **E. Hellinger** and **O. Toeplitz**

G. Valiron

1. *Lectures on the general theory of integral functions* (Toulouse, 1923).

Ch.-J. de la Vallée Poussin

1. *Cours d'analyse infinitésimale*, I (6th ed., Louvain and Paris, 1926).

2. *Intégrales de Lebesgue. Fonctions d'ensemble. Classes de Baire* (Paris 1916).

J. H. Maclagan Wedderburn

1. The absolute value of the product of two matrices, *Bull. Amer. Math. Soc.* 31 (1925), 304–308.

H. Weyl

1. *Gruppentheorie und Quantummechanik* (2nd ed., Leipzig, 1931).

2. *Singuläre Integralgleichungen mit besonderer Berücksichtigung des Fourierschen Integraltheorems*, Inaugural-Dissertation (Gottingen, 1908).

3. Bemerkungen zum Begriff des Differentialquotienten gebrochener Ordnung, *Vieteljahrsschrift d. naturforschenden Gesellsch. Zürich*, 62 (1917), 296–302.

D. V. Widder

1. An inequality related to one of Hilbert's, *Journ. L.M.S.* 4 (1929), 194–198.

F. Wiener

1. Elementarer Beweis eines Reihensatzes von Herrn Hilbert, *Math. Annalen*, 68 (1910), 361–366.

W. H. Young

1. On a class of parametric integrals and their application to the theory of Fourier series, *Proc. Royal. Soc.* (A), 85 (1911), 401–414.

2. On classes of summable functions and their Fourier series, *Proc. Royal Soc.* (A), 87 (1912), 225–229.

3. On the multiplication of successions of Fourier constants, *Proc. Royal Soc.* (A), 87 (1912), 331–339.

4. Sur la généralisation du théorème de Parseval, *Comptes rendus*, 155 (1912), 30–33.

5. On a certain series of Fourier, *Proc. L.M.S.* (2), 11 (1913), 357–366.

6. On the determination of the summability of a function by means of its Fourier constants, *Proc. L.M.S.* (2), 12 (1913), 71–88.

7. On integration with respect to a function of bounded variation, *Proc. L.M.S.* (2), 13 (1914), 109–150.

A. Zygmund

1. On an integral inequality, *Journ. L.M.S.* 8 (1933), 175–178.